519
Sch

98534

Schmitt.
Measuring uncertainty.

Date Due

**The Library
Nazareth College of Rochester, N. Y.**

MEASURING UNCERTAINTY

An Elementary Introduction to Bayesian Statistics

SAMUEL A. SCHMITT

ADDISON-WESLEY PUBLISHING COMPANY
READING, MASSACHUSETTS
MENLO PARK, CALIFORNIA · LONDON · DON MILLS, ONTARIO

This book is in the
**Addison-Wesley Series in
Behavioral Science: Quantitative Methods**

FREDERICK MOSTELLER, Consulting Editor

Copyright © 1969 by Addison-Wesley Publishing Company, Inc. Philippines copyright 1969 by Addison-Wesley Publishing Company, Inc.

All rights reserved. No part of this publication may be reproduced, stored in a retrieval system, or transmitted, in any form or by any means, electronic, mechanical, photocopying, recording, or otherwise, without the prior written permission of the publisher. Printed in the United States of America. Published simultaneously in Canada. Library of Congress Catalog Card No. 69-14299.

*To my mother
and the memory of my father*

Preface

This book is designed for an introductory course in statistics for students having a good working knowledge of high-school mathematics. I treat probability as an expression of the strength of one's knowledge, and consequently make wholehearted use of Bayes' Theorem. I show how consistent application of this technique can solve real problems. The discussion then leads naturally into the modern ideas of decision theory and information theory. Mere combinatorial problems have been avoided; the treatment of probability aims directly at statistical applications. In fact, in the Bayesian approach it is hard to draw the line between probability and statistics.

I have tried to explain basic ideas and simple techniques so thoroughly that the instructor can spend more time on the complicated aspects of practical analysis. There are a great many detailed examples and hundreds of problems, most with solutions. They make the book quite suitable for self-study. I have not talked much about experimental design or the analysis of variance, since I believe these subjects must be presented in greater detail before more homogeneous classes. And some usually highly regarded statistical concepts which have no relevance to Bayesian theory have not been mentioned at all.

Some instructors may be surprised at my gingerly treatment of the calculus. I have found that students with one or even two semesters of calculus under their belts often cannot handle it well. Fortunately, there is little opportunity to use it in elementary statistics, because most of the integrals just can't be evaluated in closed form. (More advanced students might like to derive the few formulas I merely state.) But I have used *numerical* methods of integration throughout the book in the belief that more and more problems will be done numerically to achieve greater realism. One of the notable attractions of the Bayesian approach is that numerical solutions appear almost automatically. This fits in well with the increasing availability of computers. Although most of the problems in this book do not specifically require a computer, each instructor can take advantage of the local situation to prescribe more extensive numerical work. Many of the diagrams and tables in this book were prepared—at least in part— on a computer; these can serve as challenges to the student and point the way to more sophisticated presentations.

Bayesian solutions, of course, often agree closely with the solutions attained through other methods. It is their straightforward, natural approach to inference that makes them so attractive. They try to answer the questions people really want answered. I have not argued to any extent about the merits of the various statistical schools. It just seemed high time that someone stirred the Bayesian pot on an elementary level so that practitioners, rather than theorists, could start discussions and supply feedback to one another. There still are unresolved practical problems; solutions to them will come only when a large number of applications are made, reported, and analyzed.

I have used earlier versions of this book for elementary courses at Princeton University. Whatever the statistical approach, you always run up against the heterogeneity of beginning classes. You are forced to simplify problems so that the majority of students do not get bored with the finicky details of some one subject area. There is easily enough variety and depth here for a semester or two quarters. The instructor will want to introduce more specialized material when his class is convinced it's worth while to dig in, but in the beginning, the student must be taught basic principles rather than detailed practice.

Princeton, N. J.   S.A.S.
October 1968

Acknowledgments

Statistics, like other subjects, has been developing over many years, and most of what a man knows or writes is derived from other men's hard work. My great and overwhelming debt is to Harold Jeffreys, whose approach to inference will stand out as one of the enduring landmarks of modern statistics. I first met his book when I was studying geology in school. I can't say exactly how; I may have reached for *The Earth* and pulled out *The Theory of Probability* instead, but no matter. Since then it has been my bible of statistics. My book could not have been written without it; many formulas and much of my philosophical attitude have come from it. I just hope I have done justice to the ideas.

I was also fortunate in falling in with a Bayesian crowd in my first job. Their discussions at that time and their encouragement and friendship over many years have helped me keep the faith. I must also acknowledge the aid derived from the stimulating writings of Robert Schlaifer, I. J. Good, D. V. Lindley, L. J. Savage, and others of the modern Bayesian movement; they have helped clarify many dubious points. I have gotten problem ideas from many books and articles, though usually they have had to be reshaped and simplified. The American Chemical Society, the American Council of Learned Societies, the American Statistical Association, the Athlone Press, and Taylor and Francis, Ltd., have graciously allowed me to reproduce material.

An earlier version of this book was read in full or in large measure by Don Creecy, Jack Good, John Hartigan, Fred Mosteller, Lee Neuwirth, Jack Steele, and George Vergine. I thank them for their encouragement and for the detailed critical remarks which have helped remove some of the errors from this version. I also appreciate two years of criticisms from the Princeton students who had to endure the experimentation necessary to put together a Bayesian approach at an elementary level. My boss, Dick Leibler, has constantly cheered me on; he also arranged for my taking time off work in order to write.

Fred Stephan pushed me into this project in the first place, and I thank him for a never-to-be-forgotten experience. Fred Mosteller, the editor of this series, has been extremely helpful in keeping me on the right track, although I've sometimes disregarded good advice and pursued my own stubborn path. The publishers, Addison-Wesley, have been most cooperative on all occasions. Through their excellent suggestions and continuous interest, the text has been much improved over the last manuscript version. Finally, I could not have finished this book without the wonderful cooperation of my wife and children. They have all given up many joint family activities so that I would have time for writing. They have been amazingly patient in enduring the idiosyncrasies of an author.

Contents

Introduction

Chapter 1	**Probability, the Language of Uncertainty**	1
1.1	The conditional nature of probability	2
1.2	The laws of probability	6
1.3	Probability distributions	11
1.4	Compound probabilities	21
1.5	Averages and expected values	26

Chapter 2	**More about Probability**	34
2.1	Joint and marginal distributions	35
2.2	Conditional distributions	43
2.3	Independence	50
2.4	Sampling experiments	56

Chapter 3	**Accumulating Evidence**	61
3.1	Bayes' theorem	62
3.2	Multiple observations	71
3.3	Likelihoods, odds, and factors	83
3.4	Composite alternatives	89
3.5	Prior probabilities and ultimate error	95

Chapter 4	**Continuous Variables**	100
4.1	From points to intervals	101
4.2	Probability densities	106
4.3	The rate problem revisited	114
4.4	Estimation of rates	121
4.5	Testing precise point hypotheses	129
4.6	Experimental sampling from continuous distributions	134

Chapter 5	**Describing Distributions**	**139**
5.1	The Gaussian distribution	140
5.2	The standardized variable	143
5.3	Moments	151
5.4	GAU as an approximation	158
5.5	Entropy and information	166

Chapter 6	**Inferences about the Gaussian Distribution**	**176**
6.1	Continuous likelihoods and priors	177
6.2	Inferences about μ with known σ	182
6.3	Inferences about both μ and σ	190

Chapter 7	**Nonnormality**	**203**
7.1	Two cousins of GAU	204
7.2	Probability paper	210
7.3	Long-tailed distributions	215
7.4	Transformations	225

Chapter 8	**Making Decisions**	**231**
8.1	A criterion for decision-making	232
8.2	Discrimination	240
8.3	Tests of significance and goodness of fit	253
8.4	Sequential decisions	260

Chapter 9	**Handling Several Variables**	**271**
9.1	Comparing two values	272
9.2	Increasing efficiency	283
9.3	Straight-line analysis	290

Chapter 10	A Potpourri of Applications	. . . .	**309**
10.1	Rare events	. . . .	310
10.2	Angles	. . . .	318
10.3	Reliability and failures	. . . .	323
10.4	Radar	. . . .	330
	Epilogue	. . . .	**339**
	Suggestions for Further Reading	. . . .	**341**
	Bibliography	. . . .	**345**
Appendix A	**Mathematical Details**	. . . .	**351**
A1	Sums	. . . .	351
A2	Products	. . . .	356
A3	Factorials	. . . .	356
A4	Binomial coefficients	. . . .	357
A5	The constant e	. . . .	359
A6	Numerical integration	. . . .	359
Appendix B	**Logarithms**	. . . .	**364**
B1	Using logarithms	. . . .	364
B2	Logarithms to base 10	. . . .	366
B3	$\log_{10} N!$	. . . .	368
B4	$-p \log_{10} p$	. . . .	369
Appendix C	**Random Numbers**	. . . .	**370**
C1	Flat random digits	. . . .	370
C2	Normal deviates	. . . .	371

A Gallery of Standard Distributions **375**

Table of kernels 375
Notation 376
BETA: The beta distribution 376
BIGAU: The bivariate Gaussian or bivariate normal distribution 384
BIN: The binomial distribution 385
CIRG: The circular normal distribution 385
EXPD: The exponential distribution 386
GAMMA: The gamma distribution 387
GAU: The Gaussian or normal distribution. 387
IGAM: The inverse gamma distribution 390
NEGB: The negative binomial distribution 392
POIS: The Poisson distribution 392
RECT: The rectangular distribution 393
STU: The Student or "t" distribution 393
WEIB: The Weibull distribution 395

Index **397**

Introduction

All our lives are spent trying to make sense out of what we see and hear. Facts do not speak for themselves; everything must be put into proper relation to what we already know. *Know?* Well, *believe at present.* The harsh truth is that only gradually do we approach sure knowledge. That perplexing combination of theory, analogy, persistence, and luck which is called the scientific method somehow succeeds in advancing our state.

The task of formal statistics is to set up sound procedures for evaluating certain kinds of information. Its methods apply when we can regard what we see as being random variation superimposed upon a more solid structure. Statistics then tells us how our opinion should change on the basis of data; it tells us how to estimate values of interest; it tells us what uncertainty is attached to these values; it tells us how to improve the efficiency of our investigations; it tells us how our imperfect information should be used to make decisions for action.

Heaven knows that there is more to statistical practice than such formal work! In planning large-scale surveys and experiments, statisticians and their companions in the subject field get involved with the thousand dirty details of training people, of setting up equipment, of monitoring performance, of preparing the collected data for processing. In the same way, the architect must leave his paneled study to tramp through the mud at the construction site. Yet, at the beginning, before you immerse yourself in myriad details, you have to learn basic principles for guides and for standards.

This book teaches those basic principles; you must neither underestimate nor overestimate their value. *They lead to conclusions which are correct if the conceptual model underlying them is correct.* It seems psychologically impossible to think about the world without models, but you don't take them too seriously, at least at the start. You thrash around seeing what this or that model implies. Are the implications similar? Then there's no need to worry right now about which model is right. Are the implications different? Then you see which assumptions are especially critical. Unfortunately, the techniques of formulating models are not themselves well formalized. But that's more a subject-field problem than a statistics problem. Our basic task here is to learn how to use already formulated models of random variation to evaluate the data we see.

The book has ten chapters whose content may be summarized this way:

Chapters 1–2: Probability as the measure of the uncertainty in our knowledge.

Chapters 3–4: The principles of inference using assumed models of random variation.

Chapters 5–6, 9–10: Applications of these principles to a great variety of subjects.

Chapter 7: Problems posed by hazy models.

Chapter 8: Decision-making in the face of imperfect knowledge.

Chapters 1–6 have to be read in about that order, although a few sections can be skipped at first. Chapters 7–10 can be taken more freely, but some backtracking may occasionally be necessary. Throughout I have proceeded inductively, often reserving definitions until you understand what is going on. Sometimes you may think I do too much explaining, but better that than not enough. There are hundreds of problems, many with completely worked answers.

My approach is intrinsically oriented toward computation. Ours is the age of the computer. As soon as the model of a phenomenon gets at all complicated, neat formulas disappear and numerical methods are necessary. Where formerly this meant risky approximation or a watered-down model, now we can compute most results. Numbers usually are necessary, anyway, to get a feel for the subject. Their production is not a substitute for understanding, but a vital aid to it. Most of the problems in this book can be done without a computer, but their more complex extensions demand one.

The book assumes a good working knowledge of high-school mathematics. I have put a variety of appendixes in the rear of the book to review mathematical concepts you may have forgotten. There are also suggestions for future reading. At the very end is a Gallery of Standard Distributions which gives the details of the usual models of random variation together with pertinent tables. Figures, tables, and problems are numbered according to chapter, section within chapter, and serial order within section: e.g., Problem 9.3.1 is the first problem in Section 3 of Chapter 9. References to the bibliography are given by author and year of publication: e.g., Jeffreys (1961).

CHAPTER 1

Probability, the Language of Uncertainty

*The aim of this book is to measure the uncertainty in our knowledge. To do this we need a language to describe our results, and **probability** provides the means. All statements that express uncertainty are couched in terms of probability. We say that the probability of rain is 30%. We say that the probability is .95 that the density of this evil-smelling liquid lies between 2.92 and 2.93. We say that this result is more probable than that one. When new information is acquired, we reevaluate our uncertainty by manipulating probabilities. Our first concern, then, is to settle on a terminology for talking about probabilities and groups of probabilities, and to learn the rules for calculating with them. These ideas are at the heart of the entire course.*

1.1 THE CONDITIONAL NATURE OF PROBABILITY

Our interest in probability arises from its use in statistical inference; we want to measure uncertainty and changes in uncertainty. To accomplish this we need to apply probability—our measure of uncertainty—to whatever is of concern to us. We cannot rely on methods which talk only about "long-term frequencies" or "what will happen in many trials." We get background information from extensive records, but most things we are interested in are one of a kind. We want to infer something about *this* coin, *this* drug, *this* box of fuses, the position of *this* airplane, the life expectancy of *this* person. The basis for inference is knowledge; *probability expresses the strength of our knowledge or beliefs.* Thus we will be assigning probabilities not only to events like a coin coming up heads, but also to alternative values for the cure rate of a drug or alternative identifications of a rock sample. Only in this way can we talk about accumulating information; only then can we measure the change in our knowledge.

It is most important to realize that probability cannot be isolated from the world; it is absolutely dependent on the data we have and on the assumptions we make. There is a notation which explicitly recognizes these underlying conditions. We write

$$P(A \mid H)$$

to mean "the probability of A, given the conditions H." The capital P will be used in this book solely as an abbreviation of the word *probability*; the vertical bar will always be read "given" or "given the conditions"; the information following the vertical bar describes the conditions, that is, our knowledge or assumptions. It is quite important to watch what is on which side of the bar; on some occasions the same symbol will appear on the left of the bar in one part of an equation and on the right in another. Sometimes, for the sake of brevity, the conditions will not be listed. In that case you must take extra care to understand the basis for the calculation.

We cannot always assign probabilities with great accuracy, but there are numerous ways to get reasonable values. Here are a few examples with the conditions written out in a rather wordy fashion; later we'll use more compact designations.

Example 1.1.1. P(density of this liquid is between 2.92 and 2.93 | the 10 measurements I made)

Where would such a probability come from? From the internal consistency of the measurements, of course, but only if I make some kind of assumption about the laws which describe the variation in my measurements. That assumption need not be too precise if I don't want a very precise probability. I get back in proportion to what I give.

Example 1.1.2. *P*(density of this liquid is between 2.92 and 2.93 | the 10 measurements I made and the 20 my co-worker made)

Now I have to decide whether the two of us actually measured the same thing. There's always the chance for a mistake or a systematic difference in our methods. Then I have to decide how the precisions of the two sets of measurements compare so I can weight them properly.

Example 1.1.3. *P*(I will die within the next six months | I am a 42-year-old American male)

From a Census Bureau compilation of 1959–1961, about 362 out of every 100,000 Americans alive on their 42nd birthday were dead by their 43rd. So, for a six-month period, I would take half that figure—a probability of .0018—as a starter. Yet the table included both men and women, and women live longer than men; I ought to raise the probability of my death because of that. On the other hand, I am in good health now, while some of those 42-year-olds were already sick on their birthday; I think this more than compensates for the male–female bit—perhaps we're down to .0017 now. I could quibble some more because the table is a little out of date, but life expectancy in the middle years in the United States has lately been staying rather constant.

Example 1.1.4. *P*(I will die within the next six months | I am a 42-year-old American male, am married, have three children, live in the suburbs, and do little business traveling)

Whew, this is more information, but is it useful? Some people claim that married men live longer than single men, but my children exhaust me! Then, too, my not traveling much on business should expose me to a smaller risk of accidental death, but I don't know whether my locality has an average rate to begin with. Until I can learn the effect of these factors, I'll stick with the .0017 estimated above.

Example 1.1.5. *P*(next motor off this assembly line will pass inspection | 95% of the motors off this line in the last month have passed inspection)

Having no other information, I would likely evaluate the probability as .95. This involves the judgment that past performance is a good predictor of the future—a reasonable judgment, but not an infallible one. If I knew that the line had been taken over today by trainees or that the metallurgist had complained this morning about the quality of some alloys, I would want to set a lower figure. How much lower would depend on my experience in the business; a statistics professor wouldn't have much feel for it, but the veteran foreman might. I would also change my opinion if I found out that the last 25 motors had all been bad; this information overrides any general past performance.

Example 1.1.6. P(this nickel turns up heads | fair flip)

Here we usually idealize: if the nickel in my hand were a perfect cylinder, homogeneous through and through, each of the two flat surfaces would deserve a probability of .50. But what about *my* nickel? There's a picture of Jefferson stamped on one side, Monticello on the other, and a slight nick in the edge. Is it close enough to the ideal coin for me to say that the probability of heads is $\frac{1}{2}$? This is the ever-present problem in life; the real world is infinitely complicated, and we try to approximate it by models simple enough to calculate with. I would probably say the probability is $\frac{1}{2}$, but I would be prepared to give a little in either direction if repeated flipping showed a bias. You'll learn ways to express this attitude.

Example 1.1.7. P(that fellow's nickel turns up heads | it has turned up heads 40 times in a row)

A long run of heads suggests the possibility that both sides are heads, an infrequent but not unheard of phenomenon. Did you ever inspect the coin? Has it been in plain sight the whole time? Is there any money riding on the next toss?

Example 1.1.8. P(Johnny will arrive home before 4:00 p.m. | school gets out at 3:30 p.m. and there are no extracurricular events)

Here there are no written records to go by, but a mother develops through harassing experience a good idea of the dawdling a child can exhibit.

Example 1.1.9. P(college Y will beat college H in basketball Friday night | past records of the teams and current physical condition of the players)

Some odds-makers seem to be able to produce satisfactory one-of-a-kind probabilities of this type—at least the solvent ones do. They are using performance records, but have to extrapolate mightily in a manner not at all obvious.

You can see the great variety of information used to evaluate probabilities: written records, consistency of measurements, idealization, consideration of symmetries, visceral records (usually called intuition or introspection). In those terrible cases where we have no information at all—fortunately they are rarer than you think—we may have to say that the probability is $\frac{1}{2}$, but with the proviso that we'd be willing to change our belief upon the presentation of the slightest bit of evidence. Such a fuzzy, noncommittal probability is also regularly employed when the evidence, though present, is too vague to be worth taking into account.

All these procedures ultimately depend on personal judgment. Is this idealized model pertinent? Is this information out of date? Is extrapolation reasonable here? You just can't do statistics satisfactorily unless you have a good knowledge of the subject matter to which it is applied. This poses terrible problems for an elementary book like this: the problems have to be simplified drastically for easy comprehension, and students never get to see the dirty details of real analysis.

PROBLEM SET 1.1

These problems illustrate how probability changes as your information changes. Coins and cards may not seem germane, but they are unsurpassed for illustrating elementary principles without getting entangled in details.

In Problems 1 through 4 the coin is *fair;* that is, if you know nothing else, $P(\text{heads} \mid \text{fair}) = \frac{1}{2}$. Use only the specified information.

1. A coin is tossed and lands on the table. You are too far away to see what side came up. What is $P(\text{heads} \mid \text{this information})$?
2. A coin is tossed and lands on the table. Jack Jones says, "Just what I expected." You don't know what he expected. What is $P(\text{heads} \mid \text{this information})$?
3. The previous throw of the coin came up tails. The coin is thrown again, but you do not see the result. Jack says, "Same as last time," and he has an excellent record for truthfulness. What is $P(\text{heads} \mid \text{this information})$?
4. A coin is tossed and lands on the table. You see that it is heads. What is $P(\text{heads} \mid \text{this information})$?

In Problems 5 through 11 a well-shuffled deck of 52 cards is dealt out to four players, each receiving 13. Partners sit across from one another. Each problem uses only the information specified.

5. What is $P(\text{you have the ace of spades} \mid \text{you haven't looked at your cards})$?
6. What is $P(\text{your partner has the ace of spades} \mid \text{you haven't looked at your cards})$?
7. What is $P(\text{your partner has the ace of spades} \mid \text{you don't})$?
8. What is $P(\text{your left-hand opponent has the ace of spades} \mid \text{neither you nor your partner does})$?
9. What is $P(\text{your partner has the ace of spades} \mid \text{your spades are QJ94})$?
10. What is $P(\text{your partner has the ace of spades} \mid \text{your left-hand opponent has 5 clubs and 4 hearts})$?
11. What is $P(\text{your partner has the ace of spades} \mid \text{the three of diamonds was exposed during the deal and given to him})$?

ANSWERS TO PROBLEM SET 1.1

1. It certainly came up something—I'm outlawing standing on edge. But since you don't know what it is, you have to say $\frac{1}{2}$.
2. Again $\frac{1}{2}$. You need solid information to leave this position.
3. If you were *absolutely* sure that Jack could see perfectly and was telling the truth, the answer would be 0. I would assign a small number, but not 0.
4. You saw it. The answer is 1.
5. You have 13 cards, any one of which could be the ace of spades; therefore $\frac{13}{52}$. Or, alternatively, the ace could have gone to any player; therefore $\frac{1}{4}$.

6. Same reasoning as Problem 5: $\frac{13}{52} = \frac{1}{4}$.
7. You know that the ace must be one of the 39 cards held by the other players. Your partner has 13 of them; therefore $\frac{13}{39}$. Alternatively, one of the three has it; therefore $\frac{1}{3}$.
8. There are now 26 cards unknown. That man has 13 of them; therefore $\frac{13}{26}$. Alternatively, one of the two has it; therefore $\frac{1}{2}$.
9. The ace is one of the 39 cards not in your hand. Your partner has 13 of them; therefore $\frac{13}{39} = \frac{1}{3}$.
10. There are 43 cards unplaced. Your partner has 13 of them; therefore $\frac{13}{43}$.
11. Now 51 cards are unplaced and your partner has 12 of them; therefore $\frac{12}{51}$.

1.2 THE LAWS OF PROBABILITY

This book is a book on statistics, not probability, so we'll get by with as few rules as possible. You undoubtedly know that probabilities range from 0 to 1 inclusive:

$$0 \leq P(A \mid H) \leq 1,$$

where we ordinarily interpret 0 as impossibility and 1 as certainty, under the stated conditions, of course. Here A stands for some event or alternative, H gives the background conditions, and the sign $\leq$ means *less than or equal to*. We demand, as in Problem 1.1.4, that when you know A is true, its probability is 1:

$$P(A \mid A, H) = 1.$$

["A, H" following the condition bar means that A is true and that the conditions H also apply.] We also require that either A or its opposite, $\overline{A}$, must occur:

$$P(A \mid H) + P(\overline{A} \mid H) = 1.$$

What "opposite of A" means depends on H.

People quite commonly talk about probabilities or "chances" in terms of percentages. The correspondence is the usual one:

.00 corresponds to 0%,
.25 corresponds to 25%,
1.00 corresponds to 100%.

Thus a 30% chance of rain means a probability of .30.

There is a relationship between conditional probabilities which is quite important:

$$P(B \mid A, H) = \frac{P(A, B \mid H)}{P(A \mid H)}.$$

1.2 The Laws of Probability

This formula involves these three probabilities:

i) $P(B \mid A, H)$ = probability of B being true when you know that both A and H are true,

ii) $P(A \mid H)$ = probability of A being true when you know that H is true,

iii) $P(A, B \mid H)$ = probability of both A and B being true when you know that H is true.

In (i) the truth of A is assumed; in (ii) and (iii) it is in question.

Example 1.2.1. On the Monday following a bright summer weekend we are treated to this newspaper tabulation:

Accidental deaths over the weekend

	Auto accident	Drowning
Male	49	7
Female	41	3

Here are 100 events about which we can ask various questions. A collection like this which we discuss or from which we take samples is called a *population*. Right now we are interested only in some properties of probability. Later we will be trying to infer characteristics of an unknown population from what we see in a sample.

Suppose that one of the 100 is selected by chance with the stipulation that each had an equal probability of being selected. This defines our general conditions H.

Question 1: What is $P(\text{male} \mid H)$?

Of the 100, 56 are male. $P(\text{male} \mid H) = \frac{56}{100}$.

Question 2: What is $P(\text{drowned, male} \mid H)$?

Of the 100, 7 qualify. $P(\text{drowned, male} \mid H) = \frac{7}{100}$.

Question 3: What is $P(\text{drowned} \mid \text{male}, H)$?

Of the 56 males, 7 qualify. $P(\text{drowned} \mid \text{male}, H) = \frac{7}{56}$.

Now you can see that our formula works. Let

$$A = \text{male} \quad \text{and} \quad B = \text{drowned}.$$

Then

$$P(B \mid A, H) = \tfrac{7}{56},$$
$$P(A, B \mid H) = \tfrac{7}{100},$$
$$P(A \mid H) = \tfrac{56}{100},$$

and we verify that
$$\frac{7}{56} = \frac{\frac{7}{100}}{\frac{56}{100}}.$$

In this example you could actually check the formula by counting individuals. We take the formula as a *definition* in the next examples.

Example 1.2.2. From climatic records (our H), I find that

$$P(\text{rain on July 4} \mid H) = .05, \qquad P(\text{rain on both July 4 and July 5} \mid H) = .01.$$

What is $P(\text{rain on July 5} \mid \text{rain on July 4}, H)$?

Let
$$A = \text{rain on July 4} \quad \text{and} \quad B = \text{rain on July 5}.$$

Then
$$P(B \mid A, H) = \frac{.01}{.05} = .20.$$

Note that in this example A and B occurred at different times, while in Example 1.2.1 A and B were simultaneous characteristics.

Example 1.2.3. H is the following information: The probability that a proposed tax write-off law will pass is 40%. The probability of a particular merger for tax savings is 80% if the law passes. What is $P(\text{law will pass } and \text{ merger will occur} \mid H)$?

Here
$$A = \text{law will pass} \quad \text{and} \quad B = \text{merger will occur}$$

and we rewrite the formula for conditional probability as

$$P(A, B \mid H) = P(A \mid H) \times P(B \mid A, H).$$

Thus

$P(\text{law will pass } and \text{ merger will occur} \mid H)$
$\quad = P(\text{law will pass} \mid H) \times P(\text{merger will occur} \mid \text{law will pass}, H)$
$\quad = .40 \times .80 = .32.$

We can, by the way, chain our conditional probabilities indefinitely. Suppose that after the merger there might be an increased dividend. Then

$P(\text{law will pass } and \text{ merger will occur } and \text{ larger dividend} \mid H)$
$\quad = P(\text{law will pass} \mid H) \times P(\text{merger will occur} \mid \text{law will pass}, H)$
$\quad \times P(\text{larger dividend} \mid \text{law will pass, merger will occur}, H).$

This is a little like

> "the cow with the crumpled horn,
> that tossed the dog,
> that worried the cat,
> that killed the rat,
> that ate the malt
> that lay in the house that Jack built."

Note that again in Example 1.2.4 the joint truth of A and B did *not* mean their simultaneous physical occurrence.

PROBLEM SET 1.2

Data for Problems 1 through 4. In the 1954 trials of the Salk polio vaccine, 401,974 subjects were observed. The results were as follows:

	Did not get polio	Got polio
Control (not vaccinated)	201,114	115
Vaccinated	200,712	33

Controls were used to set a standard for judging the effect of the vaccine. Suppose one of these 401,974 people is chosen by lot, each having an equal probability of being chosen (our conditions H).

1. What is P(he was vaccinated *and* got polio $\mid H$)?
2. What is P(he was vaccinated $\mid H$)?
3. What is P(he got polio $\mid$ he was vaccinated, H)?
4. Show how the answers to Problems 1, 2, and 3 fit in the formula for conditional probability.

Data for Problems 5 through 8. The 1950 U.S. census showed the following:

	Urban	Rural
Male	31%	19%
Female	33%	17%

Imagine that we picked a U.S. resident at random from that population. H is this information plus equal probability of any member of the population.

5. What is P(female $\mid$ urban, H)?
6. What is P(urban $\mid H$)?
7. What is P(female, urban $\mid H$)?

8. Show how the answers to Problems 5, 6, and 7 fit into the formula for conditional probability.

9. In Example 1.2.3 suppose

$$P(\text{larger dividend} \mid \text{law will pass } and \text{ merger will occur} \mid H) = .90.$$

What is $P(\text{law will pass } and \text{ merger will occur } and \text{ larger dividend} \mid H)$?

Data for Problems 10 and 11. In written English with punctuation and spaces removed, about 10% of the letters are t and about 5% of the letters are h. When you count letter pairs (including pairs linking the end of one word to the beginning of the next) the pair th occurs about 3.8% of the time.

10. What is $P(\text{next letter is } h \mid \text{this letter is } t, H)$?
11. What is $P(\text{previous letter was } t \mid \text{this letter is } h, H)$?

Data for Problems 12 and 13. The 365 girls in the freshman class at a women's college were found to possess the following characteristics (among others):

	Blonde	Brunette	Red-headed
Left-handed	12	13	20
Right-handed	109	166	45

A boy from the coordinate college is given a blind date with a girl selected at random from this class.

12. What is the probability that she is red-headed?
13. What is the probability that she is left-handed if you know that she is red-headed?
14. In a world series between teams C and D

$$P(\text{team } C \text{ wins first game}) = .50,$$

and for other games

$$P(\text{team } C \text{ wins} \mid \text{it lost previous one}) = .40,$$
$$P(\text{team } C \text{ wins} \mid \text{it won previous one}) = .70.$$

What is $P(\text{team } C \text{ wins first, second, fourth, and fifth games})$?

15. What happens to the formulas for conditional probability if $P(A \mid H) = 0$?

ANSWERS TO PROBLEM SET 1.2

1. $\frac{33}{401974}$

2. $\frac{200745}{401974}$

3. $\frac{33}{200745}$

4. $\frac{33}{200745} = \frac{\frac{33}{401974}}{\frac{200745}{401974}}$

5. $\frac{33}{64}$

6. $\frac{64}{100}$

7. $\frac{33}{100}$ 8. $\frac{33}{64} = \frac{\frac{33}{100}}{\frac{64}{100}}$ 9. $.40 \times .80 \times .90 = .288$

10. $P(\text{next} = h \mid \text{this} = t, H) = \dfrac{P(\text{next} = h \text{ and this} = t \mid H)}{P(\text{this} = t \mid H)} = \dfrac{.038}{.100} = .38$

That is, 38% of the t's have h's following them.

11. $P(\text{previous} = t \mid \text{this} = h, H) = \dfrac{P(\text{previous} = t \text{ and this} = h \mid H)}{P(\text{this} = h \mid H)} = \dfrac{.038}{.050} = .76$

That is, 76% of the h's are preceded by t's.

12. $\frac{65}{365}$ 13. $\frac{20}{65}$

14. $P(\text{team } C \text{ wins first, second, fourth, and fifth games} \mid H)$
$= P(\text{wins first} \mid H) \times P(\text{wins second} \mid \text{wins first}, H)$
$\times P(\text{loses third} \mid \text{wins first, wins second}, H)$
$\times P(\text{wins fourth} \mid \text{wins first, wins second, loses third}, H)$
$\times P(\text{wins fifth} \mid \text{wins first, wins second, loses third, wins fourth}, H).$

Since H says that the calculation depends only on data from one game back, this is

$P(\text{wins first} \mid H) \times P(\text{wins second} \mid \text{wins first}, H) \times P(\text{loses third} \mid \text{wins second}, H)$
$\times P(\text{wins fourth} \mid \text{loses third}, H) \times P(\text{wins fifth} \mid \text{wins fourth}, H)$
$= .50 \times .70 \times .30 \times .40 \times .70 = .00294.$

This small number is the probability of this particular sequence of results. There are other ways for team C to win the series—don't despair. [This is an example of a *Markov Chain;* there are more complicated situations in which dependence goes back more than one.]

15. $P(A \mid H)$ being 0 implies that $P(A, B \mid H) = 0$. $P(B \mid A, H)$ is then undefined.

1.3 PROBABILITY DISTRIBUTIONS

Events or alternative causes are not usually considered singly, but occur in related groups. We think of all the possible results of an investigation because we want to record our data systematically; we think of all the possible causes for a phenomenon because we want to find the one which fits our data best. Sometimes, because of ignorance, we may miss some alternatives or mention impossible ones, but the complete collection is our goal. For the time being I want to assume that we can *itemize* all the possibilities. I'm sure you know that even though all the measurements we ever make are of limited accuracy and thus *could* be itemized in advance, there is considerable mathematical advantage in imagining that they can take on *any* value within reasonable limits. There is even mathematical advantage in eliminating the limits and letting the possibilities range, conceptually at least, out to infinity. For the purposes of this chapter and the next two, however, we will work with *discrete* sets of alternatives, that is, sets we can itemize.

When we discuss such a set of alternatives we demand two features to make our work much neater. We ask that the set be *exhaustive*, that is, that every possibility be covered. We also ask that the alternatives be nonoverlapping, or *mutually exclusive*, so that just one alternative is possible at a time. People sometimes carelessly list sets of alternatives which violate these conditions, but the situation can usually be remedied.

Example 1.3.1. We are interested in the cloudiness of the sky at an airport, and define the cloudiness by estimating what fraction of the sky is obscured. We arbitrarily say that the cloudiness is

    0 if $\leq 5\%$ of the sky is obscured,
    1 if $> 5\%$ up to $\leq 10\%$ of the sky is obscured,

and so on to

   10 if $> 95\%$ of the sky is obscured.

Our *cloudiness* is thus a number from 0 to 10. Estimating the fraction of sky obscured is so rough a process that this seems a fine enough scale. If we had more precise ways of performing this measurement, we might want a scale from 0 to 100.

Example 1.3.2. We are interested in the number of inquiries directed to an airline ticket office during the period from noon to 12:30 p.m.; we will want to record data on various days and find a model that accounts for the variations.

i) The numbers from 0 on up to some reasonable upper limit are a possible set of mutually exclusive and exhaustive alternatives.

ii) The categories

        0–4 calls,   5–9 calls,   10–14 calls, etc.,

also form a mutually exclusive and exhaustive set of alternatives.

How does (ii) differ from (i)? The second choice preserves *less information* than the first. The grouped data will probably be less able to distinguish between alternative models for the variation in the number of calls. Generally the only excuses for grouping data are:

a) we can save enough in collection and processing expense to compensate for the loss in information;

b) we have so much information that we don't care;

c) we need to group to get counts large enough to form a reasonable *histogram* (bar graph). In this case remember to save the ungrouped data for your analysis.

iii) The categories starting with

$$0\text{–}5 \text{ calls}, \quad 5\text{–}10 \text{ calls}, \quad 10\text{–}15 \text{ calls}, \text{ etc.},$$

are not acceptable because 5 and 10 each appear in two groups. You must be sure that your categories are nonoverlapping.

Example 1.3.3. We are interested in the age distribution (to the nearest year) of a certain population.

i) The alternatives

$$0 \text{ years}, \quad 1 \text{ year}, \quad 2 \text{ years}, \text{ etc.},$$

form a mutually exclusive and exhaustive set of alternatives.

ii) The alternatives

$$2 \text{ years}, \quad 3 \text{ years}, \quad 4 \text{ years}, \text{ etc.},$$

are not acceptable unless no one in the population is under $1\frac{1}{2}$ years of age.

iii) The alternatives

$$(0 \text{ years, male}), \quad (0 \text{ years, female}), \quad (1 \text{ year, male}), \text{ etc.},$$

are a refinement of the original plan. We certainly would gather more information, but its value would depend on the purpose of the investigation. Often little or no additional cost is involved, and the extra data can be discarded if it is not needed. If you don't collect it now and then find later that you need it, the cost could be considerable. With modern computers available to process the data, it often is advisable to err in the direction of overdescription.

Example 1.3.4. We want to determine whether an automatic machine is in good or bad condition by sampling its output. When it is in good condition it produces, on the average, 2% defective items. When it is in bad condition it produces, on the average, 10% defective items. The two alternatives here are the two possible defect rates: .02 and .10. Note that even though a set of alternatives is to be itemized, the members do not have to be integers.

Example 1.3.5. It is becoming common to analyze the style of a writer by counting the frequencies with which he uses various words. In English the possibilities would include *a, aardvark, abacus, abandon,* etc. When there are so many possibilities you can hardly list them beforehand; instead your investigation proceeds in stages and is guided by what you see along the way.

Example 1.3.6. In a study of the effectiveness of translation from Russian to English by computer [Carroll (1966)] a scale of intelligibility from 1 to 9 was defined because no natural scale existed. Category 8, for example, reads: "Perfectly or almost clear and intelligible, but contains minor grammatical or stylistic infelicities and/or mildly unusual word usage that could, nevertheless, be easily 'corrected'." Psychologists try hard to ensure that such definitions can be employed consistently. They also aim to have the categories equally spaced in some sense, so that they form a *scale* rather than just a *ranking*.

Example 1.3.7. When nuclear fission was discovered, physicists wanted to know what products resulted from the split. Consider the uranium isotope, U^{235}, an atom with 235 protons and neutrons in its nucleus. When it splits, fragments with masses from 1 to 234 could conceivably result. The numbers from 1 to 234 formed a list of possibilities.

Example 1.3.8. When we administer an experimental drug to some subjects, the percentage cured of a particular disease will be some odd fraction. That fraction or something close to it is reported as an estimate of the cure rate. But to find out how certain or uncertain such a value is, we have to score a whole range of alternative values; we have to calculate what relative weight is to be attached to each. Here we might put forth the list of alternatives

$$.1, .2, .3, .4, .5, .6, .7, .8, .9$$

if we believe that the possibilities of curing none or all are out of the question.

If we take such a set of alternatives, assign a probability to each alternative, and have the sum of the probabilities equal to 1, then we have a *probability distribution*. Note that there are two parts: the alternatives and the probabilities. We require that the probabilities sum to 1 because if we have covered all possibilities, one of them is bound to be true. Our set is exhaustive and mutually exclusive, and 1 represents certainty. Below are some probability distributions which might have arisen in the preceding examples; I am supplying the probabilities, gratis.

Example 1.3.1 continued. After many observations and some smoothing of the observed distribution, we arrive at a distribution of cloudiness like this:

Cloudiness	0	1	2	3	4	5	6	7	8	9	10
Probability	.21	.11	.07	.02	.01	.01	.01	.02	.03	.09	.42

Note that this distribution has the largest probabilities at the ends, not in the center. It's likely to be very cloudly or very clear, but seldom half-bad. The same kind of distribution governs little girls who are very very good or horrid. You can check that the probabilities sum to 1.

Example 1.3.2 continued. After a period of observation and using a particular single-humped distribution as a model, we arrive at the following:

Number of calls	0–4	5–9	10–14	15–19	20–24	≥25
Probability	.02	.16	.31	.25	.16	.10

Our model supplied the shape; our observations told us where to put the hump. This distribution will now be used in studying staffing procedures and demands on a computerized reservation system. Check that the probabilities sum to 1.

Example 1.3.3 continued. There's not much model here; it's practically the straight returns from a mail survey of people in one branch of computing:

Age to nearest year	<25	25–29	30–34	35–39	40–44	45–49	>49
Probability	.08	.19	.16	.15	.14	.16	.12

There's obviously a bulge of recent college graduates, but one would hate to guess about other characteristics. Perhaps it is really just a summary of data, not a probability distribution. The probabilities, or relative frequencies, sum to 1.

Example 1.3.4 continued. From long consistent counts we get:

Defective rate	.02	.10
Probability	.85	.15

When the same conditions apply, we take this for the probability distribution.

Example 1.3.5 continued. Figure 1.3.1 gives a portion of a count of 272,178 words of modern Arabic prose. Half the words came from literary sources (L) and half from newspapers (N). This page includes the largest frequency (11,871) and several of the smallest (1). Of course many Arabic words do not appear in the count at all. It is not a straightforward matter to turn a count like this into a probability distribution; the approach depends on the purpose of the study. For a discussion of difficulties in counting English, see Thorndike and Lorge (1944).

Example 1.3.6 continued. When we try our mechanized translation method on a rather circumscribed range of texts and attempt to eliminate the effect of the

L	N	Tot.			
4		4	أفواه	afwāh	mouths
1		1	فوهة	fūha	mouth, crater
	3	3	فوهرر	fūhrir	Fuehrer
6001	5870	11871	في	fī	in, inside of, about, for
1		1	في أن	fī an	in what, in which
55	90	145	فيما ، فيم	fīmā-fīma	in what; amongst other matters; whilst; why
1		1	فيما بعد	fīmā ba'd	as for the following
2		2	فيمن	fīman	amongst these that...; inter alia
2		2	فيحاء	fayḥā'	roomy; adj. referring to Damascus
1		1	فاد	fāda	to accrue to
5	27	32	أفاد	afāda	to benefit (some one else)
	5	5	أفاد	afāda	to inform, denote
1	19	20	إفادة	ifāda	benefit; message
10	9	19	إستفاد	istafāda	to profit; to gather information
	9	9	إستفادة	istifāda	profit; gathering of information
	36	36	فائدة	fā'ida	benefit, profit
22	12	34	فائدة	fā'ida	percentage
2	11	13	فوائد	fawā'id	benefits; profits; percentages
4	4	8	مفيد	mufīd	profitable, advantageous
6		6	فاض	fāḍa	to be abundant, overflow
1	7	8	إستفاض	istafāḍa	to spread out
3	6	9	فيض	fayḍ	abundance, overflowing; emanation
5	1	6	فيضان	fayaḍān	flood
1		1	تفيل	tafayyala	to grow up
5		5	فيل	fīl	elephant
	1	1	أفيال	afyāl	elephants
1		1	فيلد مارشال	fīld-māršāl	field-marshal
4		4	فينة	fayna	time
8		8	فينيقي	fīnīqiyy	Phoenician
11		11	قبة	qubba	dome
1	1	2	قباب	qibāb	domes
2		2	قبح	qabbaḥa	to find fault with

Fig. 1.3.1 Reproduced with permission from J. M. Landau (1959), *A Word Count of Modern Arabic Prose* (New York: American Council of Learned Societies), p. 231.

different people who grade the translations, we might get something like this:

Intelligibility	1	2	3	4	5	6	7	8	9
Probability	.01	.02	.09	.12	.18	.11	.16	.21	.10

There is one peak at 5 and another at 7–8. The first might represent sentences whose word order is quite different from English; the second, those whose word order is much the same as English. [*Note:* This is my own distribution; Mr. Carroll is not to be blamed for it or its interpretation.]

Example 1.3.7 continued. Figure 1.3.2 shows the result of an immense number of observations on the fission products of U^{235}; the masses have been analyzed in a mass spectrometer. It turns out that most fissions yield two fragments whose sum is 234; one unit of mass is converted into energy. The observations fall nearly on a smooth curve which is bimodal (two humps) and symmetric about $\frac{234}{2} = 117$. We would take that smooth curve as our estimate of the probability distribution of a single fragment. [Note that the vertical scale is logarithmic.]

Example 1.3.8 continued. After we perform our experiment we get this probability distribution for the cure rate:

Cure rate	.1	.2	.3	.4	.5	.6	.7	.8	.9
Probability	.01	.01	.02	.05	.10	.20	.33	.20	.08

Thus the cure rate is probably close to .7, but we shouldn't be surprised if it were 0.1 or so higher or lower.

Example 1.3.8 represents a slightly special case. The cure rate p is itself part of another probability distribution:

Result of trial	No cure	Cure
Probability	$1 - p$	p

and the problem is to determine p from the data. A number like p which is fixed for one problem but which may be different in another problem with the same outward structure is called a *parameter* of the problem. [Our cure rate was something—perhaps .7—in this problem; in testing another drug it might be .2. But it's fixed within each problem.] Much of this book is devoted to estimating parameters of various problems. Locating the position of the hump in Example 1.3.2 also involved estimating a parameter.

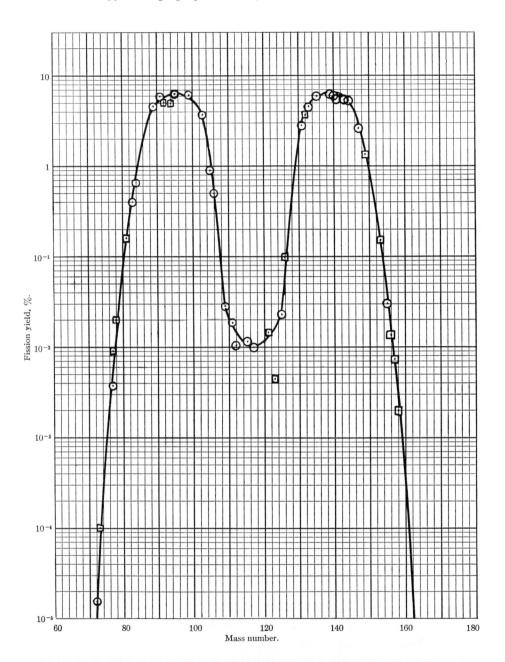

Fig. 1.3.2 Yields of U^{235} fission product chains as a function of mass. Reproduced with permission from Plutonium Project (1946), Nuclei formed in fission: Decay characteristics, fission yields, and chain relationships, *J. Am. Chem. Soc.* **68**, Figure 1, p. 2437.

PROBLEM SET 1.3

In Problems 1 through 4 tell what is wrong with the supposed probability distribution.

1.
Age to nearest year	0–4	5–9	10 and over
Probability	.1	.1	.8

2.
Age to nearest year	0–4	5–9	9 and over
Probability	.1	.1	.8

3.
Age to nearest year	0–4	5–9	10 and over
Probability	.1	.2	.8

4.
Age to nearest year	0–4	5–9	11 and over
Probability	.2	.1	.6

In Problems 5 and 6 fill in the missing probabilities.

5.
Number of successes in two trials	0	1	2
Probability	$(1-p)^2$	$2p(1-p)$	

6.
Number of successes in three trials	0	1	2	3
Probability	$(1-p)^3$	$3p(1-p)^2$		p^3

7. Here is the probability distribution of the variable X:

x (value)	A	B	C	D
$P(x)$ (probability)	$3k$	$2k$	$7k$	$8k$

What is k? What are the probabilities?

8. What is the probability distribution of the parts coming from the automatic machine of Example 1.3.4 when it is in its better state?

9. What is the probability distribution of the parts coming from the automatic machine of Example 1.3.4 when it is in its poorer state?

10. Suppose that $P(n) = (.1)(.9)^{n-1}$ for $n = 1, 2, 3, \ldots$ Show that the probabilities add up to 1.

ANSWERS TO PROBLEM SET 1.3

1. It's OK.
2. Age 9 appears in two categories.
3. The probabilities do not add up to 1.
4. Age 10 has no home, and the probabilities add up to .9.
5. The probabilities must add up to 1, so

$$P(2) = 1 - (1-p)^2 - 2p(1-p)$$
$$= 1 - (1 - 2p + p^2) - (2p - 2p^2) = p^2.$$

This is a *Binomial distribution*; see BIN in the Gallery of Distributions in the back of the book for details.

6. Again the probabilities must add up to 1, so

$$P(2) = 1 - (1 - 3p + 3p^2 - p^3) - (3p - 6p^2 + 3p^3) - p^3$$
$$= 3p^2 - 3p^3 = 3p^2(1-p).$$

This is another *Binomial distribution*.

7. $3k + 2k + 7k + 8k = 20k = 1$; $k = \frac{1}{20}$. Probabilities are $\frac{3}{20}, \frac{2}{20}, \frac{7}{20},$ and $\frac{8}{20}$. Here we were given a set of numbers: 3, 2, 7, 8, which were *proportional* to the probabilities in a probability distribution. The process of finding the factor that makes them sum to 1 is called *normalization*.

8. When the machine is in its better state, 2% of the parts are defective. Therefore:

Part from machine	Perfect	Defective
Probability	.98	.02

9. When the machine is in its poorer state, 10% of the parts are defective. Therefore:

Part from machine	Perfect	Defective
Probability	.90	.10

10. The sum of the probabilities is

$$(.1)[1 + (.9) + (.9)^2 + (.9)^3 + \cdots].$$

The series of terms in the brackets is

$$\frac{1}{1 - .9} = \frac{1}{.1}.$$

Therefore the probabilities sum to 1. This is an example of the *Negative Binomial distribution*; see NEGB in the Gallery of Distributions in the back of the book.

If you haven't seen the above series summed, it goes like this: Let

$$S = 1 + .9 + (.9)^2 + (.9)^3 + \cdots$$

Then

$$.9S = \phantom{1 + {}} .9 + (.9)^2 + (.9)^3 + \cdots$$

Subtract, and you have

$$(1 - .9)S = 1, \quad S = \frac{1}{1 - .9}.$$

1.4 COMPOUND PROBABILITIES

Now that you know what a probability distribution is, there are two more properties of probability to meet. Let's call our general conditions H, and our set of *mutually exclusive* and *exhaustive* alternatives $A_1, A_2, \ldots, A_c$. That is, there are c of them altogether; c varies from distribution to distribution.

The first formula is

$$P(A_1 \text{ or } A_2 \mid H) = P(A_1 \mid H) + P(A_2 \mid H).$$

This means that the probability of one *or* the other of the two happening—but you don't care which—is the sum of their separate probabilities.

Example 1.4.1

H:

Number of lemmings/litter	1	2	3	4	5	6
Probability	.05	.10	.20	.35	.20	.10

Then $P(4 \text{ or } 5 \text{ lemmings/litter} \mid H) = P(4 \mid H) + P(5 \mid H) = .55$. We can, of course, extend this result to any number of alternatives.

Example 1.4.2

$P(2 \text{ or } 3 \text{ or } 4 \text{ lemmings/litter} \mid H) = P(2 \mid H) + P(3 \mid H) + P(4 \mid H) = .65$

Example 1.4.3

H:

Successes in two trials	0	1	2
Probability	$(1-p)^2$	$2p(1-p)$	p^2

Then $P(\text{at most 1 success} \mid H) = P(0 \mid H) + P(1 \mid H) = 1 - p^2$.

This formula works because the A's are *mutually exclusive*. If they are not, you have to compensate for the overlap. Being *exhaustive* is not required here. However, when the A's *are* exhaustive, we see that

$$P(A_1 \text{ or } A_2 \text{ or } A_3 \text{ or } A_4 \text{ or } \cdots \text{ or } A_c \mid H) = P(\text{something will happen} \mid H) = 1.$$

This is just our old requirement that the sum of the probabilities in a probability distribution be 1.

The final property, sometimes called the *law of compound probability*, is closely related to those we have already discussed. All these elementary properties are so fundamental and intertwined that it is quite difficult, and certainly not profitable, to say which is more basic. The happy thing is that they are all consistent with each other and with our intuitive notions of probability. In Example 1.3.4, what is the probability that a perfect part will be produced by the automatic machine when we take into account the probabilities of the machine's good and bad states?

We can get a perfect part in two ways:

a) machine is good, part is perfect,

b) machine is bad, part is perfect.

These two combinations are certainly mutually exclusive—both can't be true at the same time—so we add probabilities:

$P(\text{machine good, part perfect} \mid H)$
$\qquad\qquad = P(\text{machine good} \mid H) \times P(\text{part perfect} \mid \text{machine good}, H)$
$\qquad\qquad = .85 \times .98 = .833,$

$P(\text{machine bad, part perfect} \mid H)$
$\qquad\qquad = P(\text{machine bad} \mid H) \times P(\text{part perfect} \mid \text{machine bad}, H)$
$\qquad\qquad = .15 \times .90 = .135,$

$P(\text{part perfect} \mid H) = P(\text{machine good, part perfect} \mid H)$
$\qquad\qquad + P(\text{machine bad, part perfect} \mid H)$
$\qquad\qquad = .833 + .135 = .968.$

This can be thought of as applying a set of weights to the conditional probabilities of producing a perfect part. The weights (which sum to 1) are .85 and .15. The conditional probabilities are the .98 and .90. The weighted result, .968, naturally lies between them. [*Note:* There may be many ways to divide possibilities into mutually exclusive alternatives. If you can find *any* way of doing it so the event of interest is the "or" of mutually exclusive events, the formulas work.]

1.4 Compound Probabilities 23

Example 1.4.4. In Problem 1.2.14, what is the probability that team C will win one of the first two games?

P(win one of first two) $= P$(win first, lose second) $+ P$(lose first, win second)
$= P$(win first) $\times P$(lose second | win first)
$+ P$(lose first) $\times P$(win second | lose first)
$= .50 \times .30 + .50 \times .40$
$= .35.$

Example 1.4.5. In Example 1.3.8, what is the chance that the next application of the drug will cure the disease?

P(cure) $= P$(rate $= .1) \times P$(cure | rate $= .1)$
$+ P$(rate $= .2) \times P$(cure | rate $= .2)$
$+ \cdots + P$(rate $= .9) \times P$(cure | rate $= .9)$
$= .01 \times .1 + .01 \times .2 + .02 \times .3$
$+ .05 \times .4 + .10 \times .5 + .20 \times .6$
$+ .33 \times .7 + .20 \times .8 + .08 \times .9$
$= .662.$

Example 1.4.6. In a certain city at a certain time we have these distributions:

Political affiliation	G	R
Probability	p_G	p_R

[G stands for *Good Government Party*; R stands for *Rid-the-Rascals Party*.]

If G:

Opinion of mayor	Favor	Oppose
Probability	p_{GF}	p_{GO}

If R:

Opinion of mayor	Favor	Oppose
Probability	p_{RF}	p_{RO}

If we now choose at random someone who belongs to one or the other of these parties, then

$$P(\text{favor mayor}) = p_G \times p_{GF} + p_R \times p_{RF}.$$

Example 1.4.7. A statistician has three urns (washed peanut-butter jars) in which there are colored balls. The balls are identical except for their colors.

- Urn 1: 3 green balls, 6 yellow balls, 1 white ball
- Urn 2: 4 black balls, 3 white balls, 2 green balls, 1 orange ball
- Urn 3: 6 checkered balls, 2 green balls, 2 chartreuse balls

An urn is first chosen according to this probability distribution:

Urn	1	2	3
Probability	.5	.3	.2

[I picked these probabilities arbitrarily; no theory applies.] Next a ball is chosen blindly from the chosen urn; that is, each ball in the urn has the same chance of being chosen. Then

$$P(\text{green}) = P(\text{urn 1}) \times P(\text{green} \mid \text{urn 1})$$
$$+ P(\text{urn 2}) \times P(\text{green} \mid \text{urn 2})$$
$$+ P(\text{urn 3}) \times P(\text{green} \mid \text{urn 3})$$
$$= .5 \times .3 + .3 \times .2 + .2 \times .2 = .25,$$

$$P(\text{yellow}) = P(\text{urn 1}) \times P(\text{yellow} \mid \text{urn 1})$$
$$+ P(\text{urn 2}) \times P(\text{yellow} \mid \text{urn 2})$$
$$+ P(\text{urn 3}) \times P(\text{yellow} \mid \text{urn 3})$$
$$= .5 \times .6 + .3 \times 0 + .2 \times 0 = .30,$$

$$P(\text{white}) = .5 \times .1 + .3 \times .3 + .2 \times 0 = .14.$$

The formal expression for compound probability is

$$P(B \mid H) = \sum_i P(A_i \mid H) \times P(B \mid A_i, H).$$

It holds *only* if the A's are both exhaustive and mutually exclusive. In Example 1.4.4, the A's were the outcomes of the first game, and the B was winning one game out of two. In Example 1.4.5, the A's were the values of the cure rate, and the B was curing the next patient. In Example 1.4.6, the A's were the political affiliations, and B was favoring the mayor. And in Example 1.4.7, the A's were the urns, and B was one particular color. There you have to apply the formula seven times, with B taking on each color in turn, in order to obtain the complete probability distribution. [*Note:* If you aren't familiar with the *summation sign*, $\sum$, see Appendix A1.]

PROBLEM SET 1.4

In Problems 1 through 4 use this age distribution:

H:	Age to nearest year	0–4	5–9	10 and over
	Probability	.7	.2	.1

1. What is $P(0\text{–}9 \mid H)$?
2. What is $P(5 \text{ and over} \mid H)$?
3. What is $P(3\text{–}5 \mid H)$?
4. What is $P(0\text{–}4 \text{ or } 10 \text{ and over} \mid H)$?

In Problems 5 and 6, if a fair coin is flipped 7 times, the probability distribution for the total number of heads obtained is:

H:	Number heads	0	1	2	3	4	5	6	7
	Probability	$\frac{1}{128}$	$\frac{7}{128}$	$\frac{21}{128}$	$\frac{35}{128}$	$\frac{35}{128}$	$\frac{21}{128}$	$\frac{7}{128}$	$\frac{1}{128}$

Figure 1.4.1 shows the shape of this distribution; it's characteristic of many we'll meet.

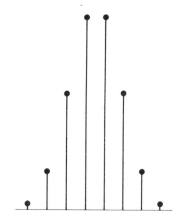

Fig. 1.4.1

5. What is $P(\text{total number of heads is an even number} \mid H)$?
6. What is $P(2 < \text{total number of heads} < 5 \mid H)$?
7. Use the data of Example 1.4.1. What is the probability of getting an even number of lemmings in a litter?
8. In Example 1.4.5 use the compound probability formula to show that $P(\text{noncure}) = .338$.
9. In Example 1.4.7 calculate the probabilities of drawing black, orange, checkered, and chartreuse balls. Check that the probabilities sum to 1.

10. Cooper is a superb tennis player. In the finals of the tournament tomorrow he will play either Jones or Smith—whoever wins their semifinal match. I estimate the probability that Jones will beat Smith as .6. I estimate that if Cooper plays Jones, Cooper's chance of winning is .65. I estimate that if Cooper plays Smith, Cooper's chance of winning is .80. What do I think is Cooper's chance of winning the tournament?

11. If a trachodon is covered by sediment soon after his death, there is about 1 chance in 10,000 that the skeleton will eventually be found by a paleontologist. If it is not covered, the bones will inevitably be dispersed by scavengers. If 1 dead trachodon in 100 gets covered quickly by sediment, what are the chances that any particular trachodon will ever be found?

12. We are going to use *stratified sampling* in Example 1.4.6. We know that $p_G = .60$ and $p_R = .40$. We do some polling and estimate that $p_{GF} = .70$ and $p_{RF} = .50$. What is our estimate of the proportion of people favoring the mayor?

ANSWERS TO PROBLEM SET 1.4

1. .9
2. .3
3. You can't tell from this data.
4. .8
5. $P(0) + P(2) + P(4) + P(6) = \frac{64}{128} = \frac{1}{2}$
6. $P(3) + P(4) = \frac{70}{128}$
7. .55
8. $.01 \times .9 + .01 \times .8 + .02 \times .7 + .05 \times .6 + .10 \times .5 + .20 \times .4 + .33 \times .3 + .20 \times .2 + .08 \times .1 = .338$
9. $P(\text{black}) = .5 \times 0 + .3 \times .4 + .2 \times 0 = .12$
   $P(\text{orange}) = .5 \times 0 + .3 \times .1 + .2 \times 0 = .03$
   $P(\text{checkered}) = .5 \times 0 + .3 \times 0 + .2 \times .6 = .12$
   $P(\text{chartreuse}) = .5 \times 0 + .3 \times 0 + .2 \times .2 = .04$
   Then $.25 + .30 + .14 + .12 + .03 + .12 + .04 = 1.00$.
10. $.6 \times .65 + .4 \times .80 = .39 + .32 = .71$
11. $.01 \times .0001 + .99 \times 0 = .000001 = 1$ in a million
12. $.60 \times .70 + .40 \times .50 = .62$

1.5 AVERAGES AND EXPECTED VALUES

Probability distributions are not just mathematical objects to play with; they are helps in explaining characteristics of the real world. We must be sensitive to every opportunity for relating a probability distribution, which is an abstract concept, to the sample counts and observations we actually see. Success in statistical analysis depends on continually juxtaposing the real world and the conceptual world. We use concepts to guide our understanding of the real world; we use the real world to steer us to better concepts and increased future understanding.

1.5 Averages and Expected Values

In this section I will discuss only distributions whose possible values are numbers. These numbers, moreover, must have a definite scale of measurement underlying them. Sometimes we convert a *ranking* of preferences like

[greatly dislike, somewhat dislike, am neutral, somewhat like, greatly like]

into $[-2, -1, 0, 1, 2]$ or $[1, 2, 3, 4, 5]$; but unless 1 and 2 are in some sense as far apart as 3 and 4 or 4 and 5, we still have just a ranking, not a scaling. Psychologists meet this problem all the time. They have techniques for trying to achieve *equal-appearing intervals* so that numerical methods can be applied. The scale of intelligibility of translations in Example 1.3.6 exhibited this. If you just have ranks, different methods are necessary. There will be some discussion of these in Chapter 7.

When we have made some measurements or counts, the most frequently calculated summary value, or *statistic*, is the *sample average*:

$$\text{sample average} = \frac{\text{total of observations}}{\text{number of observations}}.$$

If we call our observations the x's, we write

$$\bar{x} = \text{Ave}(\mathbf{x}) = \frac{\sum x}{n},$$

where $\mathbf{x}$ is the list of observations and n is their number. An average has the merit of involving only the two most obvious characteristics of a sample: the total and the number of observations. What does it mean, and when is it really useful? [We know it is *calculated* very often.]

The question can be most easily answered by forming an abstract concept to parallel our observed average. While I originally wrote

$$\bar{x} = \frac{\sum x}{n},$$

we can calculate it in a different, but equivalent, way. Suppose our observations are

$$\mathbf{x} = 0, 5, 4, 6, 4, 1, 3, 6, 6, 4, 2, 8, 0, 4, 6, 2, 2, 1, 7, 4.$$

Then

$$\bar{x} = \frac{0+5+4+6+4+1+3+6+6+4+2+8+0+4+6+2+2+1+7+4}{20}$$
$$= \frac{75}{20} = 3.75.$$

However, it is natural to count how many times each different value occurs (see

28   Probability, the Language of Uncertainty   1.5

Table 1.5.1

Value	Frequency	
0	||	(2)
1	||	(2)
2	|||	(3)
3	|	(1)
4	⌢⊓⊤	(5)
5	|	(1)
6	||||	(4)
7	|	(1)
8	|	(1)

Table 1.5.1). Then

$$\bar{x} = \frac{0 \times 2 + 1 \times 2 + 2 \times 3 + 3 \times 1 + 4 \times 5 + 5 \times 1 + 6 \times 4 + 7 \times 1 + 8 \times 1}{20}$$

$$= \frac{75}{20} = 3.75,$$

as before. Each frequency divided by n is the *relative frequency;* we can write

$$\bar{x} = \sum x \left(\frac{f_x}{n}\right),$$

where f_x means the frequency of the value x.

Now probabilities are certainly not the same thing as relative frequencies, but we often estimate them that way. This suggests that a worth-while *abstract* quantity is

$$\sum x P(x),$$

where $P(x)$ takes the place of f_x/n. We call this the *expected value* of x, and write

$$E(X) = \sum x P(x).$$

In this book E always means *expected value.* It is an operator like the $\sum$ sign. Whereas $\sum x$ means "go add up all the x's," $E(X)$ means "go to the probability distribution of the variable X and add up all the products of x by $P(x)$."

Example 1.5.1. In Problem 1.4.5, X was the number of heads in 7 tosses of a fair coin. It had the distribution:

x	0	1	2	3	4	5	6	7
$P(x)$	$\frac{1}{128}$	$\frac{7}{128}$	$\frac{21}{128}$	$\frac{35}{128}$	$\frac{35}{128}$	$\frac{21}{128}$	$\frac{7}{128}$	$\frac{1}{128}$

Therefore

$$E(X) = 0 \times \tfrac{1}{128} + 1 \times \tfrac{7}{128} + 2 \times \tfrac{21}{128} + 3 \times \tfrac{35}{128} + 4 \times \tfrac{35}{128} + 5 \times \tfrac{21}{128}$$
$$+ 6 \times \tfrac{7}{128} + 7 \times \tfrac{1}{128} = \tfrac{448}{128} = 3\tfrac{1}{2}.$$

Note that $3\tfrac{1}{2}$ is not a possible value for the variable X. It is, however, in the region of highest probabilities—right in the center, since the distribution is symmetrical.

Example 1.5.2. The distribution of lemmings/litter from Example 1.4.1 is:

x	1	2	3	4	5	6
$P(x)$	.05	.10	.20	.35	.20	.10

Hence

$$E(X) = 1 \times .05 + 2 \times .10 + 3 \times .20 + 4 \times .35 + 5 \times .20 + 6 \times .10$$
$$= .05 + .20 + .60 + 1.40 + 1.00 + .60 = 3.85.$$

Again the expected value is not a possible value of the variable, just as the average family in the United States does not really have 2.37 children. It is, however, in the part of the distribution where the largest probabilities lie.

Example 1.5.3. The distribution of cloudiness from Example 1.3.1 is:

z	0	1	2	3	4	5	6	7	8	9	10
$P(z)$	.21	.11	.07	.02	.01	.01	.01	.02	.03	.09	.42

Therefore

$$E(Z) = 0 \times .21 + 1 \times .11 + 2 \times .07 + 3 \times .02 + 4 \times .01 + 5 \times .01$$
$$+ 6 \times .01 + 7 \times .02 + 8 \times .03 + 9 \times .09 + 10 \times .42$$
$$= 5.75.$$

This expected value is right in the middle of the distribution, but here the highest probabilities are at the *ends*. It's likely to be very cloudy or very clear, but never half-bad.

Example 1.5.4. The distribution of S = number of heads in two throws of a fair coin is:

s	0	1	2
$P(s)$	$\tfrac{1}{4}$	$\tfrac{1}{2}$	$\tfrac{1}{4}$

Therefore

$$E(S) = 0 \times \tfrac{1}{4} + 1 \times \tfrac{1}{2} + 2 \times \tfrac{1}{4} = 1.$$

At last the expected value happens to coincide with a possible value of the variable!

What is the burden of all this? We have set up parallels between our models of nature and the things we actually see, as shown in Table 1.5.2.

Table 1.5.2

Model	Observation
Common possible values	
Probability	Relative frequency
Expected value (also called *population mean*)	Sample average (also called *sample mean*)

In most distributions you meet in statistics, the sample average gets closer and closer to the population mean as the number of observations gets larger and larger. A good bit of the work in this book involves estimating how close together they actually are. But this is of consequence mainly when the probability is massed in the center of the distribution, and the distribution is reasonably symmetric.

When a distribution has the general appearance of Fig. 1.5.1, the expected value marks the center, and the average provides a way of estimating it. The average is then a good single-number descriptor of the distribution.

When a distribution is massed in the center but is not symmetric (Figs. 1.5.2 and 1.5.3), the expected value is offset from the center and is not a very good single-number descriptor. In fact, no single number describes a distribution like this very well. Since we aren't so interested then in locating the expected value, the average has less importance.

And when a distribution is *U-shaped* (Fig. 1.5.4), *J-shaped* (Fig. 1.5.5), or *multimodal* (more than one hump) (Fig. 1.5.6), it's very difficult to characterize.

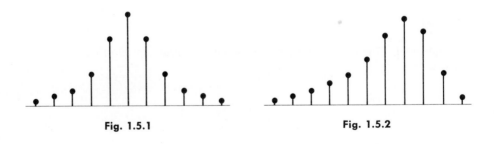

Fig. 1.5.1   Fig. 1.5.2

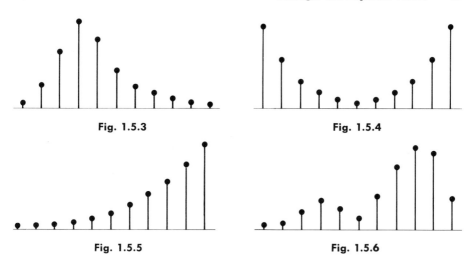

Fig. 1.5.3

Fig. 1.5.4

Fig. 1.5.5

Fig. 1.5.6

The expected value then is usually of little importance; consequently the average is not important either.

I will end with three miscellaneous remarks:

1) It is true that the sample average often estimates the expected value. [Alternatively, the *sample mean* estimates the *population mean*.] You can go the other way, too. If you know the expected value, you can use it to estimate, for planning purposes, what your sample average will be.

2) Even though a distribution is massed in the center and is symmetric, and even though the expected value marks the center, the sample average may not be the most *efficient* way of estimating the location of the center. The median, the middle sample value, is often better. This problem will be discussed at length in Chapter 7.

3) Never write Exp (X) or exp (X) for *expected value*. This symbol has a preempted meaning in mathematics, namely, e^X; it is used in place of e^x when X is a complicated expression. [If you don't know about e, see Appendix A5.] You do sometimes see a curly ε in place of E, and physicists often write $<X>$ rather than $E(X)$.

PROBLEM SET 1.5

1. Here is a distribution:

x	1	2	3	4	5
$P(x)$	.1	.2	.4	.2	.1

What is $E(X)$?

2. Here is a distribution:

y	0	100	200	300	400
P(y)	.01	.18	.50	.29	.02

What is $E(Y)$?

3. Here is a distribution:

z	−3	−2	−1	0	1	2
P(z)	.1	.2	.3	.2	.1	.1

What is $E(Z)$?

4. Here is a distribution:

u	0	1	2	3
P(u)	p	$p(1-p)$	$p(1-p)^2$	$(1-p)^3$

What is $E(U)$? What is $E(U)$ when $p = .6$? Do this directly from the probabilities and from the answer to the first part of this problem.

5. Find the expected value of p in Example 1.3.8. Compare with the discussion of Example 1.4.5.

6. Compute $E(S^2)$ in Example 1.5.4.

7. How does $E(S^2)$ from Problem 6 compare with $[E(S)]^2$ in Example 1.5.4?

8. Here is a distribution:

q	1	2	3	4	5	6	7	8	9
P(q)	.01	.16	.30	.16	.02	.07	.20	.07	.01

Calculate $E(Q)$, and comment on its meaning for this distribution.

9. Here is a distribution:

s	0	1	2	3	4	5	6	7	8	9
P(s)	.01	.03	.07	.16	.27	.19	.12	.07	.05	.03

What is $E(S)$?

Suppose we now group the values of S as (0–4) and (5–9) and assign the middle value to represent the group. We then have this distribution:

t	2	7
P(t)	.54	.46

What is $E(T)$?

10. What values could you assign to the groups in Problem 9 so that the grouped distribution would have the same expected value as the ungrouped?

ANSWERS TO PROBLEM SET 1.5

1. 3.0
4. $E(U) = 3 - 6p + 4p^2 - p^3$
   $E(U \mid p = .6) = 3 - 6(.6) + 4(.6)^2 - (.6)^3 = .624$

   Or:
u	0	1	2	3
$P(u)$	.6	.24	.096	.064

   $E(U) = .24 + .192 + .192 = .624$

6. If
s	0	1	2
$P(s)$	$\frac{1}{4}$	$\frac{1}{2}$	$\frac{1}{4}$

   then
s^2	0	1	4
$P(s^2)$	$\frac{1}{4}$	$\frac{1}{2}$	$\frac{1}{4}$

   and $E(S^2) = 0(\frac{1}{4}) + 1(\frac{1}{2}) + 4(\frac{1}{4}) = \frac{3}{2}$.

8. $E(Q) = 4.44$. This distribution is *bimodal* (two humps), so there is not too much significance in the expected value.

10. How do we make them equal? Let's write $E(S)$ as

$$.54\left[0 \times \frac{.01}{.54} + 1 \times \frac{.03}{.54} + 2 \times \frac{.07}{.54} + 3 \times \frac{.16}{.54} + 4 \times \frac{.27}{.54}\right]$$
$$+ .46\left[5 \times \frac{.19}{.46} + 6 \times \frac{.12}{.46} + 7 \times \frac{.07}{.46} + 8 \times \frac{.05}{.46} + 9 \times \frac{.03}{.46}\right]$$

The required values are those in brackets: 3.21 and 6.15. They are the *expected values* of the two five-long *conditional* distributions.

CHAPTER 2

More about Probability

So far we have talked mostly about distributions of one variable. Often, though, we meet two or more variables simultaneously. There is the problem of whether or not to record both age and sex—that makes two variables. The latitude, longitude, and altitude of a straying airplane make three. The temperature, pressure, catalyst concentration, and product yield in a chemical plant make four. Even more importantly, the simultaneous distribution of parameters and observations will arise in our study of inference. In this chapter you will learn ways of handling such multivariate distributions and ways of condensing them so the effects of a reduced number of variables can be seen.

2.1 JOINT AND MARGINAL DISTRIBUTIONS

The simultaneous distribution of several variables is called their *joint distribution*. Sometimes it is qualified as *bivariate, trivariate,* or *multivariate* to show that there are two, three, or many variables involved. A distribution of only one variable is naturally called *univariate*. The best way to begin our study of joint distributions is to look at some examples.

Example 2.1.1. According to the 1950 census, if you picked an American worker at random, this distribution would apply:

	White-collar	Manual/Service	Farm
Male	.22	.39	.11
Female	.15	.12	.01

Here the two variables are (a) type of occupation, and (b) sex. The probabilities of the combinations are given at the intersections of the coordinates: P(male, manual/service) is .39. Note that the probabilities still add up to 1. They have to, regardless of the number of variables. We could have considered this as a six-category univariate distribution; but since the structure is so evident and important, it is natural to make it bivariate.

Example 2.1.2. A fair coin is tossed three times. The results have this trivariate distribution:

First	Second	Third	Probability
T	T	T	$\frac{1}{8}$
T	T	H	$\frac{1}{8}$
T	H	T	$\frac{1}{8}$
T	H	H	$\frac{1}{8}$
H	T	T	$\frac{1}{8}$
H	T	H	$\frac{1}{8}$
H	H	T	$\frac{1}{8}$
H	H	H	$\frac{1}{8}$

The three variables are the results for each of the tosses.

Example 2.1.3. A small yacht is lost in the fog off the New Jersey coast. The Coast Guard takes bearings on the yacht's radio transmissions. An evaluation of the bearings gives this probability distribution for the yacht's position (to the nearest 10″):

		Longitude (W)			
		73° 01′ 20″	73° 01′ 10″	73° 01′ 00″	73° 00′ 50″
Latitude (N)	39° 48′ 50″	.01	.01	.01	.01
	39° 48′ 40″	.01	.15	.20	.02
	39° 48′ 30″	.01	.20	.25	.03
	39° 48′ 20″	.01	.02	.03	.03

The two variables are the latitude and the longitude. The yacht is certainly at one definite position; this distribution shows our uncertainty about it.

Example 2.1.4. Two parents, heterozygous in a given gene, are mated. The genetic composition of the child has the distribution:

		Gene from father	
		A	a
Gene from mother	A	$\frac{1}{4}$	$\frac{1}{4}$
	a	$\frac{1}{4}$	$\frac{1}{4}$

where A is the dominant allele and a the recessive one.

Example 2.1.5. A contract bridge hand is often evaluated by its high-card point count. An ace counts 4; a king, 3; a queen, 2; and a jack, 1. The joint distribution of aces and points is:

		Aces				
		0	1	2	3	4
Points	0–7	.195	.091	.000	.000	.000
	8–12	.101	.263	.082	.002	.000
	13–20	.008	.085	.130	.035	.001
	21–40	.000	.000	.002	.004	.001

That is, the probability is 26.3% that you will be dealt a bridge hand with 8–12 points and one ace. [Note that in many distributions like this the values are rounded from more exact values. Some of the .000's above are real 0's, while some of them are numbers which are closer to .000 than to .001. Rounded probabilities do not always add up to exactly 1, but they should be reasonably close.]

A joint distribution in full panoply is a splendid thing. When we finally get to statistics, you will find that our information about the results of experiments is often expressed in multivariate probability distributions. Occasionally we do stop at this point. In Example 2.1.3 (the lost yacht) the logical thing to do (after calculating the probability at more grid points) is to draw the contour lines of the joint distribution. That is, we join the points having equal probabilities; the result looks a little like a topographic map. Such a contour map conveys a tremendous amount of information quickly and directly. With plotting devices attached to computers, it can be obtained with little time and expense. The list of computed values from which the map is prepared is nowhere as readily comprehended. Contouring applies not only to geographical problems, but to any problem in which there are two numerical variables. With a little imagination you can sometimes depict surfaces of equal probability in the trivariate case.

Many times, however, we want to condense the joint distribution. There may be considerable interest, for example, in analyzing the variation of just one of the variables. Now in a joint distribution all the *combinations* of variate values are mutually exclusive. We can group the combinations into whatever categories are meaningful; the probability of each category is merely the sum of the appropriate probabilities. All we have to watch out for is that each combination must go into only one category. The result will be a new set of probabilities which add up to 1, in other words, a new probability distribution.

Example 2.1.1 continued. We might reasonably want to know the distribution of the types of jobs without caring whether a man or a woman held it. Let's then add the corresponding values in the two rows of the table:

$$.22 + .15 = P(\text{male, white-collar}) + P(\text{female, white-collar})$$
$$= P(\text{white-collar}),$$
$$.39 + .12 = P(\text{manual/service}),$$
$$.11 + .01 = P(\text{farm}).$$

Or, we might want to know what proportion of the work force was male and what proportion was female:

$$.22 + .39 + .11 = P(\text{male, white-collar})$$
$$+ P(\text{male, manual/service}) + P(\text{male, farm})$$
$$= P(\text{male}),$$
$$.15 + .12 + .01 = P(\text{female}).$$

We have obtained two new probability distributions. They are called the *marginal distributions* since their values are often written in the margins of the joint table:

	White-collar	Manual/Service	Farm		
Male	.22	.39	.11	.72	Marginal distribution
Female	.15	.12	.01	.28	of sex of worker
	.37	.51	.12		
	Marginal distribution of type of job				

You can verify that the probabilities in each marginal distribution add up to 1. Each distribution expresses the distribution of a single characteristic; the other variable has been properly eliminated. That is,

.37 is the probability that an American worker in 1950 had a white-collar job,

.72 is the probability that an American worker in 1950 was male.

Offhand, no other groupings of the joint possibilities seem meaningful.

Example 2.1.2 continued. When we toss a coin three times, we are probably more interested in the distribution of the total number of heads that come up than in the actual sequence of results. Let's group together those sequences which have 0, 1, 2, or 3 heads:

0		1		2		3	
TTT	$\frac{1}{8}$	TTH	$\frac{1}{8}$	THH	$\frac{1}{8}$	HHH	$\frac{1}{8}$
		THT	$\frac{1}{8}$	HTH	$\frac{1}{8}$		
		HTT	$\frac{1}{8}$	HHT	$\frac{1}{8}$		
	$\frac{1}{8}$		$\frac{3}{8}$		$\frac{3}{8}$		$\frac{1}{8}$

Our new probability distribution is:

Total number of heads	0	1	2	3
Probability	$\frac{1}{8}$	$\frac{3}{8}$	$\frac{3}{8}$	$\frac{1}{8}$

The probabilities obviously sum to 1. The new distribution this time is *not* a marginal distribution. We could have gotten the marginal distribution of, say, the result of the first toss. There are four combinations in which the first toss is T and four in which it is H. This leads to:

First toss	T	H
Probability	$\frac{1}{2}$	$\frac{1}{2}$

Example 2.1.3 continued. Since such a small area is involved ($1' = 1$ nautical mile, approximately) there seems to be no reason to want information separately about latitude and longitude.

Example 2.1.4 continued. There are two condensed distributions of interest here:

i) By combining

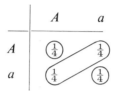

we get

Genotype	AA	Aa	aa
Probability	$\frac{1}{4}$	$\frac{1}{2}$	$\frac{1}{4}$

ii) By combining

	A	a
A	$\frac{1}{4}$	$\frac{1}{4}$
a	$\frac{1}{4}$	$\frac{1}{4}$

we get

Phenotype	Show dominant	Show recessive
Probability	$\frac{3}{4}$	$\frac{1}{4}$

Neither of these groupings forms the usual marginal distribution. Knowledge of the subject matter has to dictate your procedures.

Example 2.1.5 continued. We can get the marginal distribution of aces by adding corresponding values from each row:

$$P(0 \text{ aces}) = P(0 \text{ aces}, 0\text{--}7 \text{ points}) + P(0 \text{ aces}, 8\text{--}12 \text{ points})$$
$$+ P(0 \text{ aces}, 13\text{--}20 \text{ points}) + P(0 \text{ aces}, 21\text{--}40 \text{ points})$$
$$= .195 + .101 + .008 + .000 = .304.$$

The complete distribution is:

Number of aces	0	1	2	3	4
Probability	.304	.439	.214	.041	.002
[More exactly	.30382	.43885	.21349	.04120	.00264]

[I've added a line showing the more accurate values in case you bridge players want them. The toll exacted by rounding was not too bad.]

Similarly, the marginal distribution of the number of points is:

Number of points	0–7	8–12	13–20	21–40
Probability	.286	.448	.259	.007

PROBLEM SET 2.1

1. In Example 2.1.1, what is P(male, farm)?
2. In Example 2.1.2, what is P(third toss $= H$)?
3. In Example 2.1.2, what is P(second and third tosses both $= T$)?
4. In Example 2.1.3, what is P(yacht is at 39° 48′ 40″ or farther north)?
5. In Example 2.1.4, what is P(father will pass on a)?
6. In Example 2.1.5, what is P(2 aces *and* 8–12 points)?
7. In Example 2.1.5, what is P(0–12 points)?
8. In Problem 1.2.4, what is P(rural)?
9. Archeopteryxes lay 0, 1, 2, or 3 eggs. The fertility of an egg depends on the number of eggs laid. The joint distribution of eggs laid and eggs hatched is:

		\multicolumn{4}{c}{Eggs laid}
		0	1	2	3
	0	.10	.02	.02	.03
Eggs	1		.18	.08	.03
hatched	2			.40	.04
	3				.10

What is the marginal distribution of the number of eggs laid? What is the marginal distribution of the number of eggs hatched? Exhibit these distributions in the margins of the table.

10. Here is a table of decisive results in master chess play*:

Opening	Won by white	Won by black
Sicilian Defense	115	77
Ruy Lopez	58	52
King's Indian Defense	55	42
French Defense	43	20
Nimzo-Indian Defense	37	15
Queen's Gambit Declined	29	21
Benoni Defense	14	13
Caro-Kann Defense	14	13
Reti Opening	20	5
Gruenfeld Defense	16	9
Queen's Pawn Opening	16	9
English Opening	12	7

These are counts, not probabilities. Form the marginal *totals* instead.

11. Use the data from Problem 10. With what relative frequency did *white* win when one of the two Indian defenses was employed?

12. This is data on the number of injuries in mining accidents in the United States in 1956:

	Fatal	Nonfatal
Coal mining	448	19,816
Quarrying	50	3,754
Other mining	106	6,323

How many fatal injuries occurred altogether in mining accidents?

13. Use the data of Problem 12. How many injuries occurred in coal-mining accidents?

14. Here is the joint distribution of X and Y:

		X		
		1	2	3
Y	1	.2	.3	.1
	2	.1	.2	.1

Find the marginal distributions of X and Y.

---

* Reproduced with permission from E. Rubin (1966), The openings in decisive chess encounters, *American Statistician* **20** (December), 33.

15. Use the data of Problem 14. What is the distribution of the sum of X and Y?
16. Use the data of Problem 14. What is the distribution of $X - Y$?
17. The joint distribution of T and U is:

		\multicolumn{2}{c}{U}	
		e_1	e_2
T	d_1	$3w$	$2w$
	d_2	$5w$	$7w$

What is w?

18. The joint distribution of X, Y, and Z is:

x	y	z	$P(x, y, z)$
0	1	2	.2
0	1	5	.2
0	2	2	.1
1	1	2	.1
1	2	2	.1
1	2	5	.3

[I use a small x to mean a value of the variable X, etc.] What is the marginal distribution of (X, Y)? [You can have marginal distributions which are themselves multivariate.]

19. Use the data of Problem 18. What is the marginal distribution of Z?
20. Use the data of Problem 18. What is the distribution of $X + Y - Z$?

ANSWERS TO PROBLEM SET 2.1

2. $\frac{1}{2}$ 4. .42   6. .082   7. .734

9.
Eggs laid	0	1	2	3
Probability	.10	.20	.50	.30

Eggs hatched	0	1	2	3
Probability	.17	.29	.44	.10

10. Column totals: 429, 283
    Row totals: 192, 110, 97, 63, 52, 50, 27, 27, 25, 25, 25, 19
12. 604

14. X:
x	1	2	3
$P(x)$	.3	.5	.2

Y:
y	1	2
$P(y)$	.6	.4

[Note that I use a small x to represent a value of the variable X.]

15. We group like this:

		X		
		1	2	3
Y	1	(.2)²	.3 ³	.1 ⁴
	2	.1	.2	(.1)⁵

Thus the distribution of $X + Y$, call it W, is:

w	2	3	4	5
P(w)	.2	.4	.3	.1

17. The probabilities must sum to 1. $w = \frac{1}{17}$.

18. We sum out Z and get:

		X	
		0	1
Y	1	.4	.1
	2	.1	.4

20. Let $U = X + Y - Z$. Then:

u	−4	−2	−1	0	1
P(u)	.2	.3	.2	.2	.1

2.2 CONDITIONAL DISTRIBUTIONS

There is another way to reduce the number of variables in a joint distribution: set some of the variables equal to *fixed* values, and look at the variation in the others. We obtain *conditional distributions*.

Example 2.2.1. Consider the distribution of the labor force in Example 2.1.1:

	White-collar	Manual/Service	Farm
Male	.22	.39	.11
Female	.15	.12	.01

Suppose we are not interested in the whole distribution but just in the occupations held by men. This means that only the first row (*male*) concerns us; the second row might as well be blanked out. The first row has entries

$$.22 \quad .39 \quad .11$$

but they add up to .72, not 1.00; therefore they do not form a probability distribution. How can we get a probability distribution from them? Simple, just divide each one by .72, their total. This gives

$$\frac{.22}{.72} = .31, \quad \frac{.39}{.72} = .54, \quad \frac{.11}{.72} = .15.$$

These numbers add up to 1 and form the *conditional probability distribution*

$$P(\text{job} \mid \text{male}, H) = [.31, .54, .15].$$

The *condition* is that the worker be male.

The procedure of dividing a set of numbers by their sum so that the new numbers add up to 1 is called *normalization*. In our statistical work we will constantly be deriving sets of numbers which are *proportional* to probabilities; we then have to normalize them to get the actual probabilities. Normalization will have to become almost a reflex. This does *not* mean, however, that you can take any old set of numbers, divide by their sum, and come up with probabilities. There has to be meaning behind them.

The name *conditional* for the distribution seems obvious. We set one of the variables (sex) to a fixed value (male) and then looked at the variation in the other (type of job). Recall the definition of conditional probability,

$$P(B \mid A, H) = \frac{P(A, B \mid H)}{P(A \mid H)},$$

and put A = male and B = white-collar. The .22 is $P(A, B \mid H)$, the probability that a worker is both male and white-collar; this is the entry from the joint distribution. The .72 is $P(A \mid H)$, the marginal probability that the worker is male. The quotient, .31, is $P(B \mid A, H)$, the probability that the worker is white-collar, given that he is male.

Next, let B = manual/service, and use the same formula; finally let B = farm. We then have obtained the conditional probabilities of the various jobs under the condition that the worker is male. Why do these probabilities sum to 1? Because the B's are an exhaustive list of the jobs. You can see that we got the marginal probability (the denominator) by adding up all those joint probabilities that ever appeared in the numerator.

There are five conditional distributions associated with this joint distribution. I have shown you the first; the others are

$$P(\mathbf{job} \mid \text{female}, H) = [.54, .43, .04],$$

$(.15/.28 = .54; .12/.28 = .43; .01/.28 = .04.$ There is a little rounding trouble here.)

$$P(\mathbf{sex} \mid \text{white-collar}, H) = [.60, .40],$$

$(.22/.37 = .60; .15/.37 = .40)$

$$P(\mathbf{sex} \mid \text{manual/service}, H) = [.76, .24],$$
$$P(\mathbf{sex} \mid \text{farm}, H) = [.92, .08].$$

Example 2.2.2. Look back at the joint distribution of aces and points in Example 2.1.5. If you knew that a player had 2 aces, what could you infer about his point count? We no longer survey the entire joint distribution; we focus our attention on the column labeled "2 aces." We first calculate the marginal probability of 2 aces:

$$.000 + .082 + .130 + .002 = .214.$$

Then we normalize the column:

$$P(\text{0–7 points} \mid \text{2 aces}) = \frac{.000}{.214} = .000,$$

$$P(\text{8–12 points} \mid \text{2 aces}) = \frac{.082}{.214} = .383,$$

$$P(\text{13–20 points} \mid \text{2 aces}) = \frac{.130}{.214} = .608,$$

$$P(\text{21–40 points} \mid \text{2 aces}) = \frac{.002}{.214} = .009.$$

This is the conditional distribution and is perfectly correct on the given information. Of course, you almost always have more information and that should be taken into account, if possible. If you're a player trying to infer something about another player's hand, you have your own cards to look at and the bidding to consider. If you're a kibitzer, you've probably seen all the hands already!

Let me remind you that the way we have been calculating conditional distributions is just the way we get empirical percentages from counts we have taken: we add them up, and divide each one by the total. Many of the probabilities we use do come from counts; but often the counts are first smoothed or otherwise modified to take into account other background knowledge.

PROBLEM SET 2.2

1. In one of Mendel's experiments with peas, he encountered this joint distribution:

		Color	
		Yellow	Green
Texture	Smooth	$\frac{9}{16}$	$\frac{3}{16}$
	Wrinkled	$\frac{3}{16}$	$\frac{1}{16}$

   What is the conditional distribution of color for wrinkled peas?
2. Use the data of Problem 1 to find $P(\textbf{texture} \mid \text{green})$.
3. In Problem 2.1.9, what is $P(2 \text{ eggs hatched} \mid 3 \text{ laid})$?
4. In Problem 2.1.9, what is $P(3 \text{ eggs laid} \mid 2 \text{ hatched})$?
5. If we have one set of marginal probabilities and the corresponding conditional distributions, we can construct (or reconstruct) the joint distribution. Remember that the formula for conditional probability can be written

$$P(A, B \mid H) = P(A \mid H) \times P(B \mid A, H).$$

   Given

   $P(\text{worker is male}) = .72,$
   $P(\text{worker is female}) = .28,$
   $P(\textbf{job} \mid \text{male}) = [.31, .54, .15],$
   $P(\textbf{job} \mid \text{female}) = [.53, .43, .04],$

   what is the joint distribution?
6. Given that $P(0 \text{ aces})$ in a bridge hand is .304 and that $P(\textbf{points} \mid 0 \text{ aces}) = [.642, .332, .026, .000]$, reconstruct the first column of the table in Example 2.1.5.
7. In Problem 2.1.10, what fraction of the time was the Sicilian Defense used?
8. In Problem 2.1.10, what fraction of the time did *black* win when the English Opening was used?
9. In Problem 2.1.14, what is $P(\textbf{X} \mid Y = 2)$?
10. The weather in Shangri-la is never bad. Indeed, the usual terms are "good," "superb," and "extraordinary." Let X = weather today, and Y = weather tomorrow. The conditional probabilities are:

	Y		
	G	S	E
$P(Y \mid X = G)$	.00	.50	.50
$P(Y \mid X = S)$	.50	.25	.25
$P(Y \mid X = E)$	.25	.50	.25

If the marginal distribution of X is

x	G	S	E
$P(x)$	.28	.40	.32

what is the marginal distribution of Y?

11. Here is a table of accidental injuries, in 1000's:

		Type of accident			
		Motor vehicle	Other vehicle	Home	Work
Severity	Death	40	21	44	20
	Permanent disability	160	82	150	80
	Temporary disability	1400	2050	4160	1900

If you had to be accidentally injured, which type of accident would you prefer?

12. In a certain school, students were classified by entrance examinations into three categories:

25%: above average,   50%: average,   25%: below average.

The Statistics Department kept a record of the performance of students in its elementary course:

		Grade				
		A	B	C	D	E
Entrance rating	Above average	20%	50%	20%	10%	0%
	Average	10%	20%	35%	25%	10%
	Below average	0%	10%	15%	50%	25%

If a random student received a D, what is the probability that he had been rated *above average*?

13. Use the data of Problem 12. What is the probability that a randomly chosen student would get a C?

14. A mortality table for a short-lived species of mammal says:

of those alive at year 0, 50% die before reaching year 1;
of those alive at year 1, 10% die before reaching year 2;
of those alive at year 2, 30% die before reaching year 3;
of those alive at year 3, 80% die before reaching year 4;
of those alive at year 4, 100% die before reaching year 5.

What is the distribution of age at death? That is, what are the probabilities of dying between 0 and 1? between 1 and 2? between 2 and 3? between 3 and 4? and between 4 and 5? There is no need to worry about the endpoints of the intervals here.

15. In Problem 14, what are the life expectancies at various ages?

ANSWERS TO PROBLEM SET 2.2

1. | Color | Yellow | Green |
    |---|---|---|
    | Probability | .75 | .25 |

3. $\dfrac{.04}{.20} = .20$

5. $\begin{matrix} .72 \times [.31, .54, .15] \\ .28 \times [.53, .43, .04] \end{matrix} = \begin{Bmatrix} [.22, .39, .11] \\ [.15, .12, .01] \end{Bmatrix}$

7. $\frac{192}{712} = 27\%$

10. We reconstruct the joint distribution:

				Y	
			G	S	E
	.28	G	.00	.14	.14
X	.40	S	.20	.10	.10
	.32	E	.08	.16	.08

    Strangely enough, the marginal distribution for Y is:

y	G	S	E
P(y)	.28	.40	.32

    This is an example of a *stationary* process.

11. We have to calculate the conditional frequencies, conditional on the type of accident; that is:

	MV	OV	H	W
D	.025	.010	.010	.010
PD	.100	.038	.034	.040
TD	.875	.952	.956	.950

    Home accidents seem to be slightly preferable; there are relatively fewer deaths and permanent disabilities, though all the last three are preferable to motor vehicle accidents.

12. You first have to decide what this tabulation is. It appears to be the distributions of grades conditional on the entrance classifications. Thus we need to reconstruct the joint distribution:

	A	B	C	D	E
Above average	.050	.125	.050	.025	.000
Average	.050	.100	.175	.125	.050
Below average	.000	.025	.0375	.125	.0625

You had better check that these numbers sum to 1! Now, given that the student got a D,

$$P(\text{above average} \mid D) = \frac{.025}{.275} = .09.$$

14. The figures given are a somewhat different set of conditional probabilities, the probability that the animal will die during the coming year given that he was alive at the beginning of the year. You have to proceed by steps:

Year	Fraction alive at start of year	Died during year
0	1.000	$1.000 \times .5 = .500$
1	.500	$.500 \times .1 = .050$
2	.450	$.450 \times .3 = .135$
3	.315	$.315 \times .8 = .252$
4	.063	$.063 \times 1.0 = .063$
5	.000	

Thus:

Age at death	0–1	1–2	2–3	3–4	4–5
Probability	.500	.050	.135	.252	.063

Be sure to graph this one.

15. You have to work up from the bottom here. In the absence of more detailed information, I *define* that animals dying during a particular year average half a year of life that year. I will use E_k to mean the life expectancy of those animals who are alive at the beginning of year k. $E_4 = .5$ by my definition; we have to find E_3, E_2, E_1, and E_0. Now consider E_3. The .8 of them who will die during the year 3–4 have an expectancy of .5 years (by definition); the .2 of them who will survive year 3–4 have an expectancy of 1 (that year) + the expectancy of those alive at the beginning of year 4. Thus

$$E_3 = .5 \times .8 + (1 + .5) \times .2 = .7.$$

And we keep working back that way:

$$E_2 = .5 \times .3 + (1 + .7) \times .7 = 1.34,$$
$$E_1 = .5 \times .1 + (1 + 1.34) \times .9 = 2.156,$$
$$E_0 = .5 \times .5 + (1 + 2.156) \times .5 = 1.828.$$

The little hump at E_1 may look strange, but a similar hump shows up in the U.S. population for 1959–1961.

2.3 INDEPENDENCE

In a very tenuous sense all things on earth are related. As you lean over to turn the page in this book, the axis of rotation of the earth is shifted and the telescope at Mount Palomar goes a bit off focus. Such infinitesimal and farfetched relationships are not of importance to us. We make our approach from the point of view of information: two things are independent if knowledge of one does not in practice affect our knowledge of the other, under the stated conditions, of course. The meanderings of the stock market tell us nothing about the growth of a cancer. Our estimate of the chance of rain in the Sahara is not changed by quality control decisions in the Ford plant.

What is the role of independence in statistics? There are two aspects. First, scientists—and people in general—often hope that things are *not* independent. We wish to derive relationships between things. We want formulas for prediction, for evaluation, for condensation of data, for simplification of our theories. If everything were independent, science would not be possible. But we know that the height and weight of a man are related. The heights of brothers and sisters are related. The gross national product and tax receipts are related. The brittleness and carbon content of a steel are related. We welcome all this because discovery of relationships is one of our goals. It makes life interesting.

The type of relationship or dependence that is *not* so welcome in statistics is dependence between observations. Many statistical methods for analyzing random variations assume that observations are independent. If dependence is allowed for, calculations are generally much more difficult. If dependence is neglected when it actually is there, grossly wrong conclusions can result. I don't want to discourage you. I just hope that every time you start a statistical analysis you ask yourself, "Are my observations independent?" Better, "How can I find out if they are?" It is always *your* worry about when what method is applicable.

There are several definitions for independence between events. The one which parallels our informal definition is

A is independent of B if $P(A \mid B, H) = P(A \mid H)$.

That is, the condition B can be present or not; it makes no difference in our knowledge of A. Independence is obviously a symmetrical relationship, and we simultaneously have $P(B \mid A, H) = P(B \mid H)$. These are relationships between *two events*. Do not confuse this with the alternatives for *one variable* being mutually exclusive.

The big dividend of independence is the relationship

$$P(A, B \mid H) = P(A \mid H)P(B \mid H).$$

That is, the probability of the joint occurrence of A and B is just the product of the individual probabilities. We defined $P(A, B \mid H) = P(A \mid H)P(B \mid A, H)$, and now we use the definition of independence: $P(B \mid A, H) = P(B \mid H)$. The definition extends even farther. When you have any number of independent variables, the probability of their joint values is just the product of the probabilities of their individual (marginal) values.

Example 2.3.1. For a single observation

$$P(A \mid H) = \tfrac{1}{3} \quad \text{and} \quad P(B \mid H) = \tfrac{2}{3}.$$

Five independent observations are made: A, B, A, B, B. Then

$$P(A, B, A, B, B \mid H) = \tfrac{1}{3}\tfrac{2}{3}\tfrac{1}{3}\tfrac{2}{3}\tfrac{2}{3} = \tfrac{8}{243}.$$

This is the probability of *this particular sequence*. Do not confuse it with the probability that *some* sequence of two A's and three B's will arise. There are ten such sequences, so that probability is ten times as large.

Example 2.3.2. For one observation from the Poisson Distribution (see POIS in the Gallery of Distributions)

$$P(x \mid m) = \frac{e^{-m}m^x}{x!} \quad \text{for} \quad x = 0, 1, 2, \ldots$$

Suppose we get the four observations a, b, c, d. Then

$$P(a, b, c, d \mid m) = \frac{e^{-m}m^a}{a!} \frac{e^{-m}m^b}{b!} \frac{e^{-m}m^c}{c!} \frac{e^{-m}m^d}{d!} = \frac{e^{-4m}m^{(a+b+c+d)}}{a!b!c!d!}.$$

Two *variables* are independent if knowledge of the value of one of them does not affect knowledge of the other. We can phrase this in various ways:

a) "All the conditional distributions are the same."

b) "The conditional distributions are the same as the marginal distribution."

c) "The joint probabilities are the products of the marginal probabilities."

Example 2.3.3. Here are two bivariate distributions to illustrate all these properties:

Independent

		X		
		1	2	
Y	1	.18	.12	.3
	2	.42	.28	.7
		.6	.4	

$P(X \mid Y = 1) = [.6, .4]$
$P(X \mid Y = 2) = [.6, .4]$
$P(X) = [.6, .4]$
$P(Y \mid X = 1) = [.3, .7]$
$P(Y \mid X = 2) = [.3, .7]$
$P(Y) = [.3, .7]$

$.18 = .6 \times .3, \quad .12 = .4 \times .3$
$.42 = .6 \times .7, \quad .28 = .4 \times .7$

Dependent

		X		
		1	2	
Y	1	.2	.1	.3
	2	.4	.3	.7
		.6	.4	

$P(X \mid Y = 1) = [.67, .33]$
$P(X \mid Y = 2) = [.57, .43]$
$P(X) = [.60, .40]$
$P(Y \mid X = 1) = [.33, .67]$
$P(Y \mid X = 2) = [.25, .75]$
$P(Y) = [.30, .70]$

$.2 \neq .6 \times .3, \quad .1 \neq .4 \times .3$
$.4 \neq .6 \times .7, \quad .3 \neq .4 \times .7$

Sometimes a few of the requirements for independence between variables are satisfied, but not all. Then the variables are definitely dependent.

Example 2.3.4. The variables X and Y with the joint distribution

		X			
		0	1	2	
Y	1	.18	.10	.02	.30
	2	.42	.00	.28	.70
		.60	.10	.30	

are dependent even though

$$.60 \times .30 = .18 \quad \text{and} \quad .60 \times .70 = .42.$$

Here are some trivariate distributions. You can verify the stated degree of independence.

Example 2.3.5. X, Y, and Z are completely independent.

x	y	z	$P(x, y, z)$
0	0	0	.028
0	0	1	.042
0	1	0	.012
0	1	1	.018
1	0	0	.252
1	0	1	.378
1	1	0	.108
1	1	1	.162

Example 2.3.6. X is independent of Y and Z. Y and Z are dependent.

x	y	z	$P(x, y, z)$
0	0	0	.03
0	0	1	.00
0	1	0	.21
0	1	1	.06
1	0	0	.07
1	0	1	.00
1	1	0	.49
1	1	1	.14

Example 2.3.7. X, Y, and Z are all interdependent.

x	y	z	$P(x, y, z)$
0	0	0	.1
0	0	1	.1
0	1	0	.1
0	1	1	.3
1	0	0	.1
1	0	1	.1
1	1	0	.0
1	1	1	.2

PROBLEM SET 2.3

1. Here is a portion of the joint distribution of U and V:

		|	V		
		|	1	2	3
U	1	|	.12	.24	.04
	2	|			

   Complete it so the variables are independent.

2. Use the data from Problem 1. Complete the joint distribution so the variables are dependent. [There are a great many answers; give one.]
3. Are the variables in Problem 2.2.1 dependent or independent?
4. Demonstrate that the variables in Problem 2.1.14 are dependent.
5. Are the variables in Example 2.1.1 dependent or independent?
6. Are the variables in Example 2.1.2 dependent or independent?
7. Are the variables in Example 2.1.3 dependent or independent?
8. Are the variables in Example 2.1.4 dependent or independent?
9. Are the variables in Example 2.1.5 dependent or independent?
10. Prove that if $P(B \mid A, H) = P(B \mid H)$, then $P(A \mid B, H) = P(A \mid H)$.
11. One red ball and one green ball are in an urn. You draw them out blindfolded, one at a time. [The first ball drawn is not put back before the second is drawn.] Let

$$X = \text{identity of first ball drawn},$$
$$Y = \text{identity of second ball drawn}.$$

   What is the joint distribution of X and Y? Are they independent?

12. There are 9 "Good Government" supporters and 5 "Rid-the-Rascals" supporters in an apartment house. Two *different* persons are picked in succession at random. What is the joint distribution of the results of the selections? What is the distribution of the number of *GG* supporters selected?
13. Confirm the stated degree of dependence in Example 2.3.6.
14. What is the probability of the sequence

    heads, heads, tails, heads

   if $P(\text{heads}) = .52$ and $P(\text{tails}) = .48$ for a single trial, and the observations are independent?

15. In Problem 14, how many sequences are there with three heads and one tail in some order?
16. In Example 2.3.2, suppose $m = 2, a = 0, b = 5, c = 1, d = 1$; what is $P(0, 5, 1, 1 \mid 2)$?

17. When electronic computers first came out, people quickly realized that the computer needed a supply of random numbers. [Section 2.4 describes their use.] Nowadays random numbers are generated by a process which puts out numbers satisfying all the tests one can think of, yet which is repeatable for checking purposes. They are called *pseudo random numbers*. But at the beginning people experimented with actual physical random sources. They took *noise tubes*, vacuum tubes whose voltage fluctuated rapidly about some mean value, and measured the voltage whenever they needed a number. If the measured voltage was below the mean, the observation was called a 0; if it was above the mean, the observation was called a 1. From these random *bits*, random digits or any other desired random observation could be constructed.

Now the worry always was that the mean would drift as the tube aged, so that instead of getting a 50–50 split of 0's and 1's, they might get 55–45 or 40–60. As some insurance against this they added two outputs using *modulo 2 addition*:

$$0 + 0 = 0,$$
$$0 + 1 = 1,$$
$$1 + 0 = 1,$$
$$1 + 1 = 0.$$

Suppose that the mean of the noise tube has drifted so 0's are arising with probability .4 and 1's are arising with probability .6. If the two outputs are *independent* of each other (they are if they are far enough apart in time) their joint distribution is:

		Second		
		0	1	
First	0	.16	.24	.4
	1	.24	.36	.6
		.4	.6	

that is, the product of the marginal values. What is the distribution of the modulo 2 sum?

18. If you wanted to be extra careful, you could add, modulo 2, *two* of the pairs from Problem 17. What would be the joint distribution of the two pairs (if they were independent), and what would be the distribution of their modulo 2 sum?

ANSWERS TO PROBLEM SET 2.3

1. For independence, the marginal distribution of V must equal the conditional distributions of V given U. So $P(V) = P(V \mid U = 1) = [.3, .6, .1]$. Since $P(U = 1) = .4$, $P(U = 2) = .6$, and the bottom row of the table is $.6 \times [.3, .6, .1] = [.18, .36, .06]$.

56   More about Probability

3. Independent; the joint probabilities equal the products of the marginal probabilities.
5. Dependent   7. Dependent   8. Independent
10. We are given $P(B \mid A, H) = P(B \mid H)$. Therefore

$$P(A, B \mid H) = P(A \mid H)P(B \mid A, H) = P(A \mid H)P(B \mid H).$$

By definition

$$P(A \mid B, H) = \frac{P(A, B \mid H)}{P(B \mid H)}.$$

Substitute to get

$$P(A \mid B, H) = \frac{P(A \mid H)P(B \mid H)}{P(B \mid H)} = P(A \mid H).$$

12. We get the joint distribution by counting cases:

		First	
		GG	RR
Second	GG	$\frac{72}{182}$	$\frac{45}{182}$
	RR	$\frac{45}{182}$	$\frac{20}{182}$

There are 9 ways of picking a GG first and then 8 ways of picking a GG second. There are 5 ways of picking an RR first and then 9 ways of picking a GG second; etc. Then normalize by dividing by the sum, 182.

Number GG	0	1	2
Probability	$\frac{20}{182}$	$\frac{90}{182}$	$\frac{72}{182}$

14. $.52 \times .52 \times .48 \times .52 = .0675$ 16. $\dfrac{e^{-8} 2^7}{0!\,5!\,1!\,1!} = .000358$

The probabilities of observed sequences of observations are generally quite small.

17. $P(0) = P(0 + 0) + P(1 + 1) = .52, \quad P(1) = P(0 + 1) + P(1 + 0) = .48$

Note how much closer we are to a 50–50 split.

2.4 SAMPLING EXPERIMENTS

Throughout this book we will have occasion to test ideas and formulas by doing sampling experiments; that is, we'll try the process out and see what happens. But we don't have the time or money to do real experiments, so we *simulate* them. When each member of a class provides a simulated result, a good idea of the variability in the results can be obtained.

These sampling experiments use tables of *random numbers*. Such numbers have been generated specifically for this purpose. Appendix C1 is a table labeled "flat

2.4 Sampling Experiments

random digits." I am reserving the term *flat random* to describe values coming from a distribution in which each value has the same probability. The term *random sample* means only that the values have arisen according to *some* specified probability distribution, not necessarily the flat random one.

In the flat random table the digits are grouped in pairs purely for convenience in reading. In general, you start at some position in the table, read off as many digits as you need, and cross off the digits you have used to avoid reuse. The next time you need some digits, you start where you left off. The students in a class must not all start at the same point, for their samples would then be identical. You wouldn't see the variability in the results. Therefore the instructor must specify starting points. The direction of reading from the table—up, down, left, right—can also be varied.

Example 2.4.1. Generate a sample of ten observations from the distribution:

Color	Red	White	Blue
Probability	.09	.42	.49

These probabilities are given to hundredths; thus we'll use pairs of flat random digits, since there are 100 different pairs. We assign 9 of the pairs to mean *red*, 42 of them to mean *white*, and 49 of them to mean *blue*. A systematic assignment produces the least confusion:

$$
\begin{aligned}
&00\text{–}08 \quad \text{mean} \quad \text{red,} \\
&09\text{–}50 \quad \text{mean} \quad \text{white,} \\
&51\text{–}99 \quad \text{mean} \quad \text{blue.}
\end{aligned}
$$

Suppose I start at row 00, column 00 in the flat random table and read across: 76 60 40 33 45 33 80 40 85 91. This means our simulated observations are blue, blue, white, white, white, white, blue, white, blue, blue. [76 is in the 51–99 group; therefore it means blue; etc.] The sample can be tabulated as:

Color	Red	White	Blue
Frequency	0	5	5

We didn't happen to get any *red*'s, but that's life.

Example 2.4.2. I want ten more observations from the distribution of Example 2.4.1. I start from where I left off in the flat random digit table:

Digit pairs: 43 65 50 92 08 86 79 61 39 89
Observations: white, blue, white, blue, red, blue, blue, blue, white, blue

58   More about Probability

Example 2.4.3. I want ten observations from the distribution:

x	0	1	2	3	4	5	6	7	8	9
$P(x)$	.1	.1	.1	.1	.1	.1	.1	.1	.1	.1

This is just the flat random distribution of digits; I can read them straight from the table, starting from where I stopped before: 6 9 1 4 4 5 2 4 3 3. Do not be disturbed that one of each did not appear. There is immense variability in random processes, and one of the aims of this book is to make you realize it.

Example 2.4.4. I want ten observations simulating the rolls of a fair die. Since the possibilities are 1 to 6, I'll use single digits and discard all 0's, 7's, 8's, and 9's.

Flat random digits:	3 4 5 9 7 9 6 7 0 3 9 8 8 5 3 4 9 3 6
Observations:	3 4 5 – – – 6 – – 3 – – – 5 3 4 – 3 6

In this case I couldn't say beforehand how many digits I would need. Now, in case you're worried about the 1's and 2's, I'll take ten more observations:

Flat random digits:	0 1 5 3 7 4 9 1 6 7 0 7 8 4 7 9 0 1 2 1
Observations:	– 1 5 3 – 4 – 1 6 – – – – 4 – – – 1 2 1

Example 2.4.5. I want a sample of five workers from the distribution of Example 2.1.1. This is a bivariate distribution, but no essential change is needed in our procedure. The only thing to remember is that a *pair* of values will come up each time. I assign flat random digit pairs:

00–21	male, white-collar,
22–60	male, manual/service,
61–71	male, farm,
72–86	female, white-collar,
87–98	female, manual/service,
99	female, farm.

When I start at the next pair of digits in the table, I get:

Flat random digit pair	Worker
88	female, manual/service
17	male, white-collar
31	male, manual/service
26	male, manual/service
34	male, manual/service

This last example brings up an important problem. We got our probability distribution from a large count. Now five workers is a drop in the bucket among millions; but what if our basic population were small? Then, assuming we didn't want anyone to pop up twice, the population would be changing at each observation and so would the probabilities.

Example 2.4.6. We have just ten workers:

	White-collar	Manual/Service	Farm
Male	2	4	1
Female	2	1	0

We certainly can assign them to single digits:

　　　　　　0–1　male, white-collar,
　　　　　　2–5　male, manual/service,
　　　　　　 6 　male, farm,
　　　　　　7–8　female, white-collar,
　　　　　　 9 　female, manual/service.

If we specify any sampling that forbids duplication, we must ignore a digit when it comes up the second time. Suppose I first want to line up the ten workers to distribute Christmas bonuses:

　　Flat random digits:　　6 9 3 5 8 1 1 5 0 9 3 1 2 9 ⋯ 4
　　　　　　　Worker:　　6 9 3 5 8 1 – – 0 – – – 2 – – 4

and with nine already chosen, number 7 must be last. Or if I wanted to choose just three:

　　　　　　Flat random digits:　　8 5 8 4
　　　　　　　　　　　Worker:　　8 5 – 4

and I would have (female, white-collar), (male, manual/service), and (male, manual/service).

These two types of sampling sometimes bear the names *sampling with replacement* (or *infinite population*) and *sampling without replacement* (or *finite population*). If you have a population, small or large, and you *replace* a drawn member before making the next draw, the population is the same on each draw, and so are the probabilities. It is as if you had an infinite supply to draw from. When you don't replace the drawn member, the population changes on every draw, and its finite size will eventually make a difference. Whether or not it makes a practical difference in your calculations depends on how large your sample is compared with the size of the population.

I have been emphasizing the use of random numbers in experimental sampling because it is a method particularly well suited to an electronic computer. Sometimes you do need a million samples, and a computer is the only feasible way. On the other hand, there are less strenuous exercises where you can get by with actual physical objects. Statisticians used to—perhaps still do—pull little lotto pieces from those notorious urns. And the simplest way to sample a few bridge hands is to deal them out, if you shuffle well. But for generality remember random numbers.

PROBLEM SET 2.4

1. Draw ten observations of cloudiness from the distribution of Example 1.3.1.
2. A machine turns out 85% good parts; draw a sample of 15 of them.
3. Draw a sample of five lemming litters (Example 1.4.1).
4. Simulate the tennis tournament of Problem 1.4.10.
5. The letters of English have the probabilities:

A	B	C	D	E	F	G	H	I	J	K	L	M
.078	.013	.029	.041	.131	.029	.014	.058	.068	.002	.004	.036	.026

N	O	P	Q	R	S	T	U	V	W	X	Y	Z
.073	.082	.022	.001	.066	.065	.090	.028	.010	.015	.003	.015	.001

   Draw a sample of 60 letters from this distribution. The result is only a peculiar pseudo English because we have not provided for the dependence between successive letters. We also need spaces.

6. Assume that the length of words in English is governed by this probability distribution:

1	2	3	4	5	6	7	8	9	10	11	12
.02	.18	.28	.13	.10	.08	.07	.06	.04	.02	.01	.01

   Separate your sample from Problem 5 into "words."

7. Select a sample of six without replacement from a group of nine Good Government supporters and five Rid-the-Rascals supporters.
8. Pick a sample of ten from the distribution of Example 2.3.6.
9. Sample 20 bridge hands with 13–20 points and make a frequency count of the number of aces observed. (See Example 2.1.5.)

CHAPTER **3**

Accumulating Evidence

The analysis of data has many aspects. Theories are proposed, experiments planned and executed, data recorded, gross blunders excised, suspicious readings double-checked. In this chapter we will see how the cleaned-up observations reflect the underlying structure of the subject at hand. We will be considering alternative models for that structure, models which may be fundamentally very different, or which may differ only in the value of some numerical parameter. From the observed data we will infer the relative probabilities of the alternatives. These probabilities express our current knowledge and provide the basis for any decisions that must be made. The basis for our inferences is Bayes' Theorem, a simple extension of the ideas you have just learned. The fundamental position which we give this theorem leads to the adjective Bayesian in the book's title. In order to exhibit the ideas as clearly as possible, we will stick with discrete distributions in this chapter; we get to continuous distributions in the next one.

3.1 BAYES' THEOREM

This chapter deals with how we keep track of evidence for and against alternatives. Although an alternative could be something as grand and glorious as a new theory that casts doubt on the work of centuries, more often it is just a suggestion for a numerical value to fit into an equation or probability model. We will almost always have a set of suggested values, and will want to rate them somehow. This is the bread-and-butter problem of statistics. At least it is a procedure that we use over and over again in our iterative attempts to understand phenomena.

Our procedure is this: When we make our analysis we have at hand a probability distribution for the alternatives which expresses our accumulation of knowledge to that point. This is called the *prior distribution*, the one that comes *before* the observation or experiment. Then we make an observation intended to tell us something about the relative merits of our alternatives. On the basis of this information we modify the prior probability distribution and obtain a new one, the *posterior distribution*, the one that comes *after* the observation. If we then have another experiment to make, this posterior distribution becomes the prior distribution for the next step in the analysis. The words *prior* and *posterior* are relative words: they refer to the states before and after *any* observation.

How, then, do we modify a prior distribution on the basis of an observation? First let's look at the extreme cases; they are often a great help in clarifying ideas:

i) Suppose that the observation has the same probability under each of the alternatives we are considering. Then it is quite obvious that this observation tells us absolutely nothing that can help us distinguish between the alternatives. The posterior distribution should be exactly the same as the prior distribution.

ii) Suppose that the observation is so unique that it is possible only under one of the alternatives and impossible under all the others. In this case we would immediately acclaim that alternative as certain, and dub the rest impossible. Our posterior distribution here should have a probability of 1 for the correct alternative and 0 for all the rest.

So far so good. What about intermediate cases? If the observation is twice as probable on one alternative as on a second, we have to increase the former's relative size in forming the posterior distribution. You can't tell from this general discussion whether you should weight the two as 2 : 1, 4 : 1, or even 200 : 1. But it can be shown that we ought to do what most of us would intuitively consider the right thing: apply weights to the elements of the prior distribution which are directly proportional to the probabilities of the observation under the various alternatives. That is, of course, consistent with the treatment of the two extreme cases: in the first case the weights were all equal; in the second case one was 1 and the others were 0.

Example 3.1.1. We are trying to decide between alternatives A_1 and A_2. Suppose that the prior distribution—our current knowledge about them—is

$$\begin{array}{cc} A_1 & A_2 \\ .7 & .3 \end{array}.$$

[These numbers are out of the air; you have missed no explanation.] To get more evidence about them we perform an experiment. Suppose there are three possible outcomes to the experiment, t, u, and v, and the conditional probabilities of observing them are

$$\begin{array}{ll} P(t \mid A_1) = .8, & P(t \mid A_2) = .1, \\ P(u \mid A_1) = .2, & P(u \mid A_2) = .2, \\ P(v \mid A_1) = .0, & P(v \mid A_2) = .7, \\ \hline 1.0, & 1.0. \end{array}$$

[These numbers are also out of the air.] Do you remember that if we have a marginal distribution and the associated conditional distributions we can construct the joint distribution? We do this as follows:

$$\begin{array}{cc} A_1 & A_2 \\ \underline{.7} & \underline{.3} \\ \times & \times \\ \begin{bmatrix} .8 \\ .2 \\ .0 \end{bmatrix} & \begin{bmatrix} .1 \\ .2 \\ .7 \end{bmatrix} \end{array}$$

	A_1	A_2	
t	$.7 \times .8$	$.3 \times .1$	.59
u	$.7 \times .2$	$.3 \times .2$	.20
v	$.7 \times .0$	$.3 \times .7$	.21

$$\quad .7 \quad\quad .3$$

Now we perform the experiment:

1) If we observe t, we are interested only in the first row of the joint probability distribution. For t happened, not u or v. We want to find $P(A_1 \mid t)$ and $P(A_2 \mid t)$. By now you can do it blindfolded:

$$P(A_1 \mid t) = .56/.59, \qquad P(A_2 \mid t) = .03/.59.$$

We can rephrase the process as follows:

Weight the prior probabilities, .7 and .3 (our marginal distribution), by the conditional probabilities of the observation, .8 and .1, and get $.7 \times .8$ and $.3 \times .1$

(probabilities from the joint distribution). Now normalize:

$$\frac{.7 \times .8}{.7 \times .8 + .3 \times .1} \quad \text{and} \quad \frac{.3 \times .1}{.7 \times .8 + .3 \times .1},$$

and get the posterior probabilities, .949 and .051 (a conditional distribution).

Let's lay it out better:

Alternative	Prior		$P(t \mid \text{alt})$		Joint		Posterior
A_1	.7	×	.8	=	.56	→	.949
A_2	.3	×	.1	=	.03	→	.051
	1.0		no special sum		.59		1.000

This is an intermediate case; certainty is not achieved, but the probability distribution of the two hypotheses does change.

2) If we observe u instead of t, we have:

Alternative	Prior		$P(u \mid \text{alt})$		Joint		Posterior
A_1	.7	×	.2	=	.14	→	.7
A_2	.3	×	.2	=	.06	→	.3
	1.0		no special sum		.20		1.0

This is one of the extreme cases. Since $P(u \mid A_1) = P(u \mid A_2)$, observing u does not help us determine whether A_1 or A_2 is true. The posterior distribution is just the same as the prior distribution. Remember the well-known scientific principle that you cannot distinguish between competing hypotheses unless they predict different observable results.

3) If we observe v instead of t or u, we get:

Alternative	Prior		$P(v \mid \text{alt})$		Joint		Posterior
A_1	.7	×	.0	=	.00	→	.0
A_2	.3	×	.7	=	.21	→	1.0
	1.0		no special sum		.21		1.0

This is the other extreme case. It is not too interesting from a statistical point of view, since certainty is achieved with a bang. Any clod can draw the right conclusion here! In most practical cases the probabilities of alternatives may diminish rapidly but never actually go to 0. When they get very small you have to decide whether it's worth the bother of carrying them around anymore.

Since we've been doing just what Bayes' Theorem tells us to do, let me state it formally and then go on to a few examples.

Bayes' Theorem

If

i) the A_i's are a set of *mutually exclusive* and *exhaustive* alternatives,

ii) $P_0(A_i)$ is the prior probability of A_i,

iii) X is the observation, and

iv) $P(X | A_i)$ is the probability of the observation given that A_i is true,

then the posterior probability of A_i is

$$P(A_i | X) = \frac{P_0(A_i)P(X | A_i)}{\sum_j P_0(A_j)P(X | A_j)}.$$

I've used the new symbol $P_0(A_i)$ to mean $P(A_i | \text{all information before } X)$. Note that both A_i and X appear first on one side of the vertical bar and then on the other. The denominator is just the sum of all the numerators that will appear; it could be written simply $P(X)$, but since it is calculated by adding up all the numerators, I write it this way.

The proof is simple. The probability of the joint event (A_i, X) can be written in two ways:

$$P(A_i, X | H) = P(X | H)P(A_i | X, H) = P(A_i | H)P(X | A_i, H),$$

where H stands for all the information before X. By division,

$$P(A_i | X, H) = \frac{P(A_i | H)P(X | A_i, H)}{P(X | H)},$$

and by the rules of compound probability,

$$P(X | H) = \sum_j P(A_j | H)P(X | A_j, H).$$

Lastly, I write $P_0(A_i) = P(A_i | H)$, and drop the H's.

Example 3.1.2. I am caught up in one of those popular mass screenings for a relatively rare disease. How should I feel if I'm told I'm OK? How should I feel if I'm advised to see my doctor for a further check?

Suppose that the disease strikes about 1 person in 5000, and that there are no patent symptoms until a late stage. A reasonable description of my knowledge would be the prior probabilities:

Not diseased	Diseased
.9998	.0002

Upon inquiry I learn that this screening test has about 5% false positives and about 2% false negatives. That is, if I don't have the disease, the test will falsely say I do about 5% of the time; if I do have the disease, the test will falsely say I don't about 2% of the time. This gives me the conditional probabilities

$$P(\text{positive} \mid \text{not diseased}) = .05,$$
$$P(\text{negative} \mid \text{not diseased}) = .95,$$
$$P(\text{positive} \mid \text{diseased}) = .98,$$
$$P(\text{negative} \mid \text{diseased}) = .02.$$

i) If the test is *positive*, we have:

Alternative	Prior	$P(\text{positive} \mid \text{alt})$		Joint		Posterior
Not diseased	.9998 ×	.05	=	.049990	→	.9961
Diseased	.0002 ×	.98	=	.000196	→	.0039
	1.0000	no special sum		.050186		1.0000

This says that, even if my test is *positive*, the chances are only 1 in 256 that I do have the disease. This doesn't mean that I shouldn't check with my doctor; the consequences of the disease could be quite terrible. It just means that I should go with a calm mind.

ii) If the test is *negative*, we have:

Alternative	Prior	$P(\text{negative} \mid \text{alt})$		Joint		Posterior
Not diseased	.9998 ×	.95	=	.949810	→	.999996
Diseased	.0002 ×	.02	=	.000004	→	.000004
	1.0000	no special sum		.949814		1.000000

If the test says I'm OK, my estimate of the chance of my having the disease drops from 1 in 5000 to 1 in 237,000.

Example 3.1.3. Refer to the urns in Example 1.4.7. If I am told that the selection process has been carried through and that a green ball has been selected, what guess can I make about the urn that was chosen?

Alternative	Prior	$P(\text{green} \mid \text{alt})$		Joint		Posterior
Urn 1	.5 ×	.3	=	.15	→	.60
Urn 2	.3 ×	.2	=	.06	→	.24
Urn 3	.2 ×	.2	=	.04	→	.16
	1.0	no special sum		.25		1.00

Example 3.1.4. What if a white ball had been selected instead?

Alternative	Prior	P(white | alt)		Joint		Posterior
Urn 1	.5 ×	.1	=	.05	→	.36
Urn 2	.3 ×	.3	=	.09	→	.64
Urn 3	.2 ×	.0	=	.00	→	.00
	1.0	no special sum		.14		1.00

The selection of a green ball increased the edge for urn 1. The selection of a white one instead eliminated urn 3 and reversed the order of urn 1 and urn 2. As it happens, any other color would have determined the proper urn uniquely.

Example 3.1.5 (*Continuation of Example 3.1.4*). Suppose a white ball had been selected the first time. It was put back into its urn, the balls mixed, and another ball selected from the *same* urn. This, too, turned out to be white. What is the posterior distribution of the urns?

We can start out with the posterior distribution of Example 3.1.4. The conditional probabilities of white for the new choice are just the same as before.

Alternative	Prior	P(white | alt)		Joint		Posterior
Urn 1	.36 ×	.1	=	.036	→	.16
Urn 2	.64 ×	.3	=	.192	→	.84
	1.00			.228		1.00

I've omitted urn 3, since once a probability becomes 0, an application of Bayes' Theorem can never pull it up again.

Example 3.1.6 (*Continuation of Example 3.1.4*). The previous example discussed sampling *with replacement*. Suppose now that, after getting a white ball the first time, we do not put it back, but immediately take another ball from the same urn. It too is white. What are the posterior probabilities of the urns?

We again start with the posterior distribution of Example 3.1.4. With one white ball gone from an urn,

$$P(\text{white} \mid 1) = 0, \quad P(\text{white} \mid 2) = \tfrac{2}{9}, \quad P(\text{white} \mid 3) = 0.$$

Alternative	Prior	P(white | alt)		Joint		Posterior
Urn 1	.36 ×	0	=	.000	→	.000
Urn 2	.64 ×	$\tfrac{2}{9}$	=	.142	→	1.000
	1.00			.142		1.000

68 *Accumulating Evidence* 3.1

You could have deduced this from casual examination of the urns' contents: urn 2 is the only one with more than one white ball. But I wanted to show you that steady application of Bayes' Theorem will also lead to the proper conclusion. You need confidence for the road ahead.

PROBLEM SET 3.1

1. A coin, which you are not allowed to examine, is either a fair coin [P(heads) $= .5$] or a coin with two heads. Your initial opinion is P_0(fair) $= .90$. The coin is now flipped, and heads comes up. What is your opinion now?
2. The coin of Problem 1 is flipped again, and again heads comes up. What is your opinion now?
3. You are told that there are either 4, 5, or 6 balls in an urn. The problem is to find out how many; we assume the alternatives are equally probable to start with. The observations go as follows: On any draw from the urn, each ball has the same chance of being drawn. Initially the balls are unmarked, but after each draw you mark the drawn ball with a distinguishing number before you return it to the urn. Thus you may recognize when a once-drawn ball is drawn again. One ball is now drawn. It is labeled No. 1 and put back. What is the posterior distribution for the number of balls in the urn?
4. *Continuation of Problem 3.* A second ball, unmarked, is drawn. It is labeled No. 2 and put back. What is the probability distribution of the number of balls?
5. *Continuation of Problem 4.* A third ball is drawn; it is our No. 2. It is replaced. What is the probability distribution of the number of balls?
6. *Continuation of Problem 5.* A fourth ball is drawn; it is our No. 1. It is replaced. What is the probability distribution of the number of balls?
7. The rock strata A and B are very difficult to distinguish in the field. Through careful laboratory studies it has been determined that the only gross characteristic that might be useful is the presence or absence of a particular brachiopod fossil. In rock exposures of the size usually encountered,

$$P(\text{at least one fossil} \mid A) = .9,$$
$$P(\text{no fossil} \mid A) = .1,$$
$$P(\text{at least one fossil} \mid B) = .2,$$
$$P(\text{no fossil} \mid B) = .8.$$

The only additional information is that type A rock occurs about four times as often as type B rock in the area under study. If the fossil is observed, what is the posterior distribution of rock types?

8. *Continuation of Problem 7.* If a certain geologist always says "This rock is A" when he finds at least one fossil, and always says "This rock is B" when he doesn't find any fossils in these outcrops, what is the probability of his being right?

9. There are 11 urns. Each contains 10 balls which are identical except for color:

  urn 0 has 0 green and 10 red balls,
  urn 1 has 1 green and 9 red balls,
  urn 2 has 2 green and 8 red balls,
  $\vdots$
  urn 10 has 10 green and 0 red balls.

Each urn has probability $\frac{1}{11}$ of being selected. One urn is selected, then one ball drawn from it. The ball is green. What is the posterior probability distribution of the urns?

10. *Continuation of Problem 9.* The chosen ball is put back into its urn, and the balls are mixed well. A ball is now chosen from the *same* urn. What is the probability that it is green?

11. *Continuation of Problem 10.* Generalize the result to an arbitrary number of urns. What happens when that number becomes very large?

12. Chemist *H* and Chemist *Y* are both trying to determine the molecular weight of an important biological compound. The difficulty lies in obtaining enough of the compound for reliable analysis. Theory indicates that the molecular weight will be 212, 213, 214, or 215, but it gives no reason to prefer one value over the other. Chemist *Y* gets his experiment done first and sends the results in for publication. He reports his posterior distribution as

212	213	214	215
.1	.5	.3	.1

Some months later Chemist *H* gets his results. Just as he is about to mail in his manuscript he notices *Y*'s paper in a journal. He naturally decides to expand his paper into a review article summarizing all the evidence. (You get more Brownie points that way!) If *H*'s own results were such that

$$P(\text{results} \mid 212) = .01,$$
$$P(\text{results} \mid 213) = .07,$$
$$P(\text{results} \mid 214) = .01,$$
$$P(\text{results} \mid 215) = .01,$$

what posterior distribution combines all the evidence?

ANSWERS TO PROBLEM SET 3.1

1.

Alternative	Prior		P(heads | alt)		Joint		Posterior
Fair	.9	×	.5	=	.45	→	.82
Two heads	.1	×	1.0	=	.10	→	.18

3.

Alternative	Prior		P(unmarked | alt)		Joint		Posterior
4	$\frac{1}{3}$	×	1	=	$\frac{1}{3}$	→	$\frac{1}{3}$
5	$\frac{1}{3}$	×	1	=	$\frac{1}{3}$	→	$\frac{1}{3}$
6	$\frac{1}{3}$	×	1	=	$\frac{1}{3}$	→	$\frac{1}{3}$

The first ball gives no information.

5. If there are n balls, the probability of getting a specific ball is $1/n$.

Alternative	Prior		P(No. 2 | alt)		Joint		Posterior
4	$\frac{45}{143}$	×	$\frac{1}{4}$	=	$\frac{45}{572}$	→	.386
5	$\frac{48}{143}$	×	$\frac{1}{5}$	=	$\frac{48}{715}$	→	.329
6	$\frac{50}{143}$	×	$\frac{1}{6}$	=	$\frac{50}{858}$	→	.285

7.

Alternative	Prior		P(fossil | alt)		Joint		Posterior
Type A	.8	×	.9	=	.72	→	.947
Type B	.2	×	.2	=	.04	→	.053

9.

Alternative	Prior		P(green | alt)		Joint		Posterior
Urn 0	$\frac{1}{11}$	×	$\frac{0}{10}$	=	$\frac{0}{110}$	→	$\frac{0}{55}$
Urn 1	$\frac{1}{11}$	×	$\frac{1}{10}$	=	$\frac{1}{110}$	→	$\frac{1}{55}$
Urn 2	$\frac{1}{11}$	×	$\frac{2}{10}$	=	$\frac{2}{110}$	→	$\frac{2}{55}$
Urn 3	$\frac{1}{11}$	×	$\frac{3}{10}$	=	$\frac{3}{110}$	→	$\frac{3}{55}$
Urn 4	$\frac{1}{11}$	×	$\frac{4}{10}$	=	$\frac{4}{110}$	→	$\frac{4}{55}$
Urn 5	$\frac{1}{11}$	×	$\frac{5}{10}$	=	$\frac{5}{110}$	→	$\frac{5}{55}$
Urn 6	$\frac{1}{11}$	×	$\frac{6}{10}$	=	$\frac{6}{110}$	→	$\frac{6}{55}$
Urn 7	$\frac{1}{11}$	×	$\frac{7}{10}$	=	$\frac{7}{110}$	→	$\frac{7}{55}$
Urn 8	$\frac{1}{11}$	×	$\frac{8}{10}$	=	$\frac{8}{110}$	→	$\frac{8}{55}$
Urn 9	$\frac{1}{11}$	×	$\frac{9}{10}$	=	$\frac{9}{110}$	→	$\frac{9}{55}$
Urn 10	$\frac{1}{11}$	×	$\frac{10}{10}$	=	$\frac{10}{110}$	→	$\frac{10}{55}$

10. Using the formula for compound probability, we get

$$P(\text{green}) = P(\text{urn } 0)P(\text{green} \mid \text{urn } 0) + \cdots$$
$$= \tfrac{0}{55}\tfrac{0}{10} + \tfrac{1}{55}\tfrac{1}{10} + \tfrac{2}{55}\tfrac{2}{10} + \cdots + \tfrac{10}{55}\tfrac{10}{10}$$
$$= \tfrac{385}{550} = .70.$$

11. *Generalization:* There are $N + 1$ urns, each with prior probability $1/(N + 1)$. Urn j contains j green balls and $N - j$ red balls. A row of the table is:

Alternative	Prior		P(green | alt)		Joint		Posterior
Urn j	$1/(N + 1)$	×	j/N	=	$j/N(N + 1)$	→	$2j/N(N + 1)$

Then

$$P(\text{green}) = \sum \frac{2j}{N(N+1)} \frac{j}{N} = \frac{2N(N+1)(2N+1)}{6N(N+1)N} = \frac{2}{3} + \frac{1}{3N}.$$

This approaches $\frac{2}{3}$ as N gets large.

3.2 MULTIPLE OBSERVATIONS

In the last section you saw how a single observation modifies beliefs. This demonstrates the principle well, and is the way you would proceed if your observations came in very slowly. Again, if your observations were expensive, you might make them one at a time, calculate the posterior probabilities, and stop when the results were sharp enough. But how do you calculate when you have a great many observations? Can the procedure be shortened?

One approach is to consider the entire set of observations as one "superobservation." After all, we can look at a sequence of coin tosses as a sequence of heads or tails. Or, we can focus on sets of, say, 6 tosses, and treat the $2^6 = 64$ different sequences of 6 tosses as our basic set of possibilities for a single observation. Thus 6 observations from one point of view, 1 from another. And I could have said 103 instead of 6. If, in this spirit, we can calculate the probability of a superobservation under each of the alternative hypotheses, we are back again to the problem of 1 observation; we can then do just what we did before.

Example 3.2.1. An automatic machine in a small factory produces metal parts. Each morning the machine is started anew. Most of the time (90% by long records) it settles down to producing about 95% good parts and 5% that must be scrapped. Some mornings, for reasons not discernible by superficial inspection of the machine, it settles into a less favorable mode of operation and produces only about 70% good parts. In either case the bad parts appear to come quite haphazardly, no exceedingly long stretches of bad ones being observed. Naturally, the foreman wants to recognize the trouble, if it is there, as soon as possible so a mechanic can be called to adjust the machine. The machine is not adjusted routinely every morning because this would put it out of production for too long a time. The first dozen parts produced this morning were

$$s, u, s, s, s, s, s, s, s, u, s, u,$$

where $s =$ satisfactory and $u =$ unsatisfactory. What is the probability that the machine is in its good state?

If we *assume* that successive parts are independent (apparent "haphazardness" is no *guarantee*), we can calculate the probability of this sequence by multiplying together the probabilities of each part. Let

$$G = \text{the machine is in good condition,}$$
$$B = \text{the machine is in bad condition.}$$

Then

$$P(\text{seq} \mid G) = .95 \times .05 \times .95 \times .95 \times .95 \times .95 \times .95 \times .95$$
$$\times .95 \times .05 \times .95 \times .05 = .000079,$$

$$P(\text{seq} \mid B) = .70 \times .30 \times .70 \times .70 \times .70 \times .70 \times .70 \times .70$$
$$\times .70 \times .30 \times .70 \times .30 = .001090.$$

Alternative	Prior	$P(\text{seq} \mid \text{alt})$	Joint	Posterior
G	.90 $\times$	.000079	= .0000711 $\to$	.39
B	.10 $\times$	.001090	= .0001090 $\to$	.61
	1.00		.0001801	1.00

The odds now slightly favor the belief that the machine is not working properly. Whether anything is done depends on the relative costs of taking the machine out of production for repairs and accepting a possibly inferior rate of satisfactory parts. I will discuss decision-making in detail in Chapter 8.

Note that the conditional probabilities in this example are both quite small; if the sequence had been longer they would have been even smaller. It's just that there are so many possible sequences that each one has a small probability; yet one of them must occur! The thing that counts is the *relative size* of the two conditional probabilities. Here $P(\text{seq} \mid B)$ was 14 times as large as $P(\text{seq} \mid G)$; this was enough to overwhelm the 9 : 1 ratio of the prior probabilities.

It might have been appropriate to use a step-by-step approach to multiple observations, as we did in the last section. Our machine-shop supervisor would want to recognize trouble as soon as possible, not after a dozen expensive parts had been made. The sequence u, u, u would probably make him holler for help in a hurry. More and more people now realize the great advantages of a step-by-step, or *sequential*, analysis; I will discuss that at length in Chapter 8 also. What I want to show here is that we get absolutely the same results whether we do the analysis sequentially or in one gulp. I implicitly assumed this before, but I'm sure you want to see it demonstrated.

Example 3.2.2. We again have the automatic machine. The relevant probabilities are

$$P_0(G) = .90, \quad P(s \mid G) = .95, \quad P(s \mid B) = .70,$$
$$P_0(B) = .10, \quad P(u \mid G) = .05, \quad P(u \mid B) = .30.$$

a) Suppose the first observation is s:

Alternative	Prior	$P(s \mid \text{alt})$	Joint		Posterior
G	.9 ×	.95	= .855	→	.924
B	.1 ×	.70	= .070	→	.076
	1.0		.925		1.000

b) Suppose the second observation is u. The posterior distribution above is the new prior distribution:

Alternative	Prior	$P(u \mid \text{alt})$	Joint		Posterior
G	.924 ×	.05	= .04620	→	.67
B	.076 ×	.30	= .02280	→	.33
	1.000		.06900		1.00

c) Let's put the analysis of (a) and (b) together:

Alternative	Prior	$P(s \mid \text{alt})$		Joint	Posterior/Prior	$P(u \mid \text{alt})$		Joint		Posterior
G	.9 ×	.95	=	.855 →	.924	× .05	=	.0462	→	.67
B	.1 ×	.70	=	.070 →	.076	× .30	=	.0228	→	.33
	1.0			.925	.1000			.0690		1.00

d) So long as we go on to take a second observation, there certainly is no need to normalize after the first; normalization multiplies both lines by the same factor, so the ratio of the probabilities can't change:

Alternative	Prior	$P(s \mid \text{alt})$		$P(u \mid \text{alt})$		Joint		Posterior
G	.9 ×	.95	×	.05	=	.04275	→	.67
B	.1 ×	.70	×	.30	=	.02100	→	.33
	1.0					.06375		1.00

The only possible difference from our previous answer would be due to rounding off. If you refer back to the discussion of the probability of the sequence of 12 observations, you will recognize .95 × .05 as the $P(s, u \mid G)$ and the .70 × .30 as $P(s, u \mid B)$. It has made absolutely no difference whether we analyzed a sequence of two observations in one step or two. The complexity of the calculation will depend on whether or not they are independent. Note further that

$$P(s, u \mid G) = .95 \times .05 = .05 \times .95 = P(u, s \mid G).$$

So long as we have *independent* observations like this, we can freely interchange the order in which we consider the observations.

In this example, then, we need worry about only the *total number of satisfactory parts* and the *total number of unsatisfactory parts* that the machine has produced. However we reach some stage, the final distribution depends on only

$$P(\text{seq} \mid G) = (.95)^{\#s} (.05)^{\#u}$$

and

$$P(\text{seq} \mid B) = (.70)^{\#s} (.30)^{\#u},$$

where $\#s$ is the total number of satisfactory parts and $\#u$ is the total number of unsatisfactory parts. For any particular sequence we just plug in the observed values for $\#s$ and $\#u$. Suppose our sequence had 100 satisfactory and 23 unsatisfactory parts in some particular order. Then $\#s = 100$ and $\#u = 23$. Hence

$$P(\text{seq} \mid G) = (.95)^{100}(.05)^{23} = 7.06 \times 10^{-33},$$
$$P(\text{seq} \mid B) = (.70)^{100}(.30)^{23} = 3.05 \times 10^{-28},$$

and

$$P(G \mid \text{seq}) = \frac{.9 \times 7.06 \times 10^{-33}}{.9 \times 7.06 \times 10^{-33} + .1 \times 3.05 \times 10^{-28}},$$

$$P(B \mid \text{seq}) = \frac{.1 \times 3.05 \times 10^{-28}}{.9 \times 7.06 \times 10^{-33} + .1 \times 3.05 \times 10^{-28}}.$$

It's hardly worth while to calculate them exactly: $P(B \mid \text{seq})$ is about 4800 times the size of $P(G \mid \text{seq})$. This is about as close to certainty as we get in statistics.

Remember, however, that the conclusion depends on the model's being right. We *assumed* that successive parts were independent; there's always a chance that they weren't. I guess I should really say, "We assumed that they weren't dependent enough to matter," for I would be very surprised if there weren't some connection. In a situation like this you have to analyze sequences of results now and then just to make sure that some kind of pattern isn't showing up. The defective parts may have appeared haphazardly at one time; this does not mean they will do so forever. There is also no guarantee that the percentages of satisfactory and unsatisfactory parts will stay the same. As the machine wears, a trend may be quite noticeable. So draw your conclusions, but keep a wary eye. [The techniques of *quality control* have been developed to monitor production lines like this.]

Example 3.2.3. Suppose there are 100 people in a town, and some unknown number of them support an ordinance you are sponsoring. You question 10 different people at random and get 6 favorable replies. What you can say about the total number in the town who support your ordinance? What can you say about the accuracy of your knowledge?

The observations were taken in some definite order, of course, but it turns out—as it should, intuitively—that the order does not matter. I'll choose the convenient one,

$$S, S, S, S, S, S, O, O, O, O,$$

where S = support and O = oppose. Call the number of supporters R. We now need to find $P(\text{obs} \mid R)$, the probability that we would have gotten this sequence of observations if there were R people in the town who supported your ordinance. It has to be calculated in steps, but we'll get a general formula later. For the first observation we have

$$P(\text{first } S \mid R) = \frac{R}{100}.$$

Now after each interview there is one less person in the population, and one less supporter. Hence

$$P(\text{second } S \mid R) = \frac{R-1}{99},$$

$$P(\text{third } S \mid R) = \frac{R-2}{98},$$

$$P(\text{fourth } S \mid R) = \frac{R-3}{97},$$

$$P(\text{fifth } S \mid R) = \frac{R-4}{96},$$

$$P(\text{sixth } S \mid R) = \frac{R-5}{95}.$$

Now for the opponents.

$$P(\text{first } O \mid R) = \frac{100-R}{94}.$$

Each time an opponent is discovered there is one less person in the population, and one less opponent. Hence

$$P(\text{second } O \mid R) = \frac{99-R}{93},$$

$$P(\text{third } O \mid R) = \frac{98-R}{92},$$

$$P(\text{fourth } O \mid R) = \frac{97-R}{91}.$$

Thus

$$P(\text{seq} \mid R) = \frac{R(R-1)(R-2)(R-3)(R-4)(R-5)(100-R)(99-R)(98-R)(97-R)}{100 \times 99 \times 98 \times 97 \times 96 \times 95 \times 94 \times 93 \times 92 \times 91}.$$

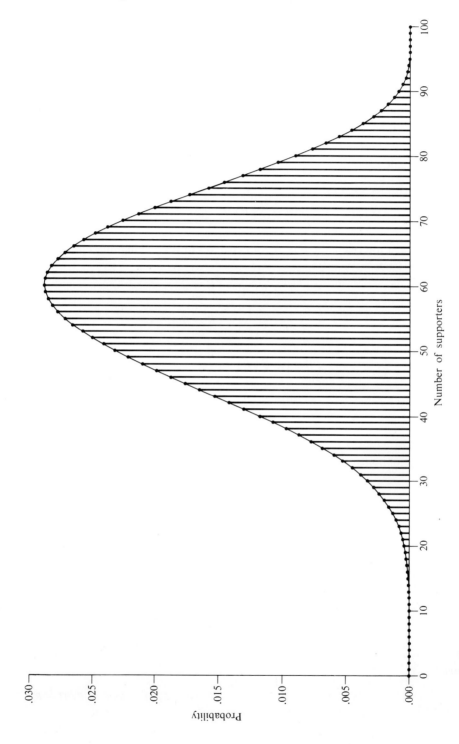

Fig. 3.2.1 Posterior distribution of number of supporters: 100 in population, 10 in sample, 6 supporters in sample.

Now if we make the assumption that any of the numbers from 0 to 100 were equally likely alternatives for R to start with, the posterior probabilities for these alternatives are proportional to $P(\text{seq} \mid R)$. [This is a general result. If a set of alternatives have equal prior probabilities, $P(\text{obs} \mid \text{alt})$ is the only thing that counts. See Problem 3.1.9.]

The numerical calculation takes a little doing, but is perfectly straightforward. Note that the posterior probabilities are 0 for $R < 6$ and $R > 96$, just as they should be. The results are shown in Fig. 3.2.1. In Chapter 5 you'll learn ways to describe this distribution without calculating everything out in full. Since the curve varies smoothly, we don't need to list every detail. Here, for example, the highest point is at $R = 60$, as you would have expected. The probabilities have decreased to half that of the maximum by the time you get out to $R = 42$ and $R = 76$.

This example illustrates the important problem of sampling from a finite population. The general solution is as follows: If

$$\begin{aligned}\text{population size} &= N \quad [\text{here } 100],\\ \text{size of category of interest} &= R \quad [\text{here } R],\\ \text{sample size} &= n \quad [\text{here } 10],\\ \text{number of category members in sample} &= r \quad [\text{here } 6],\end{aligned}$$

then

$$P(R \mid N, n, r) = \frac{\binom{R}{r}\binom{N-R}{n-r}}{\binom{N+1}{n+1}} \quad \left[\text{here } \frac{\binom{R}{6}\binom{100-R}{4}}{\binom{101}{11}}\right].$$

[Look up *Binomial Coefficients* in Appendix A4 if you don't recognize $\binom{R}{r}$.] When the sample size is small compared to the population size, the changes in probability from trial to trial can often be neglected. And if you have a nonflat prior distribution, you can introduce this information by multiplying those probabilities by the $P(R \mid N, n, r)$'s above. See Example 3.5.1.

Although this is *in principle* a complete solution to the problem of sampling from a finite population, the mathematical calculation is only a trivial part of such work. Most of the difficulties arise in *planning* and *executing* surveys, not in doing the calculations. [The same is also true of most laboratory work.] You might think, for example, that defining the population of an entire country is easy. It's not so easy. The last United Nations Statistical Yearbook had 175 footnotes (quibbles) to describe the varying practices around the world. What do you do about seasonal farming immigrants and other nomads? about refugees? about travelers in transit and passengers and crews of ships out in the harbor? about your armed forces abroad? about foreign troops stationed on your soil?

Asking questions and getting straight answers is not so easy either. Unless we, in our little problem, had taken pains to get a really random sample of 10 out of the 100, perhaps by using a random number table, we might have talked to 10 people coming from a political meeting or 10 dog lovers on their way to protest your dog-catching ordinance. Then some people won't answer; some people try to please an interviewer; some people just plain lie. Our sample of 10 was pitifully small so you could see the details of the Bayesian approach. In Chapter 5 you'll learn how to take a sample big enough to achieve the *sampling* accuracy you desire, that is, to control the purely random variation. There will still remain all these other aggravating little details which can be overcome only by experience, care, and hard work.

Example 3.2.4. Here is the slightly different problem of counting a wildlife population. Since most creatures are justifiably reticent to stand up and be counted, other methods must be used. To count the fish in a lake, you seine some fish, tag them, put them back, stir the lake thoroughly, and then make another catch. From the number of tagged and untagged fish in the second catch you try to estimate the total number of fish in the lake and estimate the uncertainty in your estimate.

Suppose there are N fish in the lake and 60 have been tagged. You now catch 100 and find that 10 of these are tagged. You naturally would guess that there are about $\frac{100}{10} \times 60 = 600$ altogether. How uncertain is this?

The probabilities are like those in Example 3.2.3, except that here N is the unknown, not R. We get

$$P(N \mid R, n, r) \propto \frac{60 \times 59 \times \cdots \times 51(N-60)(N-61)\cdots(N-149)}{N(N-1)(N-2)\cdots(N-100)}$$

when the prior probabilities are equal. [The symbol $\propto$ means *is proportional to*.] The probabilities are plotted in Fig. 3.2.2. There is here no sharp upper cutoff as there was in Fig. 3.2.1, for the number of fish could be very large. There is some upper bound on N; it's certainly smaller than the volume of the lake divided by the minimum size of the fish being considered. But your posterior probabilities are microscopic long before you get out that far.

In practice this also is a vexing problem. The biggest difficulties are the assumptions that the tagged fish are mixed in thoroughly with the untagged fish, and that each fish has an equal probability of being caught. One might suspect that fish caught once might be caught again more readily than their warier brothers. Then, too, fish, snails, flies—animals of all kinds—stay put more than most people realize. After all, a lot of humans are homebodies.

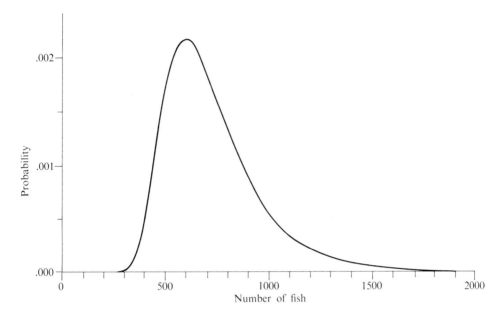

Fig. 3.2.2 Posterior distribution of number of fish: 60 fish tagged, 100 in second catch, 10 of second catch have tags.

Several helpful ideas have been advanced. You can divide the lake into several sections and tag and sample each section separately. This is a variation of the stratified sampling we used in Problem 1.4.12. The results here would also give information on the extent to which fish move around the lake. Lastly, you may be able to get additional information from the size of the total catch in relation to the effort if you have data from previous years for comparison.

Let me remind you that there is considerable bias to watch for in collections of all types. Museum specimens tend to occur in numbers inversely related to their rarity in nature. In paleontology, the number of hard parts and the ease of identification affect what is collected. In sample surveys aggressive persons may try to force themselves on the interviewer. Similarly, supporters of proposals usually write more letters to their senators than do opponents.

PROBLEM SET 3.2

1. Consider again that coin which either is fair or has two heads. Suppose 10 heads in a row have now come up. How has your opinion changed from the original $P_0(\text{fair}) = .90$?

2. Again we have the two alternatives for a coin: fair or two heads. If you said that you would assign equal posterior probabilities to the two alternatives when 20 heads in a row came up, what was your initial opinion?

3. If you take a random sample of 20 from a population of size 100 and get 15 *yes*'s and 5 *no*'s, what is the posterior probability that there are 75 in the population who think *yes*? Assume equal prior probabilities.

4. *Continuation of Problem 3.* How much less probable is 60 than 75?

5. Use the table of flat random digits to simulate production from our automatic machine in *good* condition. This means

$$P(s \mid G) = .95,$$
$$P(u \mid G) = .05.$$

Stop when you get the first unsatisfactory part, and record #*s*, the number of satisfactory parts before that point. Do this five times, and calculate

$$\frac{\sum \#s + 1}{\sum \#s + 7}.$$

6. A geneticist is investigating the linkage between two fruit-fly genes, and plans a test cross. Under hypothesis *A*, the proportions of progeny to be expected are

Type 1	Type 2	Type 3	Type 4
$\frac{9}{16}$	$\frac{3}{16}$	$\frac{3}{16}$	$\frac{1}{16}$

while under hypothesis *B*, they are

Type 1	Type 2	Type 3	Type 4
$\frac{1}{4}$	$\frac{1}{4}$	$\frac{1}{4}$	$\frac{1}{4}$

However, he really doesn't know which to expect. Unfortunately a careless lab technician lets most of the hatched flies escape, and only 17 are salvaged:

Type 1	Type 2	Type 3	Type 4
8	2	4	3

Since the genes in question do not concern wings, sight, or emotional activity, the geneticist believes that the sample is still representative and the individuals independent. What does the sample tell him?

7. A certain disease causes minks to lose patches of hair, to the great distress of their breeders. A new drug has been developed to combat it, and some small preliminary testing indicates that its cure rate is in the vicinity of $\frac{2}{3}$. Suppose we say that we have

four alternative values for the cure rate: .60, .65, .70, and .75, which are equally probable on the early evidence. [I am using a few discrete values to simplify the discussion; we will use a continuous range of values in the next chapter.] The drug is now tested on 29 animals; 20 are cured and 9 are not. What is the probability distribution of the cure rates after experimentation?

8. An archeologist working by himself in the jungle chances upon an inscription which could be in either Anahuatl or Kopacetic. From the surroundings he would have guessed that the latter was three times as probable as the former. He himself is no linguist, but has in his kit bag a table of the frequencies of the letters in the two languages:

	A	B	C	D	E
Anahuatl	.1	.4	.1	.2	.2
Kopacetic	.2	.3	.2	.1	.2

[I have transliterated into the Roman alphabet. These peoples lived in such dire poverty that they could afford only five letters in their alphabets.] He counted the letters:

A	B	C	D	E
4	7	5	0	4

If he neglects the fact that successive letters in a language are not independent, what are the posterior probabilities for the two languages?

ANSWERS TO PROBLEM SET 3.2

2.

Alternative	Prior	$P(20$ heads $\mid$ alt$)$	Joint
Fair	p ×	$(.5)^{20}$	$= .00000095p$
Two heads	$1 - p$ ×	1	$= 1 - p$

The two joint values (proportional to the posterior values) are now set equal:

$$.00000095p = 1 - p, \quad 1.00000095p = 1, \quad p = .999999.$$

3. By the formula of Example 3.2.3 we have

$$P(75) = \frac{\binom{75}{15}\binom{25}{5}}{\binom{101}{21}}.$$

The only problem is numerical evaluation. When we write out the binomial coefficients in factorials, we get

$$P(75) = \frac{75!\,25!\,21!\,80!}{15!\,60!\,5!\,20!\,101!}.$$

There is a table of $\log_{10} N!$ in Appendix B3, and we need it now.

	+		
(75)	109.3946	12.1165	(15)
(25)	25.1906	81.9202	(60)
(21)	19.7083	2.0792	(5)
(80)	118.8547	18.3861	(20)
		159.9743	(101)
	273.1482	274.4763	

$$273.1482 - 274.4763 = -1.3281 = .6719 - 2.$$

Thus $P(75) = .0470$.

6. $P(\text{sample} \mid A) = (\frac{9}{16})^8 (\frac{3}{16})^2 (\frac{3}{16})^4 (\frac{1}{16})^3 = 1.06 \times 10^{-10}$,
   $P(\text{sample} \mid B) = (\frac{1}{4})^{8+2+4+3} = (\frac{1}{4})^{17} = 5.81 \times 10^{-11}$.

Since the prior probabilities are equal, he normalizes immediately and gets

$$P(A \mid \text{sample}) = .65, \qquad P(B \mid \text{sample}) = .35.$$

Thus, for all his work, hypothesis A gets to be only twice as probable as hypothesis B. This is not much of a difference, certainly nothing to base a scientific conclusion on. It's the same as having 2 red balls and 1 green ball in an urn; the probability of drawing a red ball is twice that of drawing a green one.

7.

Alternative	Prior	$P(\text{obs} \mid \text{alt})$	Posterior
.60	.25	$(.60)^{20}(.40)^9$	.19
.65	.25	$(.65)^{20}(.35)^9$	.28
.70	.25	$(.70)^{20}(.30)^9$	.30
.75	.25	$(.75)^{20}(.25)^9$	.23

The observed cure rate for these 29 minks is $\frac{20}{29}$. But we know this is only a sample of the behavior of the drug; we want to know the range of possibilities in case we plan to give the drug to more animals.

What is realistic and unrealistic about this problem? First, it is not unusual to get preliminary results which roughly indicate the area of possible values, but which you feel are not accurate enough to include in a more precise test. The flat prior distribution is merely our expression of ignorance; that's reasonable. We would, however, want to include a wider range of alternatives. It's not that there's anything particularly wrong about using multiples of .05. The trouble is that nearby multiples on each end obviously would have had appreciable probabilities had they been included. People quite commonly overestimate the precision of frequencies. In fact, if you admit as possibilities all values between 0 and 1, the range .55 to .80 encompasses less than 90% of the total posterior probability.

The other point is that the cure rate of a drug is not a precise physical constant like the charge on an electron. There probably is no such thing as *the* cure rate of a drug for a mink; it must depend on the dose, and the weight, age, and sex of the animals. Nevertheless, you might get some information useful for general prognosis if the 29 animals you tested were fairly representative of the animals you would be treating with the drug. It would be intellectually interesting, of course, to do careful experiments to find the effect of dose, age, etc., but the cost might be high. The breeder might want to start treatments immediately to save his investment. Or it might be that a rate around $\frac{2}{3}$ is really too small to be of much value, and more effective drugs would hurriedly be sought.

8. Alternative	Prior	$P(\text{obs} \mid \text{alt})$	Joint	Posterior
Anahuatl	.25	$(.1)^4(.4)^7(.1)^5(.2)^0(.2)^4$	6.56×10^{-16}	.0049
Kopacetic	.75	$(.2)^4(.3)^7(.2)^5(.1)^0(.2)^4$	1.34×10^{-13}	.9951

Kopacetic started out 3 times as probable as Anahuatl; after the evidence was evaluated, it was 200 times as probable. It would be difficult to reach this conclusion just by gazing at the frequencies.

3.3 LIKELIHOODS, ODDS, AND FACTORS

There are several terms which are useful in talking about or arriving at posterior distributions. *Likelihood* is the name given to the probability conditional on a particular alternative—the thing I have been writing as $P(\text{sample} \mid \text{alt})$. Since most of the time our alternatives are different suggested values of an adjustable parameter, the likelihood can often be given as a mathematical expression involving that parameter. For example, the likelihood of getting 20 cures and 9 noncures in some particular sequence is $p^{20}(1 - p)^9$, where $p = P(\text{cure} \mid \text{cure rate} = p)$. Instead of having to list the values separately for all the different p's, we have neatly summarized everything. As you can imagine, however, if we actually go to evaluate the likelihood for each p, we still have to do as much work. But I'm looking ahead to Chapter 5 where you'll learn how to describe probability distributions more compactly. In any case, remember the word. I will sometimes use the letter L for the likelihood.

The other concepts come in handy when there are just two alternatives. This happens quite often, since you can always talk about *this* and *not-this*. The ratio of the prior probabilities is called the *prior odds*, and the ratio of the posterior probabilities is called the *posterior odds*. The ratio of the likelihoods is called the *factor*. In any one problem, be sure that in all these ratios the values of the two alternatives are used in the same order. That is, if you divided $P(A)$ by $P(B)$ for the odds, divide $P(\text{sample} \mid A)$ by $P(\text{sample} \mid B)$ for the factor. We have

$$\text{post}_A = \frac{\text{prior}_A \times L_A}{\text{prior}_A \times L_A + \text{prior}_B \times L_B},$$

$$\text{post}_B = \frac{\text{prior}_B \times L_B}{\text{prior}_A \times L_A + \text{prior}_B \times L_B},$$

so

$$\frac{\text{post}_A}{\text{post}_B} = \frac{\text{prior}_A}{\text{prior}_B} \times \frac{L_A}{L_B}.$$

Let me write

$$\mathcal{O} = \frac{\text{post}_A}{\text{post}_B} = \text{posterior odds},$$

$$\mathcal{O}_0 = \frac{\text{prior}_A}{\text{prior}_B} = \text{prior odds},$$

$$\mathcal{F} = \frac{L_A}{L_B} = \text{factor}.$$

Then

$$\mathcal{O} = \mathcal{O}_0 \times \mathcal{F}$$

or

posterior odds = prior odds × factor.

[I have used $\mathcal{O}$ for *odds* so it won't be confused with zero, and $\mathcal{F}$ for *factor* since F is used all over the place.]

If you have only two alternatives, you don't have to keep normalizing the distribution; knowing the ratio of the probabilities is equivalent to knowing the probabilities, since they have to sum to 1. To specify the alternatives in the expressions for odds and factors I'll write

$$\mathcal{O}\left(\frac{A}{B}\right) = \mathcal{O}_0\left(\frac{A}{B}\right) \times \mathcal{F}\left(\frac{A}{B}\right),$$

read, "the posterior odds of A with respect to B equal . . ." If the odds are $\frac{3}{1}$, the first alternative (here A) is three times as probable as the second (here B). In gambling and horse racing the quoted odds sometimes mean odds *for* and sometimes odds *against*. I will always state which way they go so you won't be confused.

Example 3.3.1. Let's recompute Example 3.1.2, where we were screening for a rare disease:

$$\mathcal{O}_0\left(\frac{\text{not diseased}}{\text{diseased}}\right) = \frac{P_0(\text{not diseased})}{P_0(\text{diseased})} = \frac{.9998}{.0002} = 4996,$$

$$\mathcal{F}\left(\frac{\text{not diseased}}{\text{diseased}} \,\bigg|\, \text{positive}\right) = \frac{P(\text{positive} \mid \text{not diseased})}{P(\text{positive} \mid \text{diseased})}$$

$$= \frac{.05}{.98} = .051 = \frac{1}{19.6},$$

$$\mathcal{F}\left(\frac{\text{not diseased}}{\text{diseased}} \,\bigg|\, \text{negative}\right) = \frac{P(\text{negative} \mid \text{not diseased})}{P(\text{negative} \mid \text{diseased})}$$

$$= \frac{.95}{.02} = 47.5,$$

$$\mathcal{O}\left(\frac{\text{not diseased}}{\text{diseased}} \,\bigg|\, \text{positive}\right) = \mathcal{O}_0 \times \mathcal{F}\left(\frac{\text{not diseased}}{\text{diseased}} \,\bigg|\, \text{positive}\right)$$

$$= 4996 \times \frac{1}{19.6} = 255,$$

$$\mathcal{O}\left(\frac{\text{not diseased}}{\text{diseased}} \,\bigg|\, \text{negative}\right) = \mathcal{O}_0 \times \mathcal{F}\left(\frac{\text{not diseased}}{\text{diseased}} \,\bigg|\, \text{negative}\right)$$

$$= 4996 \times 47.5 = 237{,}000.$$

Example 3.3.2. How do we convert odds back to probabilities? Let the two alternatives be A and B. Then

$$\mathcal{O} = \frac{p_A}{p_B} = \frac{p_A}{1 - p_A}, \qquad p_A(1 + \mathcal{O}) = \mathcal{O}.$$

Therefore

$$p_A = \frac{\mathcal{O}}{1 + \mathcal{O}}, \qquad p_B = \frac{1}{1 + \mathcal{O}}.$$

In Example 3.3.1, let $A =$ not diseased and $B =$ diseased; and let the test result be positive. Then

$$\mathcal{O}\left(\frac{A}{B}\right) = 255, \qquad p_A = \tfrac{255}{256} = .9961, \qquad p_B = \tfrac{1}{256} = .0039.$$

Example 3.3.3. This is a recalculation of Example 3.1.1. We had

$$P_0(A_1) = .7, \qquad P(t \mid A_1) = .8, \qquad P(t \mid A_2) = .1,$$
$$P_0(A_2) = .3, \qquad P(u \mid A_1) = .2, \qquad P(u \mid A_2) = .2,$$
$$ P(v \mid A_1) = .0, \qquad P(v \mid A_2) = .7.$$

Then

$$\mathcal{O}_0\left(\frac{A_1}{A_2}\right) = \frac{.7}{.3} = 2.33,$$

$$\mathcal{F}\left(\frac{A_1}{A_2}\bigg| t\right) = \frac{.8}{.1} = 8 \quad \text{[strong evidence for } A_1\text{]},$$

$$\mathcal{F}\left(\frac{A_1}{A_2}\bigg| u\right) = \frac{.2}{.2} = 1 \quad \text{[no evidence]},$$

$$\mathcal{F}\left(\frac{A_1}{A_2}\bigg| v\right) = \frac{.0}{.7} = 0 \quad \text{[absolute evidence for } A_2\text{]},$$

$$\mathcal{O}\left(\frac{A_1}{A_2}\bigg| t\right) = 2.33 \times 8 = 18.67 = \frac{.56}{.03},$$

$$P(A_1 \mid t) = \frac{18.67}{18.67 + 1} = \frac{18.67}{19.67} = .949,$$

$$P(A_2 \mid t) = \frac{1}{18.67 + 1} = \frac{1}{19.67} = .051,$$

$$\mathcal{O}\left(\frac{A_1}{A_2}\bigg| u\right) = 2.33 \times 1 = 2.33,$$

$$P(A_1 \mid u) = \frac{2.33}{2.33 + 1} = \frac{2.33}{3.33} = .7,$$

$$P(A_2 \mid u) = \frac{1}{2.33 + 1} = \frac{1}{3.33} = .3,$$

$$\mathcal{O}\left(\frac{A_1}{A_2}\bigg| v\right) = 2.33 \times 0 = 0,$$

$$P(A_1 \mid v) = \frac{0}{0 + 1} = 0,$$

$$P(A_2 \mid v) = \frac{1}{0 + 1} = 1.$$

What this tack does for you is avoid normalization. Yet the ratio of two probabilities, the odds, certainly has an intuitive meaning itself. When there are three or more alternatives, the direct analogy of odds is not apparent, though some people do keep *relative odds* and *relative factors*. An important restatement of Bayes' Theorem applicable to any number of alternatives is

posterior probs $\propto$ *prior probs* $\times$ *likelihoods*,

where $\propto$ is read *is proportional to*. This means that the same multiplier (the normalization constant) is to be applied to every alternative under consideration. If you go back to the examples in Section 3.1, you will see that this is what we did all the time: compute *prior prob* × *likelihood*, then normalize to get *posterior prob*. We could have stopped before normalizing with numbers proportional to the posterior probabilities.

This formulation will be used over and over again when we express likelihood as a function of a parameter. Both this and the odds-factor relation clearly exhibit the pertinent parts of our inferences. The prior probabilities express the information we have before we take a sample; the likelihoods express the information coming from the sample; and the posterior probabilities sum up the information afterward. Sometimes the prior odds and probabilities may be slightly hazy because of difficulty in evaluating our knowledge. Then it is useful to keep all our sampling information separated in terms of factors, relative factors, or likelihoods, so that it can be applied to whatever prior distribution or distributions are suggested.

Finally, with *independent observations*, likelihoods and factors both get multiplied. If we use logarithms we can add instead; the resulting objects are called *log-odds*, *log-factors*, and *log-likelihoods*. We will generally work with logarithms to the base 10. Log-odds and log-factors greater than 0 favor the "upper" alternative; log-odds and log-factors less than 0 favor the "lower."

Example 3.3.4. I'll rework Example 3.2.1:

$$\log \mathcal{O}_0 \left(\frac{G}{B} \right) = \log_{10} \left(\frac{.9}{.1} \right) = .9542,$$

$$\log \mathcal{F} \left(\frac{G}{B} \middle| s \right) = \log_{10} \left(\frac{.95}{.70} \right) = .1326,$$

$$\log \mathcal{F} \left(\frac{G}{B} \middle| u \right) = \log_{10} \left(\frac{.05}{.30} \right) = -.7782.$$

Then for any #s and #u,

$$\log \mathcal{F} \left(\frac{G}{B} \middle| \#s, \#u \right) = .9542 + .1326(\#s) - .7782(\#u).$$

When #s = 9 and #u = 3,

$$\log \mathcal{O} \left(\frac{G}{B} \middle| 9, 3 \right) = .9542 + 9 \times .1326 - 3 \times .7782$$

$$= .9542 + 1.1934 - 2.3346 = -.1870,$$

$$\mathcal{O} \left(\frac{G}{B} \middle| 9, 3 \right) = .65,$$

$$P(G \mid 9, 3) = \frac{.65}{.65 + 1} = .39,$$

$$P(B \mid 9, 3) = \frac{1}{.65 + 1} = .61.$$

88 Accumulating Evidence   3.3

PROBLEM SET 3.3

1. Refer to Problem 3.1.7. Express the analysis in terms of odds and factors. Recover the posterior probabilities from the odds.

2. Refer to Problem 3.2.1. Express the analysis in terms of odds and factors. Recover the posterior probabilities from the odds.

3. In Example 3.3.4, suppose we observed 100 satisfactory parts and 10 unsatisfactory ones. What are the posterior log-odds and the posterior odds?

4. Do Problem 3.2.6 using log-odds and log-factors.

5. Do Problem 3.2.8 using log-odds and log-factors.

6. Use the data from Problem 2.4.5. What is the largest factor to be obtained from a single letter in scoring English against a flat random sequence of letters?

7. There are an unknown number, N, of identical balls in an urn. They are drawn one at a time, marked with a unique number if they don't already have one, and put back. In n draws we see d different balls. What is the relative factor for each possible N?

8. In Problem 3.2.8, what is the expected sample size needed to get

$$\log \mathcal{F}\left(\frac{A}{K}\right) = 3$$

when the language is Anahuatl?

ANSWERS TO PROBLEM SET 3.3

3. $\log \mathcal{O}\left(\frac{G}{B}\middle| 100, 10\right) = .9542 + 13.26 - 7.782 = 6.4322$

   $\mathcal{O}\left(\frac{G}{B}\middle| 100, 10\right) = 2.7 \times 10^6 = 2{,}700{,}000$

6. Under the flat random hypothesis, every letter has probability $\frac{1}{26} = .0385$. The letter supplying the largest factor for English is E:

   $$\frac{.1310}{.0385} = 3.40.$$

   The letters supplying the largest factor for the flat random hypothesis are Q and Z:

   $$\frac{.0010}{.0385} = \frac{1}{38.5}.$$

7. Every time you draw an already labeled ball, the relative factor is $1/N$; the first unlabeled ball supplies a relative factor of N/N; the second unlabeled ball supplies a

relative factor of $(N-1)/N$; etc. With n draws and d different balls, you get

$$\frac{N(N-1)(N-2)\cdots(N-d+1)}{N^n} = \frac{N!}{(N-d)!}\cdot\frac{1}{N^n}.$$

Since the relative factors are unaltered if you divide by $d!$, which does not involve N, you could alternatively say

$$\binom{N}{d}\frac{1}{N^n}.$$

8. $E\left[\log \mathcal{F}\left(\frac{A}{\underline{\underline{}}K}\right) \text{ per letter} \mid A\right] = (-.3010)(.1) + (.1250)(.4) + (-.3010)(.1)$
$$+ (.3010)(.2) + (.0000)(.2) = .05.$$

Then $.05n = 3$, $n = 60$.

3.4 COMPOSITE ALTERNATIVES

The alternatives we have discussed so far have been *simple* alternatives; that is, we have been interested in scoring separately every member of the set. A *composite* alternative is a group of simple alternatives. It connotes more interest in whether or not the best answer is somewhere in this group than in what its particular value is. One possible situation is that of testing the equality of two success rates. Each of the success rates has several possibilities; the joint distribution has as many possibilities as the product of these numbers. Yet we are interested (at first) in whether or not the right answer belongs to the group of joint possibilities in which the success rates are equal, or to the group in which they are not equal.

Consider a researcher trying to determine the relative effectiveness of two drugs. To simplify things drastically, I will assume that each drug has possible cure rates of .25, .50, or .75. Now if a researcher seriously entertains the possibility that the drugs are approximately equally effective, he must ensure that his prior probability distribution reflects this attitude. Suppose he believes, on the basis of much past experience with drugs and animals of the types involved here, that it's 50–50 that the drugs have an approximately equal effect, but otherwise he has no knowledge about their cure rates. In such a case a prior distribution like the one below may be in order:

		Cure rate 2		
		.25	.50	.75
	.25	$\frac{1}{6}$	$\frac{1}{12}$	$\frac{1}{12}$
Cure rate 1	.50	$\frac{1}{12}$	$\frac{1}{6}$	$\frac{1}{12}$
	.75	$\frac{1}{12}$	$\frac{1}{12}$	$\frac{1}{6}$

Here I have arranged the nine joint possibilities in a square in order to show clearly the influence of the two variables. I have spread prior probability $\frac{1}{2}$ among the three diagonal elements where the two cure rates are equal, and have spread the other $\frac{1}{2}$ evenly over the remaining elements. This whole distribution is meant to mirror our knowledge (or ignorance). The $\frac{1}{6}$ in the upper left corner is the prior probability assigned to the alternative: cure rate 1 is .25 and cure rate 2 is .25. The $\frac{1}{12}$ in the upper right corner is the prior probability assigned to the alternative that cure rate 1 is .25 and cure rate 2 is .75, etc. The three possibilities on the diagonal make up the composite alternative that the cure rates are equal; the six off the diagonal make up the composite alternative that they are unequal.

You may legitimately complain that at the beginning of this investigation equality of the rates should hardly be considered as probable as inequality; that is, perhaps I shouldn't have used prior odds of 1. In that case, take what I call *odds* on the following pages as your *factor* for equality versus inequality; then multiply this factor by your own prior odds.

We will be making experiments with drug 1 and with drug 2. If we take care that the use of one drug on one animal has no influence on the outcome of the use of the same or other drug on another animal, the results are independent and we can multiply the probabilities. The calculations go exactly as before. Suppose that our experiments show

s successes and f failures with drug 1,
t successes and g failures with drug 2.

We calculate:

Alternative		Prior	Likelihood			
CR 1	CR 2					
.25	.25	$\frac{1}{6}$	$.25^s$	$.75^f$	$.25^t$	$.75^g$
.25	.50	$\frac{1}{12}$	$.25^s$	$.75^f$	$.50^t$	$.50^g$
.25	.75	$\frac{1}{12}$	$.25^s$	$.75^f$	$.75^t$	$.25^g$
.50	.25	$\frac{1}{12}$	$.50^s$	$.50^f$	$.25^t$	$.75^g$
.50	.50	$\frac{1}{6}$	$.50^s$	$.50^f$	$.50^t$	$.50^g$
.50	.75	$\frac{1}{12}$	$.50^s$	$.50^f$	$.75^t$	$.25^g$
.75	.25	$\frac{1}{12}$	$.75^s$	$.25^f$	$.25^t$	$.75^g$
.75	.50	$\frac{1}{12}$	$.75^s$	$.25^f$	$.50^t$	$.50^g$
.75	.75	$\frac{1}{6}$	$.75^s$	$.25^f$	$.75^t$	$.25^g$

Then we multiply *prior probability* by *likelihood* for each line, and normalize the set of nine results so they sum to 1, just as we have done many times before. The only novelty arises when we look to see how the sum of the three posterior probabilities belonging to equal cure rates compares with the sum of the other six.

3.4 Composite Alternatives

I made up some hypothetical sampling results and had the computer grind out the computations. In each case I give the four numbers

$$[s, f; t, g] = [successes_1, failures_1; successes_2, failures_2]$$

and then comment on the odds attained.

a) [0, 0; 0, 0]. This is the situation before we start. Our posterior distribution is just the prior distribution. The odds for equality versus inequality are the original 1 : 1.

b) [1, 0; 1, 0]. Here we tried each drug in one case and both outcomes were successful. The posterior distribution is now

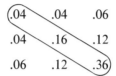

Naturally the higher cure rates get a boost. The odds for equality are now .56/.44 = 1.27. This is still practically no evidence.

c) [1, 1; 1, 1]. After each drug has had one success and one failure we have the distribution

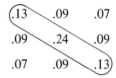

and the odds for equality are just about 1 : 1. Our four trials have not helped us to decide.

d) [0, 1; 1, 0]. The first drug failed once, the second succeeded once:

The odds for equality are 1/1.3. It is a common phenomenon that it is harder to get evidence *for* a restrictive hypothesis like equality than to get evidence *against* it. Case (b) gave 1.27 *for*; this case gives 1.30 *against*.

92 Accumulating Evidence   3.4

e) [3, 0; 2, 0].   One drug has succeeded three times, the other twice. The
                        distribution is

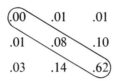

The .00 is the result of rounding; you can't get a real zero by multiplying factors different from zero. Odds for equality are now 2.4.

And now some larger samples:

f) [17, 5; 5, 17].   The posterior odds for equality are $\frac{1}{503}$; in other words it's very probable that the drugs have unequal effects.

g) [17, 5; 17, 5].   The same number of trials as (f) but with the results indicating equality. The posterior odds are 31—a further example of how hard it is to get evidence in favor of a restrictive hypothesis.

h) [22, 18; 17, 21].   There have been 78 trials in all, and the posterior odds for equality are 29.

I hope you can see that, although more calculation may be involved, there is no really new principle. The main thing to remember is that Bayes' Theorem must be applied to the *simple* alternatives. You can reach in now and then and add up some probabilities to get a score for a *composite* alternative, but you cannot multiply factors for composite alternatives.

PROBLEM SET 3.4

1. Suppose we had those 11 urns, each with 10 balls, of which from 0 to 10 were green. Repeated draws *with replacement* are made from *one* of the urns. This time we are especially interested in whether the selected urn is No. 5, the one with 5 green and 5 red balls. In fact we have some information which singles out that alternative as a favored possibility; but if it isn't that one, we have not the slightest idea which one it is. We assign half the prior probability to No. 5, and spread the other half evenly over the 10 others. In four draws we get 2 green and 2 red balls. What are the posterior odds of No. 5 versus the rest? [This is the discrete prototype of the problem of testing whether a rate equals $\frac{1}{2}$.]

2. You are given a die to experiment with and are told that it is one of the following types:

Prior probability	Type	Conditional probabilities					
		$P(1)$	$P(2)$	$P(3)$	$P(4)$	$P(5)$	$P(6)$
.70	True	$\frac{1}{6}$	$\frac{1}{6}$	$\frac{1}{6}$	$\frac{1}{6}$	$\frac{1}{6}$	$\frac{1}{6}$
.05	Loaded 1	$\frac{1}{4}$	$\frac{1}{6}$	$\frac{1}{6}$	$\frac{1}{6}$	$\frac{1}{6}$	$\frac{1}{12}$
.05	Loaded 2	$\frac{1}{6}$	$\frac{1}{4}$	$\frac{1}{6}$	$\frac{1}{6}$	$\frac{1}{12}$	$\frac{1}{6}$
.05	Loaded 3	$\frac{1}{6}$	$\frac{1}{6}$	$\frac{1}{4}$	$\frac{1}{12}$	$\frac{1}{6}$	$\frac{1}{6}$
.05	Loaded 4	$\frac{1}{6}$	$\frac{1}{6}$	$\frac{1}{12}$	$\frac{1}{4}$	$\frac{1}{6}$	$\frac{1}{6}$
.05	Loaded 5	$\frac{1}{6}$	$\frac{1}{12}$	$\frac{1}{6}$	$\frac{1}{6}$	$\frac{1}{4}$	$\frac{1}{6}$
.05	Loaded 6	$\frac{1}{12}$	$\frac{1}{6}$	$\frac{1}{6}$	$\frac{1}{6}$	$\frac{1}{6}$	$\frac{1}{4}$

You throw it 60 times and observe:

Result	1	2	3	4	5	6
Frequency	5	14	8	12	14	7

What is the posterior probability of the die's being loaded?

ANSWERS TO PROBLEM SET 3.4

1.

Alternative	Prior	Likelihood	Joint
Urn 0	.05	$(.0)^2(1.0)^2 = .0000$	.000000
Urn 1	.05	$(.1)^2(.9)^2 = .0081$	.000405
Urn 2	.05	$(.2)^2(.8)^2 = .0256$	.001280
Urn 3	.05	$(.3)^2(.7)^2 = .0441$	.002205
Urn 4	.05	$(.4)^2(.6)^2 = .0576$	.002880
Urn 5	.50	$(.5)^2(.5)^2 = .0625$	.031250
Urn 6	.05	$(.6)^2(.4)^2 = .0576$	.002880
Urn 7	.05	$(.7)^2(.3)^2 = .0441$	.002205
Urn 8	.05	$(.8)^2(.2)^2 = .0256$	.001280
Urn 9	.05	$(.9)^2(.1)^2 = .0081$	.000405
Urn 10	.05	$(1.0)^2(.0)^2 = .0000$	.000000

Then

$$\mathcal{O}\left(\frac{\text{urn 5}}{\text{others}}\right) = \frac{.03125}{.01354} = 2.31.$$

Again, if your prior odds for No. 5 *versus* the others are not $\frac{1}{1}$, 2.31 is the *factor* to be applied to those prior odds. If you had believed beforehand that it was 2 : 1 *against* No. 5, it would now be 1.15 : 1 *for* No. 5.

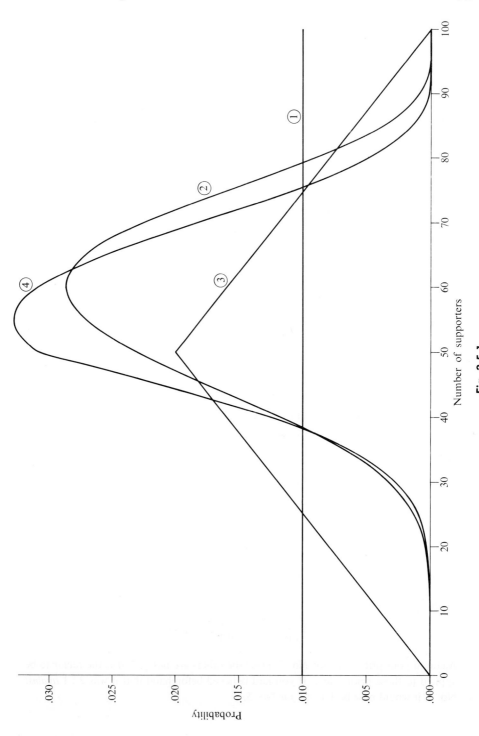

Fig. 3.5.1

3.5 PRIOR PROBABILITIES AND ULTIMATE ERROR

In this section I discuss two questions: What do I do if I don't know my prior probabilities? What are the chances of my making a mistake if I reason according to the methods of this chapter?

Evaluating prior probabilities is not intrinsically different from evaluating the probabilities which appear as likelihoods. In each case we have to make judgments; they're inescapable. However, there often are well-worn theoretical models or extensive records to suggest likelihoods, while records available for prior distributions are nowhere as plentiful. Of course no probability is just a frequency, and no probability depends on records; but all such knowledge is part of the information we must bring to bear on a problem. When we have very vague information or none at all, there seems to be no alternative but to assign equal probabilities to the stated alternatives. But the structure of the list of alternatives may itself give information. If we don't use this, we can paint ourselves into all sorts of corners. See the discussion of the "Life on Mars" paradox in Fry (1965).

Occasionally you may find it easier to say what your opinion would be if you observed such and such an experimental result than to say directly what your prior distribution is. You can then take this hypothetical posterior opinion and work your way back to the prior distribution by Bayes' Theorem. We did this in Problem 3.2.2.

One of the most useful things to do is to vary your prior probability distribution and see what happens to your analysis. If reasonable variations change your conclusions, you just haven't acquired enough experimental evidence to overcome your initial uncertainty. Additional experimentation should resolve the problem.

Example 3.5.1. In Example 3.2.3 I took a sample of 10 from a population of 100 and used a flat prior distribution. Now it has been many people's experience that about half the people will support any issue; at least you'd guess half before you'd guess an extreme result. In Fig. 3.5.1 I show the use of a simple prior probability distribution to mirror this opinion:

Curve 1 is the flat prior distribution.
Curve 2 is the curve of Fig. 3.2.3, the result of combining the flat prior with the experimental results.
Curve 3 is my new peaked prior distribution which says that numbers around 50 (50% of the population) are the most probable before I get any data.
Curve 4 combines the peaked prior distribution with the same experimental results.

Note that the highest point on curve 4 is a little higher than the highest point on curve 2; the location of the highest point is noticeably shifted towards 50. Obviously my experimental results are rather meager, and a different prior distribution has definitely affected my posterior distribution. If I had taken a sample of 20, the change would have been less noticeable.

What about the probability of making a mistake? I hope you are thoroughly convinced by now that there is no hope of avoiding all mistakes; our aim is to keep their number to a minimum by employing good statistical procedures. What can I say here that will be useful? I'll have to assume that our model and likelihoods are correct; otherwise the situation is usually too obscure. I'll discuss this simpler problem: Can I force myself into a wrong conclusion by sampling indefinitely? That is, can I achieve any result I want to if I am persistent enough?

The answer is "No!", as it should be. Consider this situation: I have two alternatives, A and B. A is the one I'm sort of attached to. But I say to myself that if I ever get a small enough factor for A versus B (large enough factor for B versus A), I'll have to give up A. How large a factor is required depends on the prior probabilities and the costs of making various decisions. Just say, for example, that if I ever got the factor for A versus B down to 1 in 1000 or less, I'd give up A. It turns out that even if I sampled day and night for the next million years, the probability that this would happen is no more than $\frac{1}{1000}$. If I set the decision point at a factor of 1 in a million or less, I'd have a probability of no more than 1 in a million of getting there. *The probability of getting strong evidence against a true hypothesis is very small.* [The proof is given in Section 8.4.]

Example 3.5.2. Take Example 3.1.1, the first one we used in our Bayesian analysis. Assume that alternative A_2 is true. What is the probability that I can get a factor of 8 or more against A_2?

Event	t	u	v
$P(\text{event} \mid A_1)$	.8	.2	.0
$P(\text{event} \mid A_2)$	.1	.2	.7
$\mathcal{F}\left(\dfrac{A_1}{A_2} \bigg| \text{event}\right)$	8.0	1.0	0.0

To get a factor of 8 or more against A_2, I have to get a t before I get a v. The u's don't count. This can be done in the following ways:

Sequence of results	$P(\text{seq} \mid A_2)$
t	.1
u, t	$.2 \times .1$
u, u, t	$(.2)^2 \times .1$
u, u, u, t	$(.2)^3 \times .1$
$\vdots$	$\vdots$

The probabilities add up to $(.1)[1 + .2 + (.2)^2 + (.2)^3 + \cdots]$, which is

$$\frac{.1}{1 - .2} = \frac{.1}{.8} = \frac{1}{8}.$$

That is, the probability of *ever* getting a factor of 8 or more against the true hypothesis A_2 is just $\frac{1}{8}$. If I were not determined to try forever, the probability would be less.

Example 3.5.3. Suppose we are given:

Event	x	y
$P(\text{event} \mid \text{true})$	.8	.2
$P(\text{event} \mid \text{false})$	.2	.8
$\mathcal{F}\left(\dfrac{\text{true}}{\text{false}} \middle| \text{event}\right)$	4.0	.25

What is the probability that I can ever get a factor of $\frac{1}{16}$ or less for the true alternative (16 or more against)? To do this, I must at some point have two more y's than x's. This can happen in the following ways:

Sequence of results	$P(\text{seq} \mid \text{true})$
y, y	.04
x, y, y, y	.0064
y, x, y, y	.0064
x, x, y, y, y, y	.001024
x, y, x, y, y, y	.001024
x, y, y, x, y, y	.001024
y, x, x, y, y, y	.001024
y, x, y, x, y, y	.001024
x, x, x, y, y, y, y, y	.00016384
$\vdots$	$\vdots$

These first nine terms add up to .05808384, still less than the theoretical $\frac{1}{16} = .0625$.

Next, if we have two alternatives, what is the expected factor for the false hypothesis? This expectation is naturally calculated assuming the true distribution is working:

$$E\left[\mathcal{F}\left(\frac{\text{false}}{\text{true}}\right)\right] = \sum_{\text{events}} \frac{P(\text{event} \mid \text{false})}{P(\text{event} \mid \text{true})} \times P(\text{event} \mid \text{true})$$

$$= \sum_{\text{events}} P(\text{event} \mid \text{false}) = 1.$$

Example 3.5.4. Assume that we sample twice from the distribution of A_2 in Example 3.5.2:

Event	P(event | A_2)	$\mathcal{F}\left(\dfrac{A_1}{A_2}\middle\| \text{event}\right)$	Product
t, t	.01	64	.64
t, u	.02	8	.16
t, v	.07	0	.00
u, t	.02	8	.16
u, u	.04	1	.04
u, v	.14	0	.00
v, t	.07	0	.00
v, u	.14	0	.00
v, v	.49	0	.00
	1.00		1.00
	[I've listed all possibilities.]		[expected factor]

Lastly, if a true hypothesis competes against several alternatives, the probability of getting a strong factor against it naturally is greater. Suppose there are k competitors, each is assigned equal prior weight, and we calculate the composite factor:

$$\frac{P(\text{event} \mid \text{true})}{\sum_{\substack{\text{false} \\ \text{alts}}} P(\text{event} \mid \text{false alts})}.$$

Then the probability of getting a factor as small as $1/a$ for the true hypothesis ($\geq a$ against it) is $\leq k/a$.

Example 3.5.5. With 10 competitors equally weighted, the probability of ever getting a factor of 100 to 1 against the true hypothesis is $\leq \frac{10}{100} = \frac{1}{10}$.

The probability can be k/a only if all competitors have equal likelihoods. This is not usually the case, so the probability of eventual error is generally smaller than this bound. There is some moderately easy discussion of all this in Kerridge (1963) and Cornfield (1966).

PROBLEM SET 3.5

1. In Example 3.5.2, what is the probability of getting a factor of 6 or more against A_2 when A_2 is true?

2. In Example 3.5.2, what is the probability of getting a factor of 64 or more against A_2 when A_2 is true? Demonstrate that the theoretical probability is reached.

3. Show that with three trials from the true distribution of Example 3.5.3,

$$E\left[\mathcal{F}\left(\frac{\text{false}}{\text{true}}\right)\right] = 1.$$

4. If a true hypothesis has 6 competitors, all equally weighted, what is the probability that you will ever get a factor of 100 to 1 against it?

ANSWERS TO PROBLEM SET 3.5

1. Because there are only a finite number of different factors, not every desired factor can be hit on the nose. The full statement is that you can achieve a factor of a against A_2 when A_2 is true with probability *less than or equal to* $1/a$. In this problem we know prob $\leq \frac{1}{6}$. It actually turns out to be $\frac{1}{8}$ again, which is $< \frac{1}{6}$.

2. By the theory, the probability is $\leq \frac{1}{64}$. We must get two t's before a v.

$$\begin{aligned}
&\text{1 way of doing it in 2 turns:} && \text{prob} = (.1)^2; \\
&\text{2 ways of doing it in 3 turns:} && \text{prob} = 2(.2)(.1)^2; \\
&\text{3 ways of doing it in 4 turns:} && \text{prob} = 3(.2)^2(.1)^2; \\
&\qquad\vdots && \qquad\vdots \\
&k-1 \text{ ways of doing it in } k \text{ turns:} && \text{prob} = (k-1)(.2)^{k-2}(.1)^2; \\
&\qquad\vdots && \qquad\vdots
\end{aligned}$$

The sum is

$$(.1)^2[1 + 2(.2) + 3(.2)^2 + 4(.2)^3 + \cdots] = \frac{(.1)^2}{(.8)^2} = \frac{1}{64}.$$

Note: There are some useful series you might like to know:

$$1 + r + r^2 + r^3 + r^4 + \cdots = \frac{1}{1-r},$$

$$1 + 2r + 3r^2 + 4r^3 + 5r^4 + \cdots = \frac{1}{(1-r)^2},$$

$$1 + 3r + 6r^2 + 10r^3 + 15r^4 + \cdots = \frac{2}{(1-r)^3}.$$

CHAPTER **4**

Continuous Variables

In Chapters 1, 2, and 3 I discussed situations in which both the variables and the set of alternative hypotheses were discrete. Such situations do exist: many variables are counted; many alternatives are disjoint by nature. The methods I described, though primarily intended for your basic initiation, have real use. But some of the problems were forced into a Procrustean bed. Genetics, for example, is almost unique in predicting simple rational fractions for its probabilities. Usually there is no theoretical backup for the rates we encounter: there is no reason to believe that the underlying cure rate of a drug is exactly $\frac{3}{4}$, or that a baseball player's built-in batting average is $\pi/10$. While we may be able to approximate rates by a discrete set of alternatives, most rates could be anything from 0 to 1. In addition, a great many variables are measured rather than counted. Measurements are always of limited accuracy, but it is advantageous here too to imagine that the variable can vary by as small an amount as we please. In this chapter we will begin the study of these continuous variables.

4.1 FROM POINTS TO INTERVALS

If we list a finite number of alternatives for a success rate and evaluate the posterior distribution, the probabilities associated with these alternatives indicate how close the "real" success rate lies to each member of the set. In Table 4.1.1 I give the results of a sampling experiment. The probability of a success was .2. I considered the nine alternatives .1, .2, ..., .9 (one of which was actually on the nose), gave them equal prior probabilities as an expression of my ignorance, and sampled 2000 times. The posterior probabilities at each stage were proportional to $p^s(1-p)^{n-s}$; I have listed the complete posterior distributions for various numbers of trials. Here n is the number of trials, s is the number of successes, and s/n is the observed success rate. Note how the observed rate varies. The underlying rate is .2, but the great god Random is at work.

The posterior probability distribution for $n = 100$ is read

$$P(.2) = .3362, \quad P(.3) = .6510, \quad P(.4) = .0128,$$

Table 4.1.1. Sampling Experiment

Posterior probabilities for 9 possibilities
[Actual success rate is .2. Prior probabilities are all .1111.]

n	s	s/n	.1	.2	.3	.4	.5	.6	.7	.8	.9
0	0		.1111	.1111	.1111	.1111	.1111	.1111	.1111	.1111	.1111
1	0	.000	.2000	.1778	.1556	.1333	.1111	.0889	.0667	.0444	.0222
2	1	.500	.0545	.0970	.1273	.1455	.1515	.1455	.1273	.0970	.0545
3	1	.333	.0982	.1552	.1782	.1745	.1515	.1164	.0764	.0388	.0109
4	1	.250	.1483	.2083	.2093	.1757	.1271	.0781	.0384	.0130	.0018
5	1	.200	.2018	.2520	.2216	.1595	.0961	.0473	.0174	.0039	.0003
6	1	.167	.2569	.2851	.2194	.1353	.0680	.0267	.0074	.0011	.0000
7	2	.286	.0993	.2204	.2543	.2092	.1313	.0620	.0200	.0034	.0001
8	2	.250	.1341	.2646	.2672	.1884	.0986	.0372	.0090	.0010	.0000
9	2	.222	.1725	.3026	.2674	.1616	.0705	.0213	.0039	.0003	.0000
10	3	.300	.0631	.2212	.2932	.2363	.1288	.0467	.0099	.0009	.0000
12	3	.250	.1106	.3065	.3110	.1841	.0697	.0162	.0019	.0001	.0000
15	3	.200	.2047	.3985	.2709	.1010	.0221	.0026	.0001	.0000	.0000
20	4	.200	.1877	.4562	.2727	.0732	.0097	.0006	.0000	.0000	.0000
30	7	.233	.0560	.4777	.3784	.0818	.0059	.0001	.0000	.0000	.0000
50	14	.280	.0011	.2546	.6074	.1326	.0043	.0000	.0000	.0000	.0000
100	26	.260	.0000	.3362	.6510	.0128	.0000	.0000	.0000	.0000	.0000
200	48	.240	.0000	.6973	.3027	.0000	.0000	.0000	.0000	.0000	.0000
500	112	.224	.0000	.9983	.0017	.0000	.0000	.0000	.0000	.0000	.0000
1000	211	.211	.0000	1.0000	.0000	.0000	.0000	.0000	.0000	.0000	.0000
2000	382	.191	.0000	1.0000	.0000	.0000	.0000	.0000	.0000	.0000	.0000

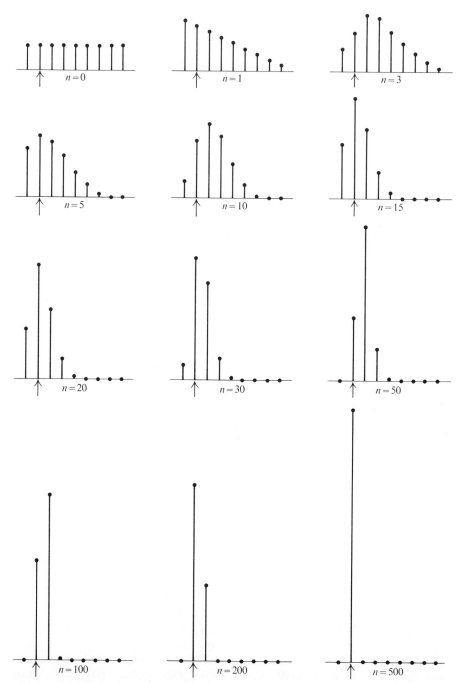

Fig. 4.1.1 Graphs of the posterior distributions in Table 4.1.1. The ↑ marks the underlying rate (.2).

and all the rest are 0 to four decimal places. If you skim the distributions, you will notice that the probability for .2 rises fairly steadily, but it does not begin to dominate until several hundred trials have taken place. *Small counts do not give a very accurate indication of the rate.* Figure 4.1.1 shows the graphs of some of the posterior distributions from Table 4.1.1.

In Table 4.1.2 I exhibit the results of another sampling experiment. This time the underlying rate is .5, not .2. The more important difference is that I have used 99 alternatives: .01, .02, .03, ... , .99, not just 9 as in Table 4.1.1. The posterior distributions are a little harder to read since they are given in double-entry form.

Table 4.1.2. Sampling Experiment

Posterior probabilities for 99 possibilities
[Actual success rate is .5. Prior probabilities are all .0101.]

	.00	.01	.02	.03	.04	.05	.06	.07	.08	.09
\multicolumn{11}{c}{10 trials, 7 successes, rate = .700}
.1										.0001
.2	.0001	.0001	.0002	.0002	.0003	.0003	.0004	.0005	.0007	.0008
.3	.0010	.0012	.0014	.0017	.0020	.0023	.0027	.0031	.0036	.0041
.4	.0047	.0053	.0059	.0066	.0074	.0082	.0091	.0100	.0109	.0119
.5	.0129	.0139	.0150	.0161	.0172	.0183	.0194	.0205	.0216	.0226
.6	.0236	.0246	.0255	.0263	.0271	.0277	.0283	.0288	.0291	.0293
.7	.0294	.0293	.0291	.0287	.0282	.0275	.0267	.0258	.0247	.0235
.8	.0221	.0207	.0192	.0176	.0160	.0143	.0126	.0109	.0093	.0078
.9	.0063	.0050	.0038	.0027	.0018	.0012	.0006	.0003	.0001	
\multicolumn{11}{c}{50 trials, 23 successes, rate = .460}
.2			.0001	.0001	.0002	.0003	.0006	.0009	.0015	.0023
.3	.0034	.0049	.0069	.0094	.0124	.0160	.0201	.0247	.0296	.0346
.4	.0397	.0445	.0488	.0524	.0552	.0569	.0574	.0569	.0552	.0525
.5	.0489	.0447	.0401	.0352	.0302	.0255	.0210	.0170	.0134	.0104
.6	.0078	.0058	.0042	.0029	.0020	.0013	.0009	.0006	.0003	.0002
.7	.0001	.0001								
\multicolumn{11}{c}{500 trials, 258 successes, rate = .516}
.4				.0001	.0005	.0023	.0077	.0215	.0489	.0909
.5	.1384	.1724	.1759	.1469	.1002	.0559	.0254	.0094	.0028	.0007
.6	.0001									
\multicolumn{11}{c}{5000 trials, 2494 successes, rate = .499}
.4								.0001	.0164	.2601
.5	.5562	.1609	.0063							

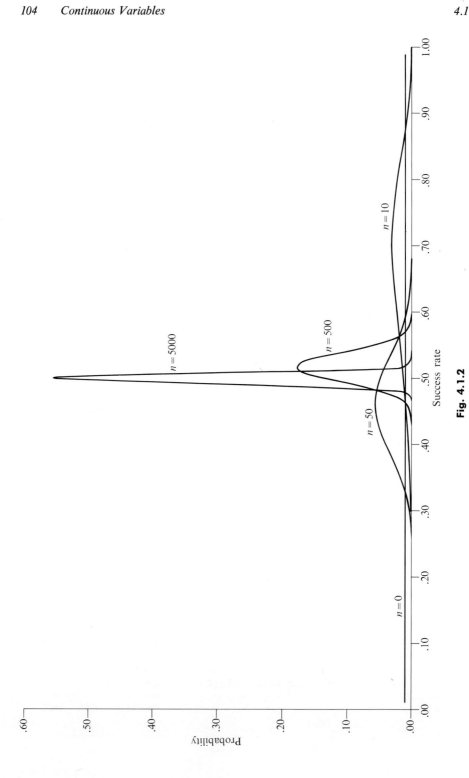

Fig. 4.1.2

The distribution for 50 trials is

$P(.22) = .0001$, $\quad P(.23) = .0001$, $\quad \ldots$, $\quad P(.46) = .0574$,

$\ldots$, $\quad P(.71) = .0001$,

and all the rest are 0 to four decimal places.

It is quite difficult to inspect tables like these; there just are too many numbers. You can see, if you look closely, that the highest point in each distribution is at the observed success rate. For example, in 50 trials there are 23 successes, and the highest point is at $\frac{23}{50} = .46$. It is much more informative to look at Fig. 4.1.2, where the curves for $n = 0, 10, 50, 500$, and 5000 are graphed.

After staring at Table 4.1.2 you have probably concluded that it won't pay to increase the number of alternatives in our set any further. You have also seen that the probabilities as plotted fall on smooth curves. This all suggests that we should go from the probability that the success rate is some *point value* to the probability that the success rate lies in some *interval*. That is, instead of asking, "What is $P(\text{rate} = .9)$?" we would ask, for example, "What is $P(\text{rate between .85 and .95})$?" To do this we need some new definitions—the subject of the next section.

PROBLEM SET 4.1

1. In Table 4.1.1, for $n = 15$, what is the probability that $.2 \leq \text{rate} \leq .4$?
2. In Table 4.1.2, for $n = 10, 50, 500, 5000$, calculate

$$\frac{\text{largest probability in distribution}}{\sqrt{n}}.$$

3. In Table 4.1.2, for $n = 10, 50, 500$, find the pairs of points whose probability is half the largest probability; multiply their distances apart by $\sqrt{n}$.
4. In Table 4.1.1, assume that you have the posterior distribution for $n = 6$. You are now told that a success has been obtained on the seventh trial. Calculate the posterior distribution for $n = 7$.
5. Make 200 trials with $p = .55$. Assume that the alternatives .50 and .55 are equally likely to start with. Are 200 trials enough to do much good in telling them apart?

ANSWERS TO PROBLEM SET 4.1

2.

n	Largest probability	$\sqrt{n}$	Quotient
10	.0294	3.162	.0093
50	.0574	7.071	.0081
500	.1759	22.36	.0079
5000	.5562	70.71	.0079

3.

n	Maximum probability	Locations of half the maximum	Distance	Distance/$\sqrt{n}$
10	.0294	.517, .848	.331	1.05
50	.0574	.378, .543	.165	1.17
500	.1759	.489, .543	.054	1.20

4.2 PROBABILITY DENSITIES

To describe the random behavior of a *continuous variable* X we use a *probability density function*, pd (x) (see Fig. 4.2.1). *Probability* is now represented by *area*, not by the ordinate of the graph (see Fig. 4.2.2). In order for these probabilities to be consistent with our discrete probabilities, we normalize by making the total area equal to 1:

$$\text{total area} = P(-\infty < X < +\infty) = P(X \text{ is something}) = 1.$$

A pd (x) is always ≥ 0. But unlike probabilities, which can be at most 1, pd (x)

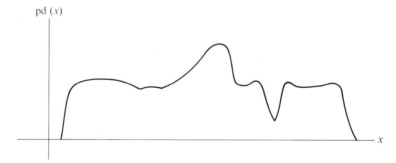

Fig. 4.2.1

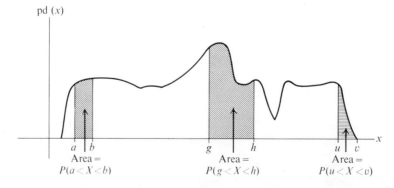

Fig. 4.2.2

has no particular upper limit. If, for example, X was restricted to lie between 0 and .000001, the average value of pd (x) would have to be 1,000,000 so that the total area would be 1. [I will always use the letter pair "pd" to stand for *probability density*, just as I always use P for *probability*.]

There is a small notational point to watch. With a continuous variable the four probabilities

$$P(a \leq X \leq b), \quad P(a \leq X < b), \quad P(a < X \leq b), \quad P(a < X < b)$$

differ only by the areas of the vertical lines at a and b; since the area of a line is 0, the four probabilities must be equal. This is not the case with a discrete variable, where the probability occurs in lumps. For example, if we have the distribution

y	1	2	3	4
$P(y)$	.4	.1	.3	.2

then

$$P(1 \leq Y \leq 4) = 1.0, \quad P(1 \leq Y < 4) = .8,$$
$$P(1 < Y \leq 4) = .6, \quad P(1 < Y < 4) = .4.$$

How do we measure these areas which represent probabilities? We can always resort to counting the squares on graph paper for a quick but crude answer. Most of the time, though, we get a numerical approximation by splitting up the area into a number of thin strips and adding up these smaller areas. There is a systematic way of doing this so that we take care of the little corners and curves at the tops of the small areas. If you have never done it, look at the article on *numerical integration* in Appendix A6. For extensive calculations you'd better consult your computer center. It has tried-and-true subroutines which produce results better and faster than you can imagine. Computers can find areas of the types discussed in this chapter in a few seconds or less.

On rather rare occasions our areas can be evaluated by the formulas of integral calculus, but for most purposes we will work numerically. Here and there I will present some formulas that were derived using calculus, but you don't need to be able to derive them in order to use them. It's necessary, though, that you *recognize* the formal expression for an integral:

$$P(a < X < b) = \int_a^b \text{pd}(x)\, dx.$$

This is *read* "the integral of pd (x) with respect to x from $x = a$ to $x = b$." It *means* "the area bounded by $x = a$, $x = b$, the x-axis, and the curve pd (x)." It is *interpreted* as $P(a < X < b)$.

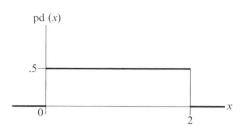

Fig. 4.2.3

Example 4.2.1. Figure 4.2.3 shows the graph of the density

$$\text{pd}(x) = \begin{cases} .5 & \text{for } 0 \le X \le 2, \\ 0 & \text{elsewhere.} \end{cases}$$

The total area is obviously $2 \times .5 = 1$. What is $P(.3 < X < .7)$? Obviously the shaded area in Fig. 4.2.4, which is $.4 \times .5 = .2$. Here we can easily generalize:

$$P(a < X < b) = .5 \times (b - a),$$

so long as both a and b are in the interval from 0 to 2.

Example 4.2.2. Figure 4.2.5 shows the graph of

$$\text{pd}(x) = \begin{cases} \tfrac{1}{2}x & \text{for } 0 \le x \le 2, \\ 0 & \text{elsewhere.} \end{cases}$$

Here

$$\text{total area} = \tfrac{1}{2} \times 2 \times 1 = 1 \quad \text{(triangular area)},$$

$$P(.3 < X < .7) = .4 \times \frac{(.15 + .35)}{2} = .1 \quad \text{(trapezoidal area)}.$$

In general, for a and b both in the interval from 0 to 2,

$$P(a < X < b) = (b - a)\frac{\left[\dfrac{a}{2} + \dfrac{b}{2}\right]}{2} = \frac{b^2 - a^2}{4}.$$

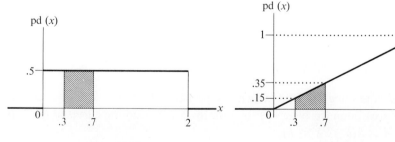

Fig. 4.2.4 Fig. 4.2.5

Example 4.2.3. What about probabilities in the unscaled, unnormalized distribution shown at the beginning of this section? What is the probability that the variable will fall in the interval shown in Fig. 4.2.6?

All we need is the ratio of the marked area to the total area. I put a square grid on the drawing (Fig. 4.2.7) and made a rough count of the squares. I got about 48 in the area between a and b and about 302 altogether. This makes the probability of the interval (a, b) about $\frac{48}{302} = .16$.

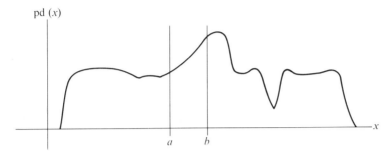

Fig. 4.2.6

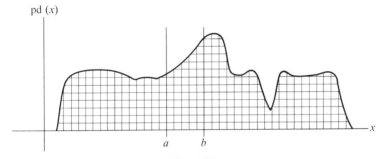

Fig. 4.2.7

Example 4.2.4. Many standard statistical distributions are well tabulated. For example, the table in the back of the book labeled "GAU (∗ | 0, 1), Areas" gives

$$\int_{-1.5}^{1.5} \frac{e^{-t^2/2}}{\sqrt{2\pi}} \, dt = .866.$$

Example 4.2.5. This is the easiest way to compute an integral—by computer!

RESEARCHER. I need to compute

$$\int_0^1 \frac{dx}{1 + x^3}.$$

COMPUTER CENTER CONSULTANT. Our integration subroutine is called ZINT. You use it this way . . .

COMPUTER. 0.83565. $0.03, please.

The cost of computing a few integrals on a computer is so low that it hardly pays to learn much about hand methods. The integral above actually has an exact form, but it is so complicated that there is a good chance that you would make a mistake in using the formula. This is not to say that you won't get a wrong answer when you use a computer. If you don't word your request properly, you may get an answer of −168.42 and a bill for $37.19. *You must check every computation for reasonableness.*

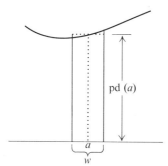

Fig. 4.2.8

There's an approximation which is extremely useful. If w is a *small* interval centered on a (Fig. 4.2.8), we approximate the little area with the curved top by a rectangle:

$$P\left(a - \frac{w}{2} < X < a + \frac{w}{2}\right) \doteq \underbrace{\text{pd}(a)}_{[\text{pd (midpoint)}]} \times \underbrace{w}_{[\text{width}]}.$$

[The symbol $\doteq$ means *is approximately equal to*.] The smaller w is, the more accurate is the approximation.

Example 4.2.6. If pd $(x) = 4x^3$ for $0 < x < 1$, then

$$P(.70 < X < .71) \doteq \underbrace{4(.705)^3}_{[\text{pd (midpoint)}]} \times \underbrace{.01}_{[\text{width}]}.$$

The value calculated from the approximation is .014016105. The exact value is .01401681.

PROBLEM SET 4.2

1. Figure 4.2.9 shows a pd that is 0 outside the interval (0, 3).
   (a) What is the value of k?
   (b) What are
   $$\wp\left(\frac{0 < X < 2}{2 < X < 3}\right) ?$$

2. Figure 4.2.10 shows a pd that is 0 outside the interval (0, 3).
   (a) What is the value of k?   (b) What is $P(.5 < X < 2.5)$?

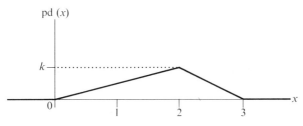

Fig. 4.2.9

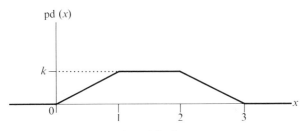

Fig. 4.2.10

3. Suppose

$$pd(x) \propto \frac{1}{1+(x-3)^2} \frac{1}{1+(x-4)^2} \frac{1}{1+(x-1)^2} \quad \text{for } -10 < x < 10,$$
$$= 0 \quad \text{elsewhere}.$$

(a) Sketch the graph.   (b) What is $P(2 < X < 4)$?

4. Figure 4.2.11 shows a pd with some areas indicated. What are

$$\wp\left(\frac{Z > 3}{Z < 3}\right)?$$

5. Let

$$pd(x) \propto \log_{10} x \quad \text{for } 1 \leq x \leq 3,$$
$$= 0 \quad \text{elsewhere}.$$

What is $P(1.5 < X < 2.5)$?

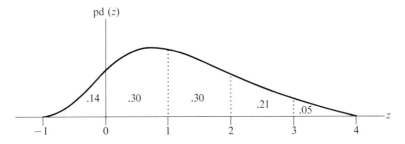

Fig. 4.2.11

6. Figure 4.2.12 shows four graphs with the areas between the curves and the horizontal axis indicated. Label each one as a

$$pd = \text{probability density function,}$$
$$\propto pd = \text{unnormalized probability density function:}$$
one that could be made into a pd by multiplying by a positive constant,
$$N = \text{neither a pd nor an } \propto pd.$$

In each case the variable is restricted to the interval (0, 2).

ANSWERS TO PROBLEM SET 4.2

1. (a) Total area $= \frac{1}{2} \times 3 \times k = \frac{3}{2}k = 1$; $k = \frac{2}{3}$
   (b) $P(0 < X < 2) = \frac{1}{2} \times 2 \times \frac{2}{3} = \frac{2}{3}$, $P(2 < X < 3) = 1 - \frac{2}{3} = \frac{1}{3}$

   $$\mathcal{O}\left(\frac{0 < X < 2}{2 < X < 3}\right) = \frac{\frac{2}{3}}{\frac{1}{3}} = 2$$

3. pd $(-1) \propto .059 \times .038 \times .2 = .00045$,   pd $(3) \propto 1.0 \times .5 \times .2 = .1$,
   pd $(0) \propto .1 \times .059 \times .5 = .00294$,   pd $(4) \propto .5 \times 1.0 \times .1 = .05$,
   pd $(1) \propto .2 \times .1 \times 1.0 = .02$,   pd $(5) \propto .2 \times .5 \times .059 = .0059$,
   pd $(2) \propto .5 \times .2 \times .5 = .05$,   pd $(6) \propto .1 \times .2 \times .038 = .00077$.

   A sketch is shown in Fig. 4.2.13. I'll use Simpson's rule (Appendix A6) to find areas; for the entire area I'll adjoin a 0 at -2 to get an odd number of points:

x	$\propto$ pd (x)	Coefficient	Product	
-2	.00000 $\times$	1	= .00000	
-1	.00045 $\times$	4	= .00180	
0	.00294 $\times$	2	= .00588	
1	.02 $\times$	4	= .08000	
2	.05 $\times$	2	= .10000	.05 $\times$ 1 = .05
3	.10 $\times$	4	= .40000	.10 $\times$ 4 = .40
4	.05 $\times$	2	= .10000	.05 $\times$ 1 = .05
5	.0059 $\times$	4	= .02360	
6	.00077 $\times$	1	= .00077	.50
			.71205	

[x-values are 1.0 apart.]

$$\text{Area } (-2 \text{ to } 6) = \frac{1.0}{3} \times .71205 = .237$$
$$\text{Area } (2 \text{ to } 4) = \frac{1.0}{3} \times .50 = .167$$

$$P(2 < X < 4) = \frac{.167}{.237} = .70$$

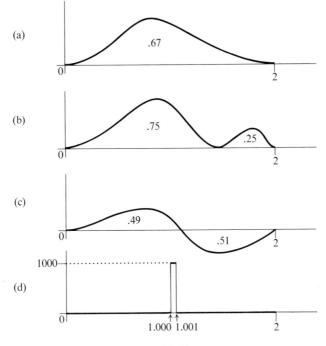

Fig. 4.2.12

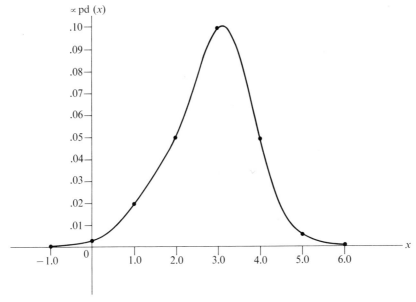

Fig. 4.2.13

5. The application of Simpson's rule goes like this:

x	$\log_{10} x$		Coefficient		Product
1.0	.0000	×	1	=	.0000
1.5	.1761	×	4	=	.7044
2.0	.3010	×	2	=	.6020
2.5	.3979	×	4	=	1.5916
3.0	.4771	×	1	=	.4771
					3.3751

$.1761 \times 1 = .1761$
$.3010 \times 4 = 1.2040$
$.3979 \times 1 = .3979$
$ 1.7780$

[x-values are 0.5 apart.]

$$\left. \begin{aligned} \text{Area (1.0 to 3.0)} &= \frac{.5}{3} \times 3.3751 = .5625 \\ \text{Area (1.5 to 2.5)} &= \frac{.5}{3} \times 1.7780 = .2963 \end{aligned} \right\} P(1.5 < X < 2.5) = \frac{.2963}{.5625} = .53$$

4.3 THE RATE PROBLEM REVISITED

I will now discuss the rate problem with p, the success rate, a continuous variable. This means that all our prior and posterior distributions will be *continuous*, although the likelihood is still *discrete*. The *discrete-discrete* case has already been encountered. We will meet the *continuous-continuous* and *discrete-continuous* cases later.

The first problem is the prior distribution. In most investigations we have little idea beforehand of the percentage of cases that will show a given characteristic. [Genetics, which can predict certain fractions like $\frac{3}{16}$, $\frac{1}{4}$, and $\frac{9}{16}$, is the outstanding exception.] We also usually have no particular reason to believe that the characteristic will be exhibited either in *all* or *none* of the cases; that is, no special emphasis is to be put on $p = 1$ or $p = 0$. Thus a reasonable prior density is the one shown in Fig. 4.3.1. This figure says that, as far as I know, any interval between 0 and 1 is as likely as any other of the same width. [Cf. Fig. 4.5.1.]

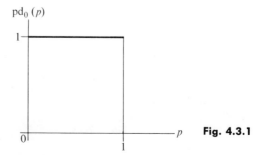

Fig. 4.3.1

Some people get very uneasy because a reasonable prior density for a continuous variable like p does not automatically correspond to a reasonable prior density for some function of p like p^2, $p/(1-p)$, or $\log p$. It would indeed be nice if there were such invariant properties, but usually there aren't. I can only say now that when the number of observations is small, the importance of the prior density is great. You cannot neglect it; you cannot avoid thinking hard about what you know, what you want to do, and which variable you ought to be considering. You should not settle for a standard, *noninformative* prior density like this if you can possibly dig up better information. Please remember that there aren't any wholly objective methods or conclusions in statistics: judgment is involved everywhere.

Our experiment consists of a series of trials in which

a) there are just two possible results: *success* and *failure*, [These are the conventional historical terms. They can mean any two contrasted events: life or death, round or smooth, rain or no rain.]

b) the probability of success, p, is the same on each trial,

c) the trials are independent.

Such trials are called *Bernoulli trials*, after James Bernoulli, an early probabilist.

If we see a particular sequence of s successes and f failures, the probability of that sequence is

$$p^s(1-p)^f$$

for any specified p. By the familiar rule

posterior prob $\propto$ *prior prob* $\times$ *likelihood,*

we have

$$\begin{aligned} \text{pd}(p) &\propto \text{pd}_0(p) \times p^s(1-p)^f \\ &= 1 \times p^s(1-p)^f \\ &= p^s(1-p)^f. \end{aligned}$$

We can normalize (make the area from $p = 0$ to $p = 1$ equal to 1) by multiplying by

$$\frac{(s+f+1)!}{s!\,f!}.$$

Often there is no need to normalize; we may be interested in the ratio of ordinates or the ratio of areas or, perhaps, only in a rough sketch to see the general shape. However, it's nice to tidy things up now and then. To prevent continual recalculation I have placed a *Table of Kernels* at the front of the *Gallery of Distributions*. The *kernel* is the expression for the probability density stripped of all multipliers not involving the variable under study. In this case we recognize from the table that $p^s(1-p)^f$ is of the form $x^a(1-x)^b$, and the table tells us that we have the BETA distribution. We then look up BETA in the Gallery to find the details.

We'll be referring to distributions like BETA so often that it's important that you get my notation straight. It's been designed to agree with our conditional outlook and the demands of present computer programming languages. A *density* will always be written with a name in *lower-case* letters. Thus

$$\text{beta}(.6 \mid 5, 10)$$

means "the ordinate of the BETA density for $p = .6$ when the parameters are $s = 5$ and $f = 10$." If I talk about the density generally without reference to a specific point, I use an * to hold the place:

$$\text{beta}(* \mid 5, 10).$$

Besides the density of a variable, we also talk about its *cumulative distribution function*. This is the area stretching all the way from the extreme left up to a specified point, in other words, $P(variable \leq certain\ value)$. Here we write the name in SMALL CAPITALS. Thus

$$\text{BETA}(.6 \mid 5, 10)$$

means P(a BETA variable with $s = 5$ and $f = 10$ is less than or equal to .6). If I speak of the cumulative distribution without reference to a particular point, I again use an *:

$$\text{BETA}(* \mid 5, 10).$$

And if I just refer to the distribution generally, I use the name in SMALL CAPITALS to attract attention.

Two last points:

$$p \sim \text{BETA}$$

means "p is distributed according to the BETA distribution." And note that PD is the generalized cumulative distribution corresponding to the generalized density pd.

That flat rectangular density in Fig. 4.3.1 is

$$\text{beta}(* \mid 0, 0),$$

the density of p when we have observed 0 successes and 0 failures. After the s successes and f failures above, the posterior density is

$$\text{beta}(* \mid s, f).$$

Thus, if we start with a flat prior density, the two parameters are always the numbers of successes and failures.

In the Gallery I have given graphs of BETA. You can see that as soon as both s and f get to be, say, at least 2 or 3, the curves have a rather similar shape. The peaks occur at different points in the interval (0, 1) and the heights of the peaks differ; yet, roughly, they all are high in the middle and decrease rapidly toward the ends. Near the peaks they give the feeling of symmetry. We shall use these characteristics later on to get approximations to the BETA distribution.

Figure 4.3.2 shows the successive posterior densities obtained in a sampling experiment. You can see how the original flat density becomes higher and narrower as the information comes in. After 500 observations it is pretty well centered about the correct value, but nearby values are by no means ruled out.

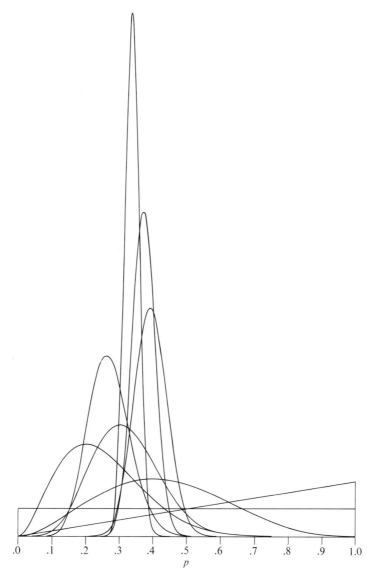

Fig. 4.3.2 Successive posterior densities for a sampling experiment with $p = \frac{1}{3}$. The nine curves correspond to $n = 0, 1, 5, 10, 20, 50, 100, 200,$ and 500.

Now a posterior probability distribution sums up all the information you have about a variable, but you may wonder what to do with it. You certainly have to keep the whole thing if you are going to do further experiments and combine the results. This is not at all difficult to do in the present case, since you need remember only the expression beta $(* \mid s, f)$. But in more complicated cases there may be quite a bit of data. The natural urge is to summarize the information; it is important to sacrifice as little information as possible. In this case you should report either s and f, or n and s, or even s/n and n; the whole distribution can be reconstructed from any of these pairs. Sets of numbers which have this property are called *sufficient statistics*. If you report s/n alone—perhaps it was .5—a reader cannot tell if you had 1 success out of 2 or 1000 successes out of 2000. The uncertainty in your knowledge is certainly different in the two cases. Remember, though, that the use of sufficient statistics depends on your model's being reasonably correct. Check periodically that your assumptions are satisfied.

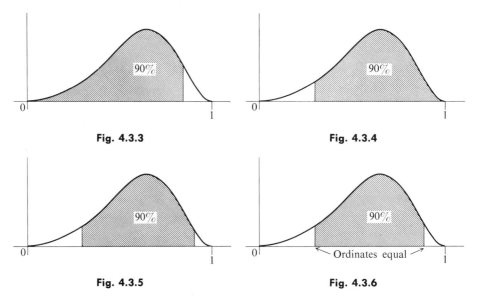

Fig. 4.3.3

Fig. 4.3.4

Fig. 4.3.5

Fig. 4.3.6

If we have a distribution like BETA which is usually high in the middle and low at the ends, we can give a good idea of where the bulk of the probability lies by quoting intervals that contain some specified large fraction of the probability. Where, for example, does 90% of the probability lie? Obviously this question does not have a unique answer. We could start at $p = 0$ and go to the right until we had accumulated 90% (Fig. 4.3.3). Or we could start at $p = 1$ and go to the left until we had accumulated 90% (Fig. 4.3.4). And in between there is an infinity of choices (Fig. 4.3.5). The criterion that makes our choice unique is this: we choose the interval such that *every ordinate inside is greater than every ordinate outside*.

That is, we start at the *mode* (the location of the largest ordinate) and go both ways, always keeping the end ordinates equal (Fig. 4.3.6). This is called a *highest density region* (HDR), since the largest ordinates are inside it. You can see quite readily that the interval is therefore the shortest one possible. [I have illustrated a BETA distribution, which is confined to the interval (0, 1). This restriction is not an essential feature.] The concept of HDR is also useful for J-shaped distributions like beta (* | 3, 0), but not for U-shaped or multimodal distributions.

Example 4.3.1. What is the 90% highest density interval for beta (* | 6, 3)? Look at the table labeled "BETA (* | s, f), Highest density regions" in the back of the book. For $s = 6, f = 3$, and column .90, we find

$$(.414, .866).$$

This means that 90% of the probability in this distribution lies between $p = .414$ and $p = .866$. Moreover, the ordinates within this interval are higher than the ordinates outside the interval. We may also say that the probability is .90 that the underlying success rate of a sample which has 6 successes and 3 failures lies between .414 and .866. This isn't very restrictive, is it? If we want to be 99% sure that the success rate is in the stated interval, we have to expand the interval to

$$(.282, .935).$$

It just takes more than 9 observations to reduce the uncertainty very much.

Example 4.3.2. Recall our sick minks. The drug cured 20 minks and didn't cure 9. What is the 95% HDR for the cure rate of the drug? We have $s = 20, f = 9$, and must interpolate because the table does not have this particular combination. The answer is roughly

$$(.51, .83).$$

Remember that a single HDR does not describe a density too well. All you know is that the indicated interval contains a given amount of probability; presumably the inner part of the interval has higher ordinates, but you really can't say more. You would need a whole series of HDR's to reconstruct the distribution that a set of sufficient statistics does so easily. However, the HDR provides an interpretation that is not apparent from the bare mathematical form. An HDR is sometimes called a *Bayesian confidence interval* or a *credible region*.

Another way to describe a distribution is to quote the odds that the variable is above (or below) a certain critical point. Here we might not particularly care exactly what p is, so long as we are fairly sure that it is high (or low) enough. The procedure is simple: we compare the probabilities, i.e., areas, above and below the critical point. Generally, this means calculating an integral, but there is a table of *right tail/left tail* for BETA in the back of the book. The *right tail* is the area to the right of a point; the *left tail*, the area to the left. The left tail is the *cumulative probability* which I mentioned earlier.

Example 4.3.3. We observe 0 defects in 50 parts. If we assume that this is a sample of Bernoulli trials, can we get some idea of an upper limit for the defect rate?

We know that the posterior density of p is

$$\text{beta} (* \mid 0, 50)$$

if a flat prior is assumed. We look at the table labeled "BETA $(* \mid s, f)$, right tail/left tail." The table gives the values of p for which this ratio is a nice round value. The line

s	f
0	50

is read as follows:

At $p = .002$, the right tail is 10 times as large as the left tail.
At $p = .046$, the right tail is $\frac{1}{10}$ the size of the left tail, etc.

Thus we can report results like this:

The odds are $\frac{10}{1}$ that $p < .046$.
The odds are even that $p < .014$.
The odds are $\frac{10}{1}$ *against* p's being less than .002.

These tables and the ones of HDR's are decidedly limited in size. In the next chapter we will learn convenient ways to approximate our BETA distributions. But although calculation of complete tables like these is burdensome, there is no difficulty or expense in calculating the import of the *particular* results *you* obtain in an experiment. It is nonsensical to spend a lot of time and effort on your laboratory work and begrudge the nickels and dimes it takes to evaluate your results properly. Learn to use your computer!

PROBLEM SET 4.3

1. What is the 90% HDR for the defect rate in Example 4.3.3?
2. In Example 4.3.2, what are the odds that the cure rate is greater than $\frac{2}{3}$?
3. In Example 4.3.2, what is the 95% HDR for the failure rate of the drug?
4. A sample of 25 electrical resistors is taken at well-spaced intervals from the production line. One was found to be defective. If this is a random sample from the production line and we had only vague information before, what is the 90% HDR for the defect rate?

4.4   Estimation of Rates   121

5. There were 60 girls present in the A–Ch section of the Freshman Orientation Lecture. Ten of them were redheads. If this is a representative sample of the freshman class as a whole, what is the 99% HDR for the fraction of freshman girls that are red-headed (for that instant in time, of course)?

6. The standard drug for treating Disease X cures about 70% of the cases. A new drug has been developed and proves very successful in animal experimentation. It is now tested on 60 humans and cures 50. If this sample of patients is representative of the people on whom the drug would be used, what are the odds that the cure rate of the new drug is at least 75%?

7. What BETA distribution approximately expresses the following information: the mode is near $p = \frac{1}{5}$, and 95% of the probability lies between $\frac{1}{10}$ and $\frac{1}{3}$?

ANSWERS TO PROBLEM SET 4.3

2. In the table:

s	f	$\frac{1}{1}$
20	8	.704
20	10	.659

The answer is thus very close to $\frac{1}{1}$. This means it is as likely to be above $\frac{2}{3}$ as below it.

4. Here $s = 1$ and $f = 24$. This combination is not in our table, but by rough graphical interpolation we get a 90% HDR of (.004, .15). Note that this is *not* symmetric about $\frac{1}{25} = .04$.

5. The table gives 6.9% to 31.1%. But note that while minks and resistors came from populations so large that we had no qualms about regarding them as infinite, this freshman class is considerably smaller. Here we already know definitely the hair color of 60 of them. If this is a large fraction of the class, we should use the finite population analysis demonstrated in Example 3.2.3.

7. We want to look for (s, f)-combinations where $s = f/4$, since $\frac{1}{5} = s/(s + f)$. These requirements are approximately satisfied by a combination around (8, 32) or (10, 40). You sometimes can persuade people to give you probabilistic information in the form stated in this problem; this is how we can turn it into mathematical form.

4.4   ESTIMATION OF RATES

By now you are probably thinking to yourself: "I don't need all these odds and intervals; I need a value of p to plug into some equations I have. Do I use s/n or don't I?" The answer is that s/n is as satisfactory as anything else you might use when n is large. For smaller n, however, you can make a little money by using a slightly different value. The reasoning is quite basic, and goes as follows.

What is the probability that the next try will be a success? This is that old "compound probability" problem; remember drawing colored balls from urns? We first choose a p under the guidance of our probability distribution for p. Then we make a trial with our chosen p as the probability of a success. Earlier we calculated the compound probability of success as

$$\sum p\, P(p).$$

In the continuous case, of course, we calculate

$$\int p\, \text{pd}\,(p)\, dp.$$

In either case I have written down the expression for the *expected value*, or *mean*, of p. Do you remember?

The expected value of a BETA variable has been calculated once for all to be

$$\frac{s+1}{s+f+2} = \frac{s+1}{n+2}.$$

Example 4.4.1. In curing our minks we had 20 successes and 9 failures. Our estimate of p is

$$\frac{20+1}{20+9+2} = \frac{21}{31} = .68.$$

You can see that this value is always somewhat more moderate—closer to .5—than s/n. Is this reasonable?

i) $s = 0, f = 0$: $E(p) = \frac{1}{2}$. This seems right in view of our flat prior density.
ii) $s = 1, f = 0$: $E(p) = \frac{2}{3}$. $s = 0, f = 1$: $E(p) = \frac{1}{3}$. I said at the start specifically that we had no reason to believe p was exactly 0 or 1. Certainly we wouldn't be convinced of either on the strength of one vote. [If we think $p = 0$ and $p = 1$ are definite possibilities, we need to use the different kind of prior distribution described in Section 4.5.]
iii) As $n \to \infty$,

$$\frac{s+1}{n+2} \to \frac{s}{n}.$$

That is, as n gets larger, the difference between these two possible estimates disappears.

Example 4.4.2. For the minks

$$\frac{s}{n} = \frac{20}{29} = .69.$$

The estimate s/n is called the *maximum likelihood estimate* because it is the value at which the likelihood, $p^s(1 - p)^{n-s}$, is a maximum. When we use a flat prior, the posterior density is just the normalized likelihood; then the maxima of the likelihood and the posterior density coincide.

We can also look at the estimation problem from the point of view of *decision theory*, which will be covered more extensively in Chapter 8. Decision theory is concerned with minimizing the loss you expect to incur when you make a decision with incomplete knowledge. What kind of a loss is involved in trying to estimate p? Obviously it is somehow measured by how far you are from the true value. With a continuous variable like p, you're sure to be wrong, that is, never right on the nose. There was a chance to be exactly right when we had discrete alternatives. Calculations have been carried out and the results are these: Let $\hat{p}$ be the estimate and let p_0 be the "true" value (for this problem). Then

a) if the loss is $\propto (\hat{p} - p_0)^2$, that is, to the square of the error, the expected loss is minimized by choosing $\hat{p} = E(p)$;

b) if the loss is $\propto |\hat{p} - p_0|$, that is, to the magnitude of the error, the expected loss is minimized by choosing $\hat{p}$ = median (p). [The median is the point where 50% of the probability lies to the left and 50% to the right.]

Most of the ordinary usages of statistics are concerned with the squares of errors; that is, we are very worried about very large errors. Using the absolute value of the error de-emphasizes the importance of large errors somewhat. Since the median lies between the mean and the mode, all three are about equal for large n.

This discussion was based on the use of the first power of p in your equations. If you have to give a guess for p^2, you do not use $\hat{p}^2$, the square of your estimate for p, unless n is large. Instead you calculate

$$E(p^2) = \sum p^2 P(p) \quad \text{or} \quad \int p^2 \, \text{pd}(p) \, dp,$$

and similarly for other functions of p. For example, the probability that the next two trials are both successes is

$$\text{not} \quad E^2(p) \doteq \left(\frac{s+1}{s+f+2}\right)^2 \quad \text{but} \quad E(p^2) = \frac{(s+1)(s+2)}{(s+f+2)(s+f+3)},$$

which is larger. A little calculation also shows that if you have observed s successes without any failures, the probability that the next m trials will all be successes is

$$\frac{s+1}{m+s+1}.$$

Thus, even if we observed a long string of successes we would never quite believe that $p = 1$. This is due to our choice of prior distribution in which we gave no credence to the possibility that p might be exactly 0 or 1. I will discuss some very different prior distributions in Section 4.5.

Undoubtedly, you need *point estimates*. The great danger is that you will glibly quote an isolated value and then forget the uncertainty associated with it. You have to use the entire posterior distribution to combine evidence or to make statements about the probabilities of intervals. It is, in fact, a Bayesian principle to keep information in the form of a distribution until the last possible moment. Yet estimates and HDR's are different in a significant way: HDR's describe the density; a point estimate must also take into account the consequences of the particular choice.

I have been using a flat prior density in order to show no preference for any interval or any particular point. It was chosen to express our complete ignorance of the situation. But we are not always completely ignorant; sometimes we have information, perhaps from a previous experiment, which we feel is relevant to the case at hand. If it's not too vague and tenuous, we want to use it in our prior distribution. [If it's really vague or half-remembered, we're better off with our flat prior.] The easiest kind of prior to use is the posterior BETA density from a previous experiment, say beta $(* \mid x, y)$. Then if we obtain s successes and f failures in the new experiment, we have

$$\text{pd}(p) \propto \text{beta}(p \mid x, y) \times p^s(1-p)^f$$
$$\propto p^x(1-p)^y \times p^s(1-p)^f = p^{s+x}(1-p)^{f+y},$$

so

$$\text{pd}(p) = \text{beta}(p \mid s+x, f+y).$$

We can reword this as follows: If you have two experimental results bearing on the same p, you combine them by adding together the successes and adding together the failures. The final inference is based on the grand total of successes and the grand total of failures.

Example 4.4.3. Our information is as follows:

first experiment: 10 successes, 5 failures;
second experiment: 21 successes, 16 failures;
combined results: 31 successes, 21 failures.

Then

$\hat{p}_1$, the first estimate of p, is $\frac{11}{17}$,
$\hat{p}_2$, the estimate of p from the second experiment alone, is $\frac{22}{39}$,
$\hat{p}$, the estimate of p from the combined information, is $\frac{32}{54}$.

Note that

$$\frac{32}{54} \neq \frac{(\frac{11}{17} + \frac{22}{39})}{2};$$

that is, you do *not* average the individual estimates of p when you consolidate information.

Example 4.4.4. Recall that in Problem 3.2.5 I asked you to simulate the production of an automatic machine until you got a bad part, to do this five times, and then to calculate

$$\frac{\sum \#s + 1}{\sum \#s + 7}.$$

[We were then using $\#s$ for what we have lately called s.] This is just the use of the above combining formula: The total number of successes is $\sum s$; the total number of failures is 5. Thus you calculated

$$\frac{\sum s + 1}{\sum s + \sum f + 2}.$$

But your information may be from a different kind of source. If it can be put into the BETA form without too much mangling (see Problem 4.3.7), the solution goes as above. Sometimes, however, this would involve too much distortion of the information. All is not lost, you just are forced to apply your general procedure rather than bask in the luxury of a neat solution. Simply use the familiar

posterior density $\propto$ prior density $\times$ likelihood.

The likelihood for the new experiment is $p^s(1 - p)^f$, as before. Your prior is something—either an expression involving p, or a graph, or a reasonably detailed list of values for a selection of points. When you multiply the prior by the likelihood, *point by point*, you get your unnormalized posterior density. It may or may not have a compact mathematical form, but that is not important; you can always normalize by numerical integration.

Example 4.4.5. In Fig. 4.4.1 prior, likelihood, and posterior are all given as graphs with arbitrary vertical scales. Here I picked nine values of p, multiplied the prior and likelihood together at these points (Table 4.4.1), and drew a quick sketch. Please don't quibble about my decimals or the accuracy of my sketch. I expect you or your computer to do the problem more accurately and thoroughly. Many computers can also draw the sketch for you. [Remember that the numbers given are in arbitrary units; the posterior distribution must eventually be normalized.]

126 Continuous Variables 4.4

Table 4.4.1

Prior	× Likelihood	= Posterior
9.5	0	0
7.8	2	15.6
6.6	5	33
5.3	8	42.4
4.3	8.9	38
3.5	8	28
2.3	5	11.5
1.3	2	2.6
.0	0	0

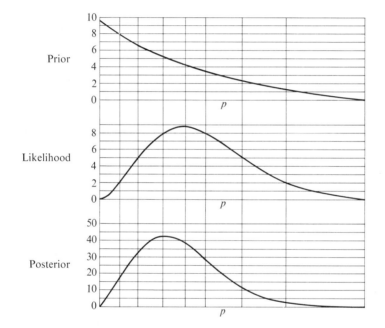

Fig. 4.4.1

One caution about combining information: If the other information comes from somebody else's experiment, make sure that his prior is acceptable to you. Maybe you will find it preferable to extract his likelihood function alone, but this involves knowing his prior density. [It's not always safe to assume it was flat.] You also must take pains that the same information is not used twice.

PROBLEM SET 4.4

1. In Problem 4.3.4 we observed 1 defect in 25. What is $E(p)$?
2. In Problem 4.3.6 we observed 50 cures out of 60. What is $E(p)$?

3. What are $E(p)$, mode (p), $E^2(p)$, and $E(p^2)$ for BETA $(*\mid 20, 12)$?
4. What are $E(p)$, mode (p), $E^2(p)$, and $E(p^2)$ for BETA $(*\mid 4, 12)$?
5. Suppose the prior density of p is 0 at $p = .34$. How many successes are required to make the posterior density at this point at least 0.01?
6. You are trying to make an inference about a rate p. You observe 6 successes and 2 failures. You have other *sure* information that p cannot be less than .6, but the exact value is not predicted. Sketch the posterior density of p. What is $E(p)$?
7. The prior density is $\propto 1 - |p - \frac{1}{2}|$ for $0 \le p \le 1$, and the likelihood is $\propto p^3(1-p)^2$. Sketch the posterior density.
8. You are investigating the fraction of houseflies killed within two hours by a new insecticide. On Thursday you observe that 199 out of 251 die within two hours. On Friday, in a new setup, you observe that 156 out of 218 die within two hours. You are quite confident that the experimental conditions are essentially the same on the two days. The only prior information you have is that the insecticide must be good for something or the manufacturer would not have paid you for testing it. What is the posterior distribution of the fraction of houseflies killed by the insecticide within two hours under your laboratory conditions?
9. Because of genetic differences, some people can taste phenyl thiocarbamide and others cannot. Three students from a psychology class were sent out to get data. Their results were as follows:

> student 1: 109 persons tested, 23% could not taste;
> student 2: 43 persons tested, 21% could not taste;
> student 3: 71 persons tested, 28% could not taste.

If we assume that the experiments were conducted equivalently and no one got interviewed twice, what is the estimate of the fraction of people who cannot taste this chemical?

10. Show that

$$\frac{s+1}{n+2} < \frac{s}{n} \quad \text{for} \quad s > \frac{n}{2}, \qquad \frac{s+1}{n+2} > \frac{s}{n} \quad \text{for} \quad s < \frac{n}{2}.$$

ANSWERS TO PROBLEM SET 4.4

1. $\frac{2}{27} = .074$
3. $E(p) = \frac{21}{34} = .618$; mode $(p) = \frac{20}{32} = .625$; $E^2(p) = .618^2 = .382$; and $E(p^2) = (21 \times 22)/(34 \times 35) = .388$
6. The prior density, likelihood, and posterior density are sketched in Fig. 4.4.2. We calculate $E(p)$ by numerical integration. We first find the integral of $p^6(1-p)^2$ from .6 to 1; then the integral of $p \times p^6(1-p)^2$ for the same interval. The calculations are shown in Table 4.4.2.

Table 4.4.2

p	$p^6(1-p)^2$	Coefficient	Product
.6	.0465 × .16 = .0074	1	.0074
.7	.1175 × .09 = .0106	4	.0424
.8	.2625 × .04 = .0105	2	.0210
.9	.5310 × .01 = .0053	4	.0212
1.0	1.0000 × .00 = .0000	1	.0000
			.0920

$$\frac{.1}{3} \times .0920 = .00307$$

p	$p \times p^6(1-p)^2$	Coefficient	Product
.6	.6 × .0074 = .0045	1	.0045
.7	.7 × .0106 = .0074	4	.0296
.8	.8 × .0105 = .0084	2	.0168
.9	.9 × .0053 = .0048	4	.0192
1.0	1.0 × .0000 = .0000	1	.0000
			.0701

$$\frac{.1}{3} \times .0701 = .00234$$

$$E(p) = \frac{.00234}{.00307} = .763$$

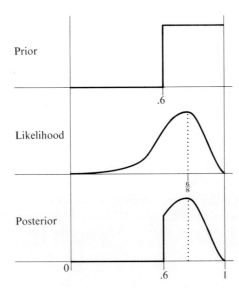

Fig. 4.4.2

8. I will assume that we can combine the observations. [If this is not true, we cannot use the technique just described.] There are 355 successes in all, and 114 failures:

$$\text{pd}(p) = \text{beta}(p \mid 355, 114), \qquad E(p) = \tfrac{356}{471} = .758.$$

Unfortunately, it is often true that experimental setups cannot be reestablished exactly; extreme care must be used. As a consumer of insecticides [I spray, not eat!] I must point out that the kill rate achieved in the laboratory is not necessarily that achieved by the casual purchaser.

9. The first difficulty is that the students have reported their results in percentages; we have to reconvert to actual numbers before we can combine:

	Not taste	Taste
1	25	84
2	9	34
3	20	51
	54	169

$$E(p) = \tfrac{55}{225} = .244$$

Now let me point out that there was no attempt made to define the population under study. Who was interviewed: students, faculty, passersby, townspeople? Was there any attempt made to pick a random sample? If you don't define a population and sample properly from it, the numbers you have collected apply to the people you have sampled but to no more. Next, the literature on phenyl thiocarbamide shows that generally more women than men can taste this chemical. The experiment would have been more interesting had the results been separated by sex. Then, too, the percentage of nontasters varies according to ethnic origin; American Indians, for example, are nearly 100% tasters. Was there any opportunity for exploring this? Lastly, was the intended population large enough to be treated as infinite? See Boyd (1952), page 278.

4.5 TESTING PRECISE POINT HYPOTHESES

Most of the time we are interested in estimating rates and their uncertainties. In relatively rare circumstances we might want to ask, "Is $p = \tfrac{1}{2}$ (or $\tfrac{3}{4}$, 0, 1, or some other specified number)?" Not just close—that we can handle by areas—but right on the nose. We cannot use the ordinary prior density here because the line (ordinate) at a point has 0 area. If we actually contemplate that p is some precise number specified in advance, we have to put a *lump* of probability on that point to show that we have some interest in it. Our distribution then becomes a *mixture* of discrete and continuous. But everything proceeds quite naturally.

Example 4.5.1. We want to estimate p, but believe there is a distinct possibility that p may be exactly 0 or 1. A possible prior density is shown in Fig. 4.5.1. Here

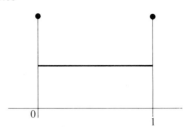

Fig. 4.5.1

I put a lump of .25 on $p = 0$, a lump of .25 on $p = 1$, and spread the rest out evenly from 0 to 1. How you make the allocation depends on your prior opinion, but it isn't that critical. Now, as soon as we have gotten both a success and a failure, the estimation problem will be back to what it was before: both the lumps will have disappeared. But if, say, no failures are observed, the lump at $p = 1$ becomes increasingly important. Our estimate of the chances of an uninterrupted sequence of successes will get closer and closer to 1. After 50 successes, we will have

likelihood: $P(50, 0 \mid 1) = 1$, $P(50, 0 \mid p) = p^{50}$,

posterior: $P(1 \mid 50, 0) = \dfrac{.25 \times 1}{.25 \times 1 + .50 \times \int p^{50} \, dp} = \dfrac{51}{53}$

and

pd $(p \mid 50, 0) = \tfrac{2}{53}$ beta $(p \mid 50, 0)$.

Then

$E(p) = \tfrac{51}{53} + \tfrac{2}{53} \tfrac{51}{52} = .999$ instead of $\tfrac{51}{52} = .981$.

Example 4.5.2. Concern over $p = 0$ or $p = 1$ is apt to be short-lived, since either can be definitely disproved by a single observation. Suppose, however, we want to know whether $p = v$, where v is some intermediate value. What are

$$\mathfrak{O}\left(\dfrac{p = v}{p \neq v}\right) = \dfrac{P(p = v)}{P(p \neq v)}?$$

Again we have to give v a lump of probability and spread the rest out over the interval 0 to 1. We calculate the posterior distribution, and then take the ratio of the lump on v to the integral of the alternatives. This is very much like the composite alternative problem I discussed in Section 3.4. If I put half on $p = v$ and spread out the other half evenly, it turns out that

$$\mathfrak{O}\left(\dfrac{p = v}{p \neq v} \,\bigg|\, s, f\right) = \text{beta}(v \mid s, f).$$

Thus if $s = 25, f = 12$, and $v = \tfrac{1}{2}$,

$$\mathfrak{O}\left(\dfrac{p = .5}{p \neq .5} \,\bigg|\, 25, 12\right) = \text{beta}(.5 \mid 25, 12) = .512.$$

4.5 Testing Precise Point Hypotheses

That is, it's 2 to 1 that $p \neq \frac{1}{2}$. Again, if your prior odds were not $\frac{1}{1}$, you should treat beta $(v \mid s, f)$ as the *factor* to be applied to those prior odds.

In the BETA section of the Gallery of Distributions there is a chart that can be used for $v = \frac{1}{2}$. It really is just a contour map of beta $(.5 \mid s, f)$. If you have other v's in mind, you'll have to calculate them yourself.

Example 4.5.3. A 2×2 (read 2-by-2) classification scheme such as the one below is called a 2×2 *contingency table*.

	Vocabulary test score		
	High	Low	
Girls	50	31	81
Boys	39	39	78
	89	70	159

A natural question is: Are the classifications independent or not? Do girls get higher scores on vocabulary tests? We must assign some lump of probability to the hypothesis of independence and spread the rest over all the alternatives. The hitch is that the alternatives to independence are not that clear-cut, nor is it obvious how they should be weighted. It is fair to say that controversy exists. A solution by Jeffreys (1961) is

$$\mathcal{F}\left(\frac{\text{independence}}{\text{dependence}}\right) = \frac{(\text{smallest marginal total} + 1)! \prod (\text{other marginal total})!}{(\text{total})! \prod (\text{table entry})!}$$

Here $\prod$ means *product*. [This should not be confused with π, which has the value 3.1416.] See Appendix A2 if you aren't familiar with this sign.

In the contingency table above, the smallest marginal total is 70. Therefore

$$\mathcal{F}\left(\frac{\text{independence}}{\text{dependence}}\right) = \frac{71! \, 89! \, 81! \, 78!}{159! \, 50! \, 31! \, 39! \, 39!}.$$

To calculate this I use logarithms. There is a table of log-factorials in Appendix B3. From it I get

(71!)	101.9297	(159!)	282.4693
(89!)	136.2177	(50!)	64.4831
(81!)	120.7632	(31!)	33.9150
(78!)	115.0540	(39!)	46.3096
	473.9646	(39!)	46.3096
			473.4866

so

$$473.9646 - 473.4866 = .4780 = \text{log-factor}.$$

Thus we get a factor of 3 for independence.

PROBLEM SET 4.5

1. After fitting a model to data, a statistician carefully examines the residuals, the variation unexplained by the model. When the observations occurred in a definite time sequence, he wonders if any of this unexplained variation could be variation due to changes as time progressed. A quick way of investigating this (but one which ignores some information) is the *sign test*. You look to see whether the signs of the residuals tend to persist. If time had no effect you would expect that the sign would change from one residual to the next about half the time and would stay the same half the time. Suppose we have 20 observations. We try to explain the data by a model, and find these differences between the observations and the predictions of the model:

    $-.19, .66, .54, 1.22, .32, -.31, .78, .31, 1.78, .84, -.06,$
    $-.11, -1.92, -.87, -.30, -.56, -.35, -.20, -.43, -1.04.$

    If p is the fraction of times two adjacent residuals have the same sign, what is

    $$\mathcal{F}\left(\frac{p = .5}{p \neq .5}\bigg|\text{ these residuals}\right)?$$

2. Describe how a psychologist might use the sign test to evaluate the performance of a subject trying to follow a straight line with a pencil. Consider the independence of his observations. What are the arguments for and against a flat distribution for the alternatives to $p = \frac{1}{2}$?

3. There were 24 fatal accidents in your locality last year. In the first six months of this year there were 18. Has the rate increased?

4. A poll is taken to find the support for a proposed ordinance. The political affiliation of the respondent is also noted. Analyze the resulting contingency table:

	Support	Oppose	
GG Party	24	17	41
RR Party	21	10	31
	45	27	72

5. Analyze the following contingency table giving the results of the Salk vaccine test:

	No polio	Polio	
Control	201,114	115	201,229
Vaccinated	200,712	33	200,745
	401,826	148	401,974

6. A sociologist is trying to find out whether cigar smoking is associated with being an executive. He plans to sample both executives and production workers. His first 11

4.5 *Testing Precise Point Hypotheses*

samples all turn out to be executives, 10 of whom smoke cigars. What factor does this give toward the association of cigar smoking with level of employment?

ANSWERS TO PROBLEM SET 4.5

1. Let
$$1 = \text{same sign on two adjacent residuals,}$$
$$0 = \text{different sign.}$$

The sequence of changes is 0, 1, 1, 1, 0, 0, 1, 1, 1, 0, 1, 1, 1, 1, 1, 1, 1, 1, 1, so $s = 15$, $f = 4$. Hence

$$\mathcal{F}\left(\frac{p = .5}{p \neq .5} \bigg| 15, 4\right) = \text{beta}(.5 \mid 15, 4) = \frac{20!}{15!\,4!}(.5)^{19}.$$

Then
$$\text{log-factor} = 18.38861 - 12.1165 - 1.3802 - 19(.3010) = -.8296,$$
so
$$\text{factor} = .148 = 1/6.75.$$

This simple test gives a factor of 6.75 for there being some time effect. Presumably we would now want to examine the *magnitudes* of the residuals to get some further clues. Of course, a factor of 6.75 is not an awful lot, but experience has shown that there usually is some kind of time effect. This means that the prior odds for an effect were also greater than 1.

3. We are testing whether 24 and 18 are in the ratio 2 to 1, since the time periods are 12 months and 6 months. This is the same as testing $p = \frac{2}{3}$. Thus

$$\mathcal{F}\left(\frac{\text{no change}}{\text{change}}\right) = \text{beta}(\tfrac{2}{3} \mid 24, 18)$$
$$= \frac{43!}{24!\,18!}\left(\frac{2}{3}\right)^{24}\left(\frac{1}{3}\right)^{18}$$
$$= \frac{2.3}{1}.$$

There is a chance that accidents could be regarded as independent. This would not be true of injuries or deaths, which tend to occur in bunches. In this problem seasonal variation may play a part: in northern areas the first six months of the year are the iciest months.

4. The factor for independence of the categories—that is, that political affiliation did not affect support of the ordinance—is

$$\frac{28!\,45!\,31!\,41!}{72!\,24!\,17!\,21!\,10!} = \frac{4}{1}.$$

This is mild evidence in favor of no connection.

4.6 EXPERIMENTAL SAMPLING FROM CONTINUOUS DISTRIBUTIONS

In Section 2.4 we studied the drawing of samples from discrete distributions; now we look at continuous ones. Special tables of random values have been prepared for the more common types. In Appendix C2 you will find random values from the *Gaussian* or *normal distribution,* a workhorse which will be discussed in the next chapter. The *rectangular distribution* of Fig. 4.6.1 has its own obvious answer. The flat curve says that all values between 0 and 1 are equally probable; therefore we just form a random number between 0 and 1 to as many decimal places as we like and, presto, we have the random value. The four random digits 4, 9, 7, and 1, for example, form the value .4971. This, then, is our sample. [But we seldom need random values to four decimal places!]

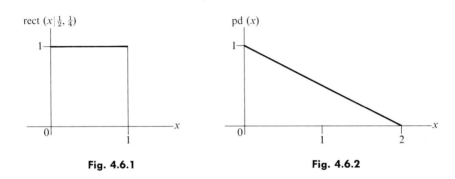

Fig. 4.6.1   Fig. 4.6.2

We still want a general method for handling the general distribution. You probably can guess that we must form a discrete distribution to mimic the continuous one to some satisfactory degree of approximation. When you are going to do extensive experimentation on the computer, you must consider space and speed; special programming procedures have been developed. But the two related methods described below will work perfectly well for any sampling you're likely to do at this stage.

One way of handling the general distribution is to divide the range of the variable into small intervals and determine the area, that is, probability, belonging to each. We then allocate random numbers to each interval and, when the interval comes up, take some point in the interval, say the midpoint, as the value of the random variable. If we had the density shown in Fig. 4.6.2, we might divide the interval (0, 2) into 20 small intervals and approximate the distribution by a discrete distribution with 20 categories. These calculations are shown in Table 4.6.1. If we now pull the random digits 5872 out of the hat, we get the random value .75 from the distribution. The random digits 1454 correspond to the random value .15.

4.6 Experimental Sampling from Continuous Distributions

Table 4.6.1

Interval	Probability	Number allocation	Random value
.0– .1	.0975	0000–0974	.05
.1– .2	.0925	0975–1899	.15
.2– .3	.0875	1900–2774	.25
.3– .4	.0825	2775–3599	.35
.4– .5	.0775	3600–4374	.45
.5– .6	.0725	4375–5099	.55
.6– .7	.0675	5100–5774	.65
.7– .8	.0625	5775–6399	.75
.8– .9	.0575	6400–6974	.85
.9–1.0	.0525	6975–7499	.95
1.0–1.1	.0475	7500–7974	1.05
1.1–1.2	.0425	7975–8399	1.15
1.2–1.3	.0375	8400–8774	1.25
1.3–1.4	.0325	8775–9099	1.35
1.4–1.5	.0275	9100–9374	1.45
1.5–1.6	.0225	9375–9599	1.55
1.6–1.7	.0175	9600–9774	1.65
1.7–1.8	.0125	9775–9899	1.75
1.8–1.9	.0075	9900–9974	1.85
1.9–2.0	.0025	9975–9999	1.95

Example 4.6.1. Draw a sample of 15 from the distribution we have been discussing. Use the midpoint of each interval as the random value. Find the sample mean and the sample median (Table 4.6.2).

Table 4.6.2

Random digits	Random value	Random digits	Random value
0870	.05	8781	1.35
2534	.25	4465	.55
2403	.25	9006	1.35
7283	.95	7146	.95
6269	.75	3237	.35
2863	.35	4832	.55
0950	.05	1977	.25
9882	1.75		9.75

Mean: .650   Median: .55
The population mean is .667.
The population median is .586.

In case you've forgotten, the sample mean, or average, is the total divided by the number of observations. Here

$$\frac{9.75}{15} = .650.$$

To find the median, the middle value, you have to arrange the observations in order:

.05, .05, .25, .25, .25, .35, .35, (.55,) .55, .75, .95, .95, 1.35, 1.35, 1.75,

If there are an *even* number of observations, there is no middle value. Then one usually takes the average of the two values nearest the center.

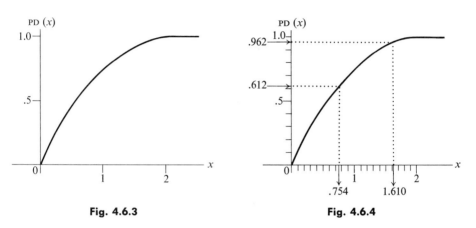

Fig. 4.6.3   Fig. 4.6.4

There is another way of looking at random sampling which introduces some new features. Let's graph not the density but the *cumulative distribution*, the area from $-\infty$ to the given point. Remember that lower-case letters are used for naming densities and small capitals for cumulative distributions. So PD $(x) = P(X \leq x)$. The cumulative distribution function starts way over at the left, where it is 0, and ends up way over at the right, where it is 1. In between, it either stays constant or rises; it never decreases. [*Monotonic increasing* is the technical term.] For the distribution we have been discussing, it looks like Fig. 4.6.3.

We now use our table of random digits to get a random value for PD (x). The x that corresponds to that value is the random sample for the population. Suppose I pick the random digits 6, 1, and 2. I then want the x such that PD $(x) = .612$. Figure 4.6.4 shows graphically that $x = .754$. Similarly, if I choose the random value .962 for PD (x), then $x = 1.610$.

If you think for a moment you can see that what we did now is just what we did before. Look back at the table of the 20 intervals: 6120 corresponds to the interval .7–.8; 9620 corresponds to the interval 1.6–1.7. To get PD (x) we had to add up areas. To get the number allocation I had to add up those same areas. However, this second method automatically interpolates within the intervals.

Naturally, I couldn't see three decimal places on my graph. To get good results you need an equation for PD (x). To go from pd (x) to PD (x) or vice versa with explicit formulas, you almost always need the methods of calculus:

$$\text{PD}(x) = \int_{-\infty}^{x} \text{pd}(y)\, dy, \qquad \text{pd}(x) = \frac{d}{dx} \text{PD}(x).$$

In the present case

$$\text{pd}(x) = 1 - \frac{x}{2}, \qquad \text{PD}(x) = \frac{x(4-x)}{4} \qquad \text{for } 0 < x < 2.$$

When I gave a value a for PD (x), I had to solve the quadratic equation

$$x^2 - 4x + 4a = 0$$

to get x. The solution is

$$x = 2(1 - \sqrt{1-a}).$$

If you can't get explicit formulas, you have to use numerical methods.

PROBLEM SET 4.6

1. Figure 4.6.5 shows a PD. What is pd (x)?
2. Figure 4.6.6 shows a pd. Secure a sample of 12 observations from this distribution. Calculate the sample mean and median. [Show details.]
3. Suppose

$$\text{PD}(x) = \begin{cases} 0 & \text{for } x < 0, \\ x^3 & \text{for } 0 < x < 1, \\ 1 & \text{for } 1 < x. \end{cases}$$

The population mean is $\frac{3}{4}$. What is the population median? Draw a large graph of PD (x). Obtain a sample of 15. Calculate your sample mean and median.

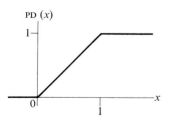

Fig. 4.6.5

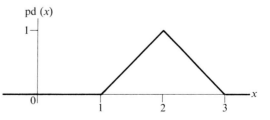

Fig. 4.6.6

4. Draw a sample of 15 from GAU $(* \mid 0, 1)$. Find the sample mean and median.
5. Draw a sample of 15 from GAU $(* \mid 1, 4)$. Find the sample mean and median.
6. Draw a sample of 15 from the distribution whose density is flat between -5 and $+5$ and is 0 elsewhere.

ANSWERS TO PROBLEM SET 4.6

1. Outside the interval $(0, 1)$, PD (x) does not change; therefore pd $(x) = 0$ there. Inside $(0, 1)$, PD (x) changes at a constant rate; therefore pd (x) is constant throughout $(0, 1)$. All this means that
$$\text{pd}(x) = \text{rect}(x \mid \tfrac{1}{2}, \tfrac{1}{4}).$$
See Fig. 4.6.1.
3. PD (median) $= \tfrac{1}{2} =$ median^3, median $= \sqrt[3]{\tfrac{1}{2}} = .794$.
5. Use the table in Appendix C2 and the relation

random value from GAU $(* \mid \mu, \sigma^2) = \mu + \sigma[\text{random value from GAU}(* \mid 0, 1)]$.

CHAPTER **5**

Describing Distributions

We have been deriving posterior distributions for variables of interest. Sometimes tables for these distributions were available; others had to be evaluated numerically on the spot. It's nice to have some quick, crude methods for describing distributions so we can get at least a feel for the range of the uncertainty in a variable. Tables are always limited in size, computers are not always available, and the computation of a few cases does not let us survey the general picture. I'll treat two important topics here: first, the Gaussian distribution, which provides excellent approximations to the more central regions of many distributions; second, the use of moments to describe distributions of variables and combinations of variables. These two topics take care of distributions whose variables have numerical values. I will add a third topic, entropy, which can be used to describe distributions whose variables are nonnumerical. With these techniques at hand, we'll be prepared to move rapidly through a great variety of situations.

5.1 THE GAUSSIAN DISTRIBUTION

The *Gaussian distribution*, GAU, is the most widely used and most widely abused distribution in statistics; here we concentrate on its simpler uses. GAU is also called the *normal distribution*. To avoid the insidious connotation that it is the "normal" thing, I will use the more common international name.

GAU is a function of two parameters: μ and σ^2. [These are the Greek letters "mu" and "sigma."] The density is

$$\text{gau}(x \mid \mu, \sigma^2) = \frac{1}{\sigma\sqrt{2\pi}} e^{-\frac{1}{2}\left(\frac{x-\mu}{\sigma}\right)^2} \quad \text{for} \quad \begin{cases} -\infty < x < +\infty, \\ -\infty < \mu < +\infty, \\ 0 < \sigma < +\infty. \end{cases}$$

When an expression in an exponent gets complicated like this, we usually use the alternative notation

$$\text{gau}(x \mid \mu, \sigma^2) = \frac{1}{\sigma\sqrt{2\pi}} \exp\left[-\frac{1}{2}\left(\frac{x-\mu}{\sigma}\right)^2\right].$$

Remember that exp (Y) means e^Y, *not* expected value.

Let's take it apart, piece by piece. Consider first the simpler density

$$\text{gau}(x \mid 0, 1) = \frac{1}{\sqrt{2\pi}} \exp(-\tfrac{1}{2}x^2).$$

Its graph is shown in Fig. 5.1.1. I've drawn this with the same scale on both axes so you can compare it with those graphs of BETA. Usually you see it with the vertical scale magnified three to five times (Fig. 5.1.2). A table of the ordinates is given in the GAU section of the Gallery.

Note that although the distribution goes off to $-\infty$ and $+\infty$, the density decreases very rapidly the farther you get from 0. It becomes almost invisibly small by the time you get to -3 or $+3$. This is due to the $-x^2$ in the exponent. Since x occurs as a square, the distribution is symmetric about the point 0, which is the *expected value* or *mean*. As in other symmetric distributions, the same point is also the *median* (the *mode*, too, if the distribution is single-humped).

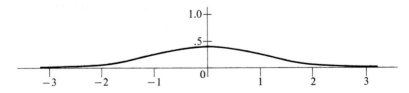

Fig. 5.1.1

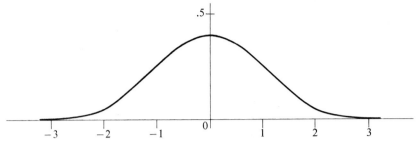

Fig. 5.1.2

All the tables for GAU are tables of GAU (* | 0, 1). There are tables both of areas (probabilities) and ordinates. Since we will be using them quite a bit, both for GAU itself and for approximating other distributions, it is worth while to acquire some fluency in reading them. Look first at the table labeled "GAU (* | 0, 1), Areas." If X is the variable and z is the point being referenced:

the *left tail* is $P(X < z) = $ GAU $(z \mid 0, 1)$;
the *right tail* is $P(X > z) = 1 - $ GAU $(z \mid 0, 1)$;
the *center* is $P(-z < X < z) = 2$ GAU $(z \mid 0, 1) - 1$;
the *both tails* is $P(X < -z$ or $X > +z) = 2[1 - $ GAU $(z \mid 0, 1)]$.

Let's read some values. If $X \sim $ GAU (* | 0, 1) (read "X is distributed according to . . ."),

$P(X < -2.4) = .00820$ (left tail),
$P(X < -.85) = .198$ (left tail, interpolated),
$P(X > -2.4) = .99180$ (right tail).

Note that the first and third results say

left tail + right tail = 1,

as they have to. The other columns of the table tell us that

$P(X < 1.7) = .9554$ (left tail; second column),
$P(X > 3.0) = .00135$ (right tail; second column),
$P(-1.0 < X < 1.0) = .683$ (center),
$P(X < -1.0$ or $X > 1.0) = .317$ (both tails).

Note that the last two results say

center + both tails = 1,

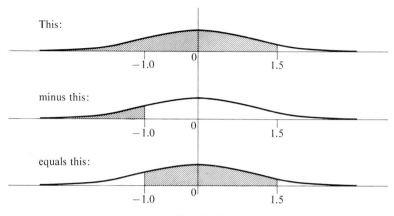

Fig. 5.1.3

as they have to. There are also simple relationships connecting *center* and *both tails* with the individual tails; I'm sure you can discover them for yourself.

You can find other tables of GAU ($z \mid 0, 1$) which are given in much greater detail. For our purposes we will use simple linear interpolation. This suffices save in those cases where we are near some critical value; then the more accurate tables may come in handy.

You can get other areas from this same table. How about $P(-1.0 < X < 1.5)$? This is not tabulated directly but can be gotten by a little subtraction (see Fig. 5.1.3). That is,

$$P(-1.0 < X < 1.5) = P(X < 1.5) - P(X < -1.0)$$
$$= \text{left tail } (1.5) - \text{left tail } (-1.0).$$

The same relation is true for *any* distribution, but you seldom have the extensive tables available that you do for GAU.

Sometimes you want nice round values for the probabilities. In this case consult "GAU ($* \mid 0, 1$), Special Areas." For example.

$$P(-2.576 < X < +2.576) = .99, \qquad P(|X| > 4.417) = .00001.$$

There is also a table of the ordinates, that is gau ($x \mid 0, 1$):

$$\text{gau } (0.0 \mid 0, 1) = .3989,$$
$$\text{gau } (2.5 \mid 0, 1) = .0175,$$
$$\text{gau } (-3.2 \mid 0, 1) = .0024$$

(since the distribution is symmetric about 0). We actually will not use this table much. For when we calculate the likelihood of a normal sample, it's a lot easier to add the x^2's in the exponents than to multiply all the individual terms directly.

PROBLEM SET 5.1

These are just to make sure you can use the tables. For all these problems, $X \sim \text{GAU}(* \mid 0, 1)$.

1. $P(X < 2.15)$
2. $P(X < -.05)$
3. $P(X > -1.2)$
4. $P(X > .45)$
5. z such that $P(|X| < z) = .75$
6. $P(|X| < 2.8)$
7. $P(|X| > 3.0)$
8. z such that $P(|X| > z) = .1$
9. $\text{gau}(-1.4 \mid 0, 1)$
10. $\text{gau}(1.75 \mid 0, 1)$
11. $P(-2.0 < X < 1.6)$
12. $P(0.4 > X > 0.2)$

ANSWERS TO PROBLEM SET 5.1

1. .984
3. .885
6. .99489
8. 1.645
9. .1497
11. $.9452 - .0228 = .9224$

5.2 THE STANDARDIZED VARIABLE

What about our two parameters? For the *unit normal density* gau $(* \mid 0, 1)$, 0 was the mean, the center of symmetry. For any GAU, the first parameter, μ, specifies the mean. [The Greek letter μ ("mu") is often used for the mean of a distribution even if it isn't GAU.] So long as the second parameter of GAU is the same, the *shape* of two GAU curves with different means is the same; they are just displaced right or left with respect to each other (Fig. 5.2.1).

The second parameter controls the shape of the distribution. A large σ^2 lowers the curve and spreads it out wide; a small σ^2 raises and concentrates it. The actual *scale factor* is σ. When σ is doubled (σ^2 is multiplied by 4) the curve is half

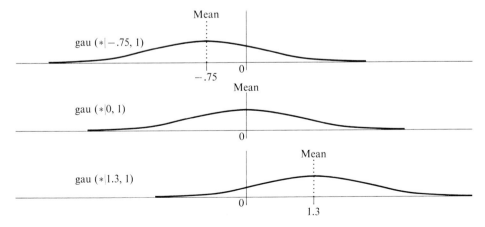

Fig. 5.2.1

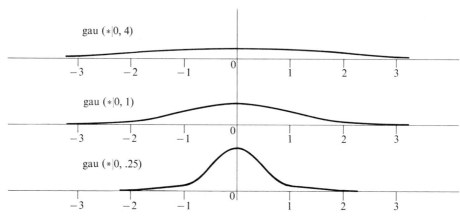

Fig. 5.2.2

as high and spreads out twice as far. When σ is halved (σ^2 is multiplied by $\frac{1}{4}$) the curve is twice as high and twice as concentrated (Fig. 5.2.2). Confusion between σ and σ^2 is frequent; please keep your eyes and ears open.

The three curves in Fig. 5.2.2 all have $\mu = 0$, but we can use any combination of μ and σ^2. A change in μ shifts the location of the mean, and a change in σ^2 changes the shape of the curve; the two act independently. See, for example Fig. 5.2.3.

Let's go back to the defining equation

$$\text{gau}(x \mid \mu, \sigma^2) = \frac{1}{\sigma\sqrt{2\pi}} \exp\left[-\frac{1}{2}\left(\frac{x - \mu}{\sigma}\right)^2\right]$$

to see why these relations are true.

1) If μ is increased by some quantity h, then we have the same ordinate at $x + h$ that we formerly had at x, since $(x + h) - (\mu + h) = x - \mu$. This means that the entire curve is shifted over h.

2) If σ changes, $x - \mu$, the distance of a point from the mean, has to change in the same proportion to maintain the same exponent. [Roughly, the exponent

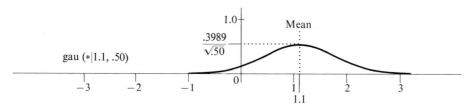

Fig. 5.2.3

5.2 The Standardized Variable

determines the shape.] The $1/\sigma$ at the front of the formula means that all the ordinates are also multiplied by a new quantity: if σ is doubled,

$$\frac{1}{\sigma_{\text{new}}} = \frac{1}{2\sigma_{\text{old}}}.$$

You can see immediately that the key to the GAU curves is the *standardized variable*

$$\frac{x - \mu}{\sigma}.$$

For then we have

$$\text{gau}(x \mid \mu, \sigma^2) = \frac{1}{\sigma} \text{gau}\left(\frac{x - \mu}{\sigma} \mid 0, 1\right).$$

We can find the ordinate for any GAU curve by looking up the ordinate of the *standardized variable* in the GAU $(* \mid 0, 1)$ table of ordinates and dividing the tabular entry by σ. If $\mu = 0$ and $\sigma^2 = 1$, we naturally have $(x - \mu)/\sigma = x$.

Example 5.2.1. What is gau $(1.8 \mid 1, .25)$? Here

$$\mu = 1, \qquad \sigma = \sqrt{.25} = .5,$$

[Remember, the second parameter is σ^2, not σ.]

$$\frac{x - \mu}{\sigma} = \frac{1.8 - 1}{.5} = \frac{.8}{.5} = 1.6.$$

[We can say that 1.8 lies 1.6 σ's to the right of the mean.] Our table gives gau $(1.6 \mid 0, 1) = .1109$, so

$$\text{gau}(1.8 \mid 1, .25) = \frac{.1109}{.5} = .2218.$$

We could have evaluated the formula directly:

$$\text{gau}(1.8 \mid 1, .25) = \frac{1}{.5\sqrt{2\pi}} \exp\left(-\tfrac{1}{2} \times 1.6^2\right)$$

$$= \frac{2}{\sqrt{2\pi}} \exp(-1.28),$$

$$\log_{10} 2 = .3010,$$

$$\log_{10} \frac{1}{\sqrt{2\pi}} = -.3991,$$

$$\log_{10} \exp(-1.28) = -1.28 \log_{10} e$$

$$= -1.28 \times .4343 = -.5559,$$

and therefore

$$\log_{10} \text{gau}(1.8 \mid 1, .25) = .3010 - .3991 - .5559$$
$$= -.6540 = .3460 - 1,$$
$$\text{gau}(1.8 \mid 1, .25) = .2218.$$

Example 5.2.2. What is gau $(-3.4 \mid 1.9, 2.66)$? Here

$$\mu = 1.9, \qquad \sigma = \sqrt{2.66} = 1.63,$$

[To get σ from the second parameter, take the square root!]

$$\frac{x - \mu}{\sigma} = \frac{-3.4 - 1.9}{1.63} = \frac{-5.3}{1.63} = -3.25,$$

[-3.4 is 3.25 σ's to the *left* of the mean.]

so

$$\text{gau}(-3.25 \mid 0, 1) = .0020,$$

$$\text{gau}(-3.4 \mid 1.9, 2.66) = \frac{.0020}{1.63} = .00123.$$

Example 5.2.3. What is gau $(0.6 \mid -0.1, .18)$? Now

$$\frac{x - \mu}{\sigma} = \frac{0.6 - (-0.1)}{.424} = \frac{.7}{.424} = 1.65,$$

[0.6 is 1.65 σ's to the right of the mean.]

so

$$\text{gau}(1.65 \mid 0, 1) = .1023,$$

$$\text{gau}(0.6 \mid -0.1, .18) = \frac{.1023}{.424} = .242.$$

Although these manipulations of the ordinates are perfectly correct, we will need the product of many ordinates for the likelihood of a sample; then we usually add the exponents rather than work with each ordinate separately. The standardized variable plays a more important role in evaluating areas: If

$$X \sim \text{GAU}(* \mid \mu, \sigma^2),$$

then

$$P(X < z) = \text{GAU}(z \mid \mu, \sigma^2)$$
$$= \text{GAU}\left(\frac{z - \mu}{\sigma} \mid 0, 1\right).$$

That is, the left tail of the general GAU distribution can be obtained from the unit normal tables by looking up the standardized variable. There is *no* multiplier here as there was in the formula for the ordinate.

Example 5.2.4. $X \sim \text{GAU}(* \mid -4, 9)$; what is $P(X < 0)$? We see that we want a left tail area:

$$\mu = -4,$$
$$\sigma = \sqrt{9} = 3 \quad \text{(remember that square root)},$$
$$\frac{0 - (-4)}{3} = \frac{4}{3} = 1.33.$$

[0 is 1.33 σ's to the right of the mean, $0 = -4 + 1.33 \times 3$.] So

$$P(X < 0) = \text{GAU}(1.33 \mid 0, 1)$$
$$= \text{left tail}(1.33) = .908.$$

Remember our notation: GAU (1.33 | 0, 1) is the cumulative area; gau (1.33 | 0, 1) is the ordinate at 1.33.

Example 5.2.5. $X \sim \text{GAU}(* \mid 0, 100)$; what is $P(X > -1)$? In this case we want the right tail area:

$$\mu = 0, \quad \sigma = \sqrt{100} = 10, \quad \frac{-1 - 0}{10} = -.1,$$
$$P(X > -1) = \text{right tail}(-.1) = .540.$$

We can write it also as

$$P(X > -1) = 1 - \text{GAU}(-.1 \mid 0, 1).$$

Example 5.2.6. $X \sim \text{GAU}(* \mid 1.1, .50)$; what is $P(-.5 < X < 1.5)$? Here we will have to take the difference of two areas:

$$\mu = 1.1, \quad \sigma = \sqrt{.50} = .707,$$
$$\frac{-.5 - 1.1}{.707} = \frac{-1.6}{.707} = -2.26, \quad \frac{1.5 - 1.1}{.707} = \frac{.4}{.707} = .57.$$

[$-.5$ is 2.26 σ's to the left of the mean; $-.5 = 1.1 - 2.26 \times .707$. Also 1.5 is .57 σ's to the right of the mean; $1.5 = 1.1 + .57 \times .707$.] The desired probability is then

$$\text{GAU}(.57 \mid 0, 1) - \text{GAU}(-2.26 \mid 0, 1) = .716 - .012 = .704.$$

Example 5.2.7. $X \sim \text{GAU}(* \mid 0, \sigma^2)$; what is σ if $P(-1 < X < 1) = .95$? This is a center area since $1 - 0 = 0 - (-1)$:

$$\text{center of GAU}(* \mid 0, 1) = .95 \quad \text{when} \quad z = 1.96.$$

Thus

$$\frac{1-0}{\sigma} = 1.96 \quad \text{and} \quad \sigma = .51.$$

Example 5.2.8. $X \sim \text{GAU}(* \mid \mu, 5)$ and $P(X > 3) = .074$; what is μ? We have been given a right tail area:

$$\text{right tail of GAU}(* \mid 0, 1) = .074 \quad \text{when} \quad z = 1.45.$$

Thus

$$\frac{3-\mu}{2.24} = 1.45, \quad \mu = 3 - 2.24 \times 1.45 = -.25.$$

Example 5.2.9. Suppose we are monitoring the length of parts produced by an automatic machine and have definite knowledge of the proper length (μ) and the random variation (σ) to be expected. We want to see whether the machine is working properly and set upper and lower *control limits* of $\mu \pm 3\sigma$. Our *control chart* looks like Fig. 5.2.4. Now we start checking our product and plot successive values of the length; if the items are independent and under control, we expect to see something like Fig. 5.2.5 most of the time. There is variation, but all of it is contained within acceptable limits.

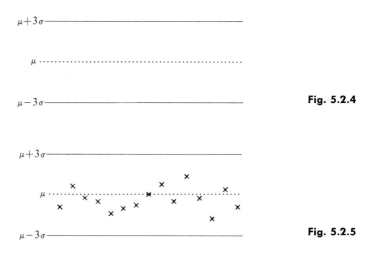

Fig. 5.2.4

Fig. 5.2.5

Now suppose there is a trend—the machine is wearing or loosening or becoming contaminated; we might see something like Fig. 5.2.6. We would then stop the machine to see what was the matter. The trouble is that this will inevitably happen even if things are really going all right, just because of random variation. The probability depends on how long the sequence of parts is.

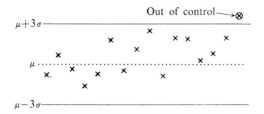

Fig. 5.2.6

What is the probability that a series of 100 independent items under control will *not* exceed the $(\mu \pm 3\sigma)$-limits? From our GAU table, center (3) = .99730. This means that

$$P(\mu - 3\sigma < \text{length} < \mu + 3\sigma) = .99730 \quad \text{for one part.}$$

For 100 of them,
$$\text{prob} = (.99730)^{100} = .763.$$

This means that the probability of going out at some time or other during this sequence is $1 - .763 = .237$.

PROBLEM SET 5.2

1. What is gau $(1.1 \mid 1.1, .50)$?
2. What is gau $(-1.1 \mid 1.1, .50)$?
3. What is μ if gau $(1 \mid \mu, 4) = .0470$?
4. What is μ if gau $(0 \mid \mu, 2) = .00735$?
5. What is the ordinate at a point m σ's from the mean?
6. What is the ordinate at a point 3 σ's from the mean when $\sigma = 2.5$?
7. $X \sim$ GAU $(* \mid 0, 25)$; what is $P(X < -2)$?
8. $X \sim$ GAU $(* \mid 0, 25)$; what is $P(X > 3)$?
9. $X \sim$ GAU $(* \mid 1, 2)$; what is $P(0 < X < 5)$?
10. $X \sim$ GAU $(* \mid 0, 25)$; what is $P(-9.80 < X < 9.80)$?
11. $X \sim$ GAU $(* \mid 0, 1)$; what is $P(|X| < 4.0)$?
12. $Y \sim$ GAU $(* \mid 4, \sigma^2)$; what is σ if $P(2 < Y < 6) = .890$?
13. What is the area of any GAU distribution contained between $\mu - \sigma$ and $\mu + \sigma$?
14. What is the area of any GAU distribution contained between $\mu - 2\sigma$ and $\mu + 2\sigma$?
15. *Continuation of Example 5.2.9.* What is the probability that no one of 500 items will leave the $(\mu \pm 3\sigma)$-limits?
16. *Continuation of Example 5.2.9.* What is the probability that all of 1000 items are less than $\mu + 3\sigma$?

ANSWERS TO PROBLEM SET 5.2

1. .565

3. $\sigma = 2$; we must find a z such that $\frac{1}{2}$ gau $(z \mid 0, 1) = .0470$; $z = \pm 1.7$ makes gau $(z \mid 0, 1) = .0940$, so

$$\frac{1 - \mu}{2} = \pm 1.7, \qquad \mu = -2.4 \text{ or } 4.4.$$

5. $\dfrac{1}{\sigma}$ gau $(m \mid 0, 1)$ 7. left tail $(-.4) = .345$

9. left tail (2.83) − left tail $(-.71) = .998 - .239 = .759$

11. .9999367   13. .683   15. $(.99730)^{500} = .259$

5.3 MOMENTS

The expected value of a numerical variable is a definite point in its distribution. Sometimes it is the exact center, sometimes not; in any case it still is a definite point. If we have two distributions with exactly the same shape and one is shifted to the left or right of the other, lining up their expected values removes the shift; we can then see that they are the same except for the shift. Calculation of the expected value is thus the first basic step in describing a distribution.

But what else can we do? If we lined up all distributions so that their expected values coincided, there still would be a bewildering variety: tall ones, short ones, thin ones, fat ones, two-headed ones, distributions gnarled with age. After all, the only requirement levied on a distribution is that it lie above the axis and have its area or sum of probabilities be 1. How can we introduce a little order?

Now the expected value was the expected value of the *first power* of the variable. A promising approach is to look at the expected value of other *powers* of the variable. This has the most meaning if we measure from the one point in the distribution that we have already nailed down. So we calculate

$$E\{[X - E(X)]^2\}, \qquad E\{[X - E(X)]^3\}, \qquad E\{[X - E(X)]^4\}, \quad \text{etc.}$$

These are the second, third, fourth, etc., *moments about the mean*. We don't bother with the first, since

$$E\{[X - E(X)]\} = 0.$$

We will have no trouble performing these calculations. $E(X)$ is just a number which we throw into a formula, and that formula has another "E".

5.3 Moments

Example 5.3.1. From

y	1	2	3	4
P(y)	.4	.3	.2	.1

we get

$$E(Y) = 1 \times .4 + 2 \times .3 + 3 \times .2 + 4 \times .1 = 2.$$

What is $E\{[Y - E(Y)]^2\}$? For each y we calculate

$$[y - E(Y)]^2 = [y - 2]^2$$

and associate it with the probability belonging to y:

y	1	2	3	4
P(y)	.4	.3	.2	.1
$[y - E(Y)]^2 = [y - 2]^2$	1	0	1	4

Then $E\{[Y - E(Y)]^2\} = 1 \times .4 + 0 \times .3 + 1 \times .2 + 4 \times .1 = 1.0$. This is just

$$\sum [y - E(Y)]^2 P(y).$$

Example 5.3.2. For the distribution given in Example 5.3.1, what is

$$E\{[Y - E(Y)]^3\},$$

the third moment about the mean?

y	1	2	3	4
P(y)	.4	.3	.2	.1
$[y - E(Y)]^3 = [y - 2]^3$	-1	0	1	8

$$E\{[Y - E(Y)]^3\} = (-1) \times .4 + 0 \times .3 + 1 \times .2 + 8 \times .1 = .6.$$

This is just

$$\sum [y - E(Y)]^3 P(y).$$

In general, then,

$$E\{[Y - E(Y)]^k\} = \sum [y - E(Y)]^k P(y)$$

for a discrete distribution, and

$$E\{[Y - E(Y)]^k\} = \int [y - E(Y)]^k \, \text{pd}(y) \, dy$$

for a continuous distribution. You can calculate moments as high as you want (though sometimes you start to get ∞). But remember that theoretical concepts in statistics usually acquire significance only if they can be related to what we see in our observations. Can we set up any parallels?

We already had:

Model	Observation
$E(X) = \sum x P(x)$ or $\int x \text{ pd}(x) dx$	Ave $(\mathbf{x}) = \bar{x} = \dfrac{1}{n} \sum x_i$

Now we add:

$E\{[X - E(X)]^k\} = \sum [x - E(X)]^k P(x)$ or $\int [x - E(X)]^k \text{ pd}(x) dx$	$\dfrac{1}{n} \sum [x_i - E(X)]^k$

That is:

population moments about mean	sample moments about mean

What benefit comes from this correspondence? The answer is that only the first few moments are of much practical value. The higher sample moments are so greatly affected by the observations farthest from $\bar{x}$ that their fluctuations are extreme. I mentioned before that the sample average, $\bar{x}$, is not always the best estimator of the population expected value, $E(X)$. It turns out that the higher sample moments are even poorer estimators of the corresponding population values. The one moment that is of great value is the second. It has a special name—*variance*—and a special symbol—$D^2(X)$. That is,

$$D^2(X) = E\{[X - E(X)]^2\}.$$

The square root of the variance, $\sqrt{D^2(X)} = D(X)$, is called the *standard deviation* or *standard error*. The variance is important because

i) it's the simplest descriptor of a distribution beyond $E(X)$,

ii) squared quantities are easy to manipulate mathematically,

iii) the variance of a GAU variable is its second parameter, σ^2. That is, if $X \sim$ GAU $(* \mid \mu, \sigma^2)$, then $D^2(X) = \sigma^2$, $D(X) = \sigma$.

For this last reason, the variance is often written σ^2 rather than D^2.

The fact that the variance of a Gaussian variable is σ^2 lets us immediately relate an arbitrary distribution and GAU. First, however, there are a number of important properties of $E(X)$ and $D^2(X)$ which must be learned. They concern what happens when a variable has a constant added to it, when a variable is multiplied by a constant, and when two variables are added together. They apply to any variables, not just Gaussian ones. In the following rules X and Y are variables, and a, b, and c are constants.

Rule 1. When you add a constant to a variable or multiply by a constant, do the same thing to $E(X)$:

$$E(X + c) = E(X) + c,$$
$$E(X - c) = E(X) - c,$$
$$E(cX) = cE(X),$$
$$E(X/c) = E(X)/c,$$
$$E(a + bX) = a + bE(X).$$

Example 5.3.3. In Example 5.3.1 set $Z = Y + 3$:

z	4	5	6	7
$P(z)$	.4	.3	.2	.1

$E(Z) = 4 \times .4 + 5 \times .3 + 6 \times .2 + 7 \times .1 = 5.0 = E(Y) + 3.$

Example 5.3.4. In Example 5.3.1 set $W = 2Y - 1$:

w	1	3	5	7
$P(w)$	.4	.3	.2	.1

$E(W) = 1 \times .4 + 3 \times .3 + 5 \times .2 + 7 \times .1 = 3.0 = 2E(Y) - 1.$

Rule 2. When you add or subtract variables—whether they are independent or dependent—add or subtract the expected values:

$$E(X + Y) = E(X) + E(Y),$$
$$E(X - Y) = E(X) - E(Y),$$
$$E(aX + bY) = aE(X) + bE(Y),$$

Example 5.3.5. Take the two distributions in Example 2.3.3. In both cases

$E(X) = 1 \times .6 + 2 \times .4 = 1.4, \qquad E(Y) = 1 \times .3 + 2 \times .7 = 1.7.$

What is $E(X - Y)$? In the independent distribution, the distribution of $X - Y$ is

−1	0	1
.42	.46	.12

and the expected value of the difference is

$$(-1) \times .42 + 0 \times .46 + 1 \times .12 = -.30.$$

In the dependent distribution, the distribution of $X - Y$ is

	−1	0	1
	.4	.5	.1

and the expected value of the difference is

$$(-1) \times .4 + 0 \times .5 + 1 \times .1 = -.3.$$

Lastly,
$$E(X) - E(Y) = 1.4 - 1.7 = -.3.$$

Rule 3. When you add a constant to a variable, the variance is unchanged:

$$D^2(X + c) = D^2(X - c) = D^2(X).$$

This is because we *defined* the variance so that a shift right or left did not matter.

Example 5.3.6. In Example 5.3.3:

z	4	5	6	7
$P(z)$	.4	.3	.2	.1
$[z - E(Z)]^2 = [z - 5]^2$	1	0	1	4

The same squared values appear that appeared in Example 5.3.1; the probabilities are also the same. Therefore $D^2(Z) = D^2(Y)$.

Rule 4. When you multiply a variable by a constant, the variance is multiplied by the square of the constant:

$$D^2(cX) = c^2 D^2(X), \qquad D^2(a + bX) = b^2 D^2(X).$$

Example 5.3.7. In Example 5.3.4:

w	1	3	5	7
$P(w)$	.4	.3	.2	.1
$[w - E(W)]^2 = [w - 3]^2$	4	0	4	16

$$D^2(W) = 4 \times .4 + 0 \times .3 + 4 \times .2 + 16 \times .1 = 4.0 = 4D^2(Y).$$

Rule 5. If two variables are independent, the variance of their sum or difference is the *sum* of their variances:

$$D^2(X+Y) = D^2(X-Y) = D^2(X) + D^2(Y),$$
$$D^2(aX+bY) = a^2 D^2(X) + b^2 D^2(Y).$$

Example 5.3.8 (*Continuation of Example 5.3.5*)

$$D^2(X) = (1-1.4)^2 \times .6 + (2-1.4)^2 \times .4 = .240,$$
$$D^2(Y) = (1-1.7)^2 \times .3 + (2-1.7)^2 \times .7 = .210.$$

In the independent case the variance of the difference is

$$(-1+.3)^2 \times .42 + (0+.3)^2 \times .46 + (1+.3)^2 \times .12 = .45$$

and this is $D^2(X) + D^2(Y)$. Now we can see that the formula does *not* apply to the dependent case. There the variance of the difference is

$$(-1+.3)^2 \times .4 + (0+.3)^2 \times .5 + (1+.3)^2 \times .1 = .41 \neq .24 + .21.$$

In our study of GAU we reached the concept of the *standardized variable* $(X-\mu)/\sigma$ by analyzing the formula for the density. We found that it was $\sim$GAU $(* \mid 0, 1)$. What happens to the analogous

$$\frac{X - E(X)}{D(X)}$$

for an arbitrary distribution?

$$E\left[\frac{X-E(X)}{D(X)}\right] = \frac{1}{D(X)} E[X - E(X)] = \frac{E(X) - E(X)}{D(X)} = 0,$$
$$D^2\left[\frac{X-E(X)}{D(X)}\right] = \frac{1}{D^2(X)} D^2[X - E(X)] = \frac{D^2(X)}{D^2(X)} = 1.$$

Example 5.3.9. Use Examples 5.3.4 and 5.3.7. There

$$E(W) = 3, \qquad D^2(W) = 4.$$

Define

$$Q = \frac{W - E(W)}{D(W)} = \frac{W-3}{2}.$$

We find that

q	-1	0	1	2
$P(q)$	.4	.3	.2	.1

$$E(Q) = (-1) \times .4 + 0 \times .3 + 1 \times .2 + 2 \times .1 = 0,$$
$$D^2(Q) = (-1)^2 \times .4 + 0^2 \times .3 + 1^2 \times .2 + 2^2 \times .1 = 1.0.$$

PROBLEM SET 5.3

1. Figure 5.3.1 shows the probability density of X. Sketch the probability density of $Y = X - 3$.
2. Figure 5.3.2 shows the probability density of U. Sketch the probability density of $V = 2U + 1$.

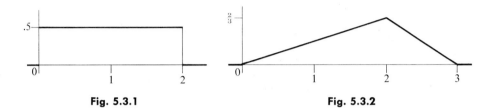

Fig. 5.3.1   Fig. 5.3.2

3. X has the distribution:

x	-1	0	1	2
$P(x)$	.1	.5	.2	.2

Find the second and third moments about the mean.

4. *Continuation of Problem 3.* Exhibit the distribution of $Y = X - 7$ and calculate $E(Y)$.
5. *Continuation of Problems 3 and 4.* Find $D^2(Y)$ and relate it to $D^2(X)$.
6. What is the standardized variable in Problem 3?
7. Show that $D^2(X) + D^2(Y) \neq D^2(X - Y)$ in Example 2.3.4.
8. Show that $E(XY) = E(X)E(Y)$ in Example 2.3.5.
9. $X \sim$ GAU $(* \mid 15, 26)$; what is $E(X)$?
10. $X \sim$ BETA $(* \mid 5, 8)$; what is $E[(X + .1)/5]$?
11. In Example 5.3.1, find the fourth moment about the mean.
12. Prove that $E(a + bX) = a + bE(X)$ for a discrete variable.
13. In Example 2.3.7, show that

$$E(2X + 3Y - Z) = 2E(X) + 3E(Y) - E(Z).$$

14. $X \sim$ GAU $(* \mid 2, 3)$, $Y \sim$ GAU $(* \mid -2, 3)$, and X and Y are independent. What are $E(X - Y)$ and $D^2(X - Y)$?
15. $X_1, X_2, \ldots, X_n$ are n independent variables, all with $E(X_i) = \mu$ and $D^2(X_i) = \sigma^2$. What is $E(\sum x_i)$?
16. For the conditions of Problem 15, what is $E(\bar{x})$?
17. For the conditions of Problem 15, what is $D^2(\sum x_i)$?

18. For the conditions of Problem 15, what is $D^2(\bar{x})$?
19. Show that $D(a + bX) = |b|D(X)$.
20. $X \sim \text{GAU}(*\,|\,5, 20)$, $Y \sim \text{GAU}(*\,|\,10, 12)$, and X and Y are independent. $Z = aX + (1-a)Y$ with $0 < a < 1$. What value of a makes $D^2(Z)$ the smallest?
21. Show that $D^2(X) = E(X^2) - [E(X)]^2$.

ANSWERS TO PROBLEM SET 5.3

1. See Fig. 5.3.3.

Fig. 5.3.3

3. .85 and .30
5. $D^2(Y) = .85 = D^2(X)$
6. $\dfrac{X - .5}{\sqrt{.85}}$
7. $E(X) = .7$, $D^2(X) = .81$; $E(Y) = 1.7$, $D^2(Y) = .21$

$x - y$	-2	-1	0	1
$P(x - y)$	.42	.18	.38	.02

$E(X - Y) = -1.0$, $D^2(X - Y) = .88 \neq .81 + .21$

9. 15

11.

y	1	2	3	4
$P(y)$	.4	.3	.2	.1
$(y - 2)^4$	1	0	1	16

$E\{[Y - E(Y)]^4\} = 1 \times .4 + 0 \times .3 + 1 \times .2 + 16 \times .1 = 2.2$

14. $E(X - Y) = 2 - (-2) = 4$; $D^2(X - Y) = 3 + 3 = 6$
15. The rule $E(X + Y) = E(X) + E(Y)$ extends to $E(X + Y + Z) = E(X) + E(Y) + E(Z)$, etc. Therefore
$$E(\sum x_i) = n\mu.$$

17. The result about addition of variances extends to any number of independent variables. Therefore
$$D^2(\sum x_i) = n\sigma^2.$$

20. $D^2(Z) = 20a^2 + 12(1-a)^2$
$= 32a^2 - 24a + 12$
$= 32(a^2 - \frac{3}{4}a + \frac{9}{64}) + 12 - \frac{9}{2}$ (complete the square)
$= 32(a - \frac{3}{8})^2 + \frac{15}{2}$

This is a minimum when $a = \frac{3}{8}$. You will find that the minimizing value can be calculated as

$$\frac{\frac{1}{20}}{\frac{1}{20} + \frac{1}{12}} = \frac{12}{32} = \frac{3}{8}.$$

21. $D^2(X) = E\{[X - E(X)]^2\} = E(X^2) - 2E(X)E(X) + [E(X)]^2$
$= E(X^2) - 2[E(X)]^2 + [E(X)]^2$
$= E(X^2) - [E(X)]^2$

5.4 GAU AS AN APPROXIMATION

It is an empirical fact that many continuous distributions are approximated fairly well by a Gaussian curve. Several times I've pointed out densities with that high-in-the-middle-and-very-low-at-the-ends shape. Someone has said, "The center of every distribution is Gaussian." How do you proceed? Just use

$$\text{GAU}\ (*\mid E(X), D^2(X)).$$

Before you do too many calculations, please note the implied conditions. We're talking about *one-humped* curves and, in particular, about their *central areas*, that is, areas not too far out in the tails. As you can imagine, the approximation also is better when the density is *symmetric*, since the Gaussian density is symmetric. But there is altogether no guarantee of the accuracy; you'd be very foolish to do a lot of approximating without checking the accuracy you're getting.

Example 5.4.1. What GAU distribution approximates BETA $(*\mid 4, 7)$? We match the expected value and variance:

$$E[\text{beta}\ (*\mid 4, 7)] = \frac{4+1}{4+7+2} = \frac{5}{13} = .385,$$

$$D^2[\text{beta}\ (*\mid 4, 7)] = \frac{(4+1)(7+1)}{(4+7+2)^2(4+7+3)}$$

$$= \frac{40}{13^2 \times 14} = .0169.$$

That is,
$$\text{gau}\ (p\mid .385, .0169) \stackrel{?}{=} \text{beta}\ (p\mid 4, 7).$$

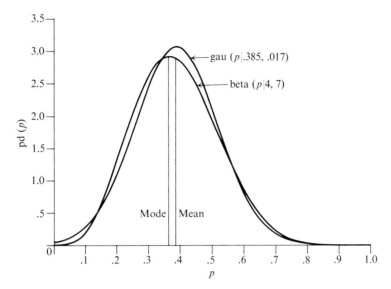

Fig. 5.4.1 beta $(p \mid 4, 7)$ approximated by gau $(p \mid .385, .017)$.

You can judge how close this is from Fig. 5.4.1. The BETA curve is somewhat skew; the GAU curve is not, and it has a longer tail going off to the left.

90% HDR   for BETA $(* \mid 4, 7) = [.169, .595]$ (from the BETA table),
90% HDR   for GAU $(* \mid .385, .0169) = .385 \pm 1.645\sqrt{.0169}$
$$= .385 \pm .214 = [.171, .599].$$

This is certainly quite satisfactory.

99% HDR   for BETA $(* \mid 4.7) = [.055, .794]$,
99% HDR   for GAU $(* \mid .385, .0169) = .385 \pm 2.576\sqrt{.0169}$
$$= .385 \pm .335 = [.050, .720].$$

This is not quite so good; we're too far out in the tails. We can also compare ordinates:

$$\text{beta }(.5 \mid 4, 7) = 1.93,$$
$$\text{gau }(.5 \mid .385, .0169) = \frac{1}{.13} \text{ gau }(.885 \mid 0, 1) = 2.08.$$

This is reasonable. But

$$\text{beta }(.05 \mid 4, 7) = .0173$$

while

$$\text{gau }(.05 \mid .385, .0169) = \frac{1}{.13} \text{ gau }(-2.58 \mid 0, 1) = .110.$$

Here there's a factor of 6.4 between the two!

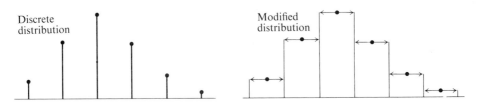

Fig. 5.4.2

It's just not possible to match all the characteristics of BETA with GAU: they are different curves. The approximation in the center seems good. You might think you could do better by putting the center of the GAU distribution at the *mode* of BETA. This would improve things at the center, where they aren't bad, but the discrepancies in the tails would become worse. Part of the trouble with the tails here is that BETA lies between 0 and 1, while GAU goes out to $\pm\infty$.

While I'm talking about BETA, let me show you a simpler notation: Call

$$\hat{p} = \frac{s+1}{s+f+2} = \frac{s+1}{n+2}.$$

Then

$$D^2(p) = \frac{\hat{p}(1-\hat{p})}{n+3} \quad \left(\text{or, for large } n, \doteq \frac{\hat{p}(1-\hat{p})}{n}\right).$$

This is a handier, more easily remembered form for the variance.

We can also use GAU to approximate a discrete distribution, though a little care must be taken. The probabilities must not have 0's scattered among them, for GAU does not go down to 0 and then up again. And since probability corresponds to area in GAU, we must imagine the modification shown in Fig. 5.4.2. Each variate value extends over to points halfway between it and its neighboring values. If we want to include a variate value in a probability we are calculating, we must include the whole bar which now represents it. The height of this bar is the original probability if the spacing between values is 1; otherwise it is prob/spacing.

Example 5.4.2. Consider a trivial case just to illustrate the mechanism:

s	0	1	2	3	4	5	6	7	8
bin $(s \mid 8, \frac{1}{2})$	.004	.031	.109	.219	.274	.219	.109	.031	.004

Look at Fig. 5.4.3. We calculate

$$E(s) = 8 \times \tfrac{1}{2} = 4,$$
$$D^2(s) = 8 \times \tfrac{1}{2} \times \tfrac{1}{2} = 2,$$
$$D(s) = 1.414.$$

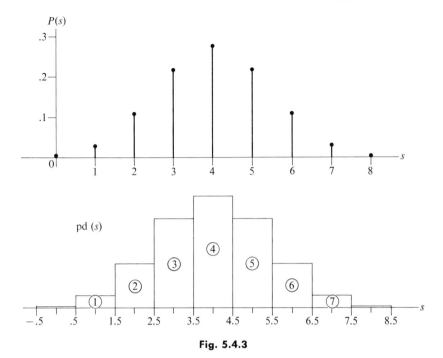

Fig. 5.4.3

What is $P(3 \leq s \leq 5)$?
This is $P(3) + P(4) + P(5) = .219 + .274 + .219 = .712$. We approximate by finding the area under the approximating GAU distribution from 2.5 to 5.5. This includes the 3-bar, 4-bar, and 5-bar:

$$\frac{5.5 - 4}{1.414} = 1.06, \quad \frac{2.5 - 4}{1.414} = -1.06,$$

left tail (1.06) − left tail $(-1.06) = .855 - .145 = .710$.

What is $P(s < 2)$?
The exact answer is $P(0) + P(1) = .004 + .031 = .035$. We approximate by finding the area under the GAU curve from $-\infty$ to 1.5. [There is always a conflict in choosing the best way to handle the end values. If I use $-.5$ as the left-hand boundary of the approximating distribution, I know that the total area will be less than 1 since I am neglecting the tails. If I use $-\infty$ (and $+\infty$ at the other end), the end cases are treated specially. It's a personal choice, and I choose to go out to $\pm\infty$.]

$$\frac{1.5 - 4}{1.414} = -1.77, \quad \text{left tail } (-1.77) = .038.$$

What is $P(s = 4)$?

Exactly, it's .274. Approximately, it's the area under the GAU curve from 3.5 to 4.5:

$$\frac{4.5 - 4}{1.414} = .35, \quad \frac{3.5 - 4}{1.414} = -.35,$$

left tail (.35) − left tail (−.35) = .637 − .363 = .274.

That came out very well, but it was the center of a symmetric distribution.

When you are asked to find a probability like this, always look carefully at the "<" or "≤" signs to make sure that you know which variate values to include. A "≤" or "≥" includes the endpoint; a "<" or ">" does not. Then make sure that your approximating area includes the entire set of relevant bars. I find it safest to list the variate values in a row and draw a ring around the ones I want; this automatically shows what the boundaries of the area should be:

$3 \le s \le 5$:   0   1   2 (3   4   5) 6   7   8
                         2.5   5.5

$s < 2$:   (0   1) 2   3   4   5   6   7   8
         −∞   1.5

$s = 4$:   0   1   2   3 (4) 5   6   7   8
                          3.5   4.5

$1 < s < 7$:   0   1 (2   3   4   5   6) 7   8
                    1.5   6.5

Remember again that the quality of the approximation is best when the distribution is symmetric and we don't go too far from the center.

I want to consider again briefly the problem of sampling from a finite population (Example 3.2.3). I will give you the formulas for the mean and variance and show you how this finite problem grades into the Bernoulli trials problem where the population is infinite. Remember that N was the size of the population; R, the number of supporters in the population; n, the size of the sample; and r, the number of favorable replies. Thus r corresponds to the s of BETA, and n corresponds to $s + f$. I get

$$E(R \mid N, n, r) = \frac{N(r + 1) + r - (n - r)}{n + 2},$$

$$D^2(R \mid N, n, r) = \frac{(N - n)(N + 2)(r + 1)(n - r + 1)}{(n + 2)^2(n + 3)},$$

and for R/N, the *fraction* of supporters, I get

$$E\left(\frac{R}{N}\bigg| N, n, r\right) = \frac{r+1}{n+2} + \frac{1}{N}\left(\frac{2r-n}{n+2}\right),$$

$$D^2\left(\frac{R}{N}\bigg| N, n, r\right) = \frac{(N-n)(N+2)}{N^2}\frac{(r+1)(n-r+1)}{(n+2)^2(n+3)}.$$

The relations become clearer if I put

$$\hat{p}_\infty = \frac{r+1}{n+2}, \qquad D_\infty^2 = \frac{(r+1)(n-r+1)}{(n+2)^2(n+3)},$$

which are the BETA values. Then

$$\hat{p}_N = E\left(\frac{R}{N}\right) = \hat{p}_\infty + \frac{1}{N}\left(\frac{2r-n}{n+2}\right),$$

$$D_N^2 = D^2\left(\frac{R}{N}\right) = \frac{(N-n)}{N}\frac{(N+2)}{N} D_\infty^2 \doteq \left(1 - \frac{n}{N}\right)D_\infty^2.$$

As N gets large, the difference disappears, as it should.

Example 5.4.3. With $N = 100$, $n = 10$, $r = 6$, and a flat prior, what is $P(50 \leq$ number of supporters ≤ 70)?

We want the area from 49.5 to 70.5:

$$48 \quad 49 \;(\; 50 \quad 51 \quad \ldots \quad 69 \quad 70 \;)\; 71 \quad 72$$
$$\qquad\qquad 49.5 \qquad\qquad\qquad\qquad 70.5$$

We get

$$E(R) = \frac{100 \times 7 + 6 - 4}{12} = \frac{702}{12} = 58.5,$$

$$D^2(R) = \frac{90 \times 102 \times 7 \times 5}{12 \times 12 \times 13} = 172,$$

$$D(R) = \sqrt{172} = 13.1,$$

$$\frac{49.5 - 58.5}{13.1} = -.69,$$

$$\frac{70.5 - 58.5}{13.1} = .92,$$

GAU $(.92 \mid 0, 1) -$ GAU $(-.69 \mid 0, 1) = .821 - .245 = .576.$

That is the approximate value. Table 5.4.1 gives the exact calculation. The approximation is off by 3%.

Table 5.4.1

R	P(R)	R	P(R)
50	.0230	61	.0287
51	.0240	62	.0286
52	.0249	63	.0282
53	.0258	64	.0279
54	.0265	65	.0272
55	.0272	66	.0265
56	.0277	67	.0257
57	.0282	68	.0248
58	.0285	69	.0237
59	.0287	70	.0226
60	.0288	Total	.557

PROBLEM SET 5.4

1. Figure 5.4.4 shows beta $(p \mid 10, 6)$ together with its approximating GAU curve. Show how the expected value and variance are calculated. Compare the 95% HDR's for the two curves. Point out a value of p for which the ratio of the two ordinates is about 5 to 1.

2. Draw beta $(p \mid 1, 6)$ and its GAU approximation on the same axes.

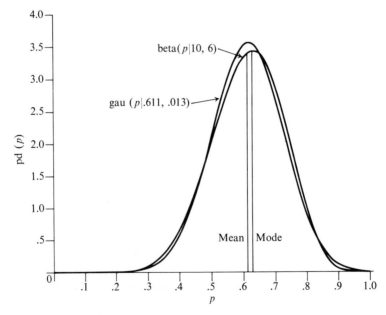

Fig. 5.4.4 beta $(p \mid 10, 6)$ approximated by gau $(p \mid .611, .013)$.

3. The 90% HDR for BETA $(* \mid 16, 32)$ is $(.230, .448)$. Determine what probability the GAU approximation assigns to that interval.

4. Use a GAU approximation to find

$$\mathcal{F}\left(\left.\begin{array}{c}p = .5 \\ \overline{p \neq .5}\end{array}\right| 19 \text{ successes, 14 failures, flat prior}\right).$$

5. $p \sim$ BETA $(* \mid 20, 50)$; at what point z are

$$\mathcal{O}\left(\frac{p > z}{p < z}\right) = \frac{100}{1}?$$

What is the corresponding point in the GAU approximation?

6. $p \sim$ BETA $(* \mid 75, 39)$; what are

$$\mathcal{O}\left(\frac{|p - \frac{2}{3}| > .02}{|p - \frac{2}{3}| < .02}\right)?$$

7. $s \sim$ BIN $(* \mid 25, .2)$; what is $P(3 \leq s \leq 6)$?
8. $s \sim$ BIN $(* \mid 25, .4)$; what is $P(8 < s < 11)$?
9. A fair coin is tossed 100 times. What range of values contains about 90% of the probability distribution of the total number of heads? [The *binomial distribution*, BIN, applies.]
10. A fair die is thrown repeatedly. What is the approximate probability that the 10th deuce will occur no earlier than the 55th nor later than the 65th toss? [The *negative binomial distribution*, NEGB, applies.]
11. In Example 5.4.3, calculate $P(55 \leq R \leq 65)$ both exactly and with the help of GAU.
12. In Example 5.4.3, approximate $P(R = 40)$. The exact value is .0118.
13. Check the finite population results for consistency. Show that

   i) if $n = N$,
   $$E(R) = r, \qquad D^2(R) = 0;$$
   ii) if $n = 0$,
   $$E(R) = N/2.$$

ANSWERS TO PROBLEM SET 5.4

1. $E(p) = .611$, $D^2(p) = .01251$, $D(p) = .112$

   From the table 95% HDR for BETA is $(.392, .823)$.

   95% HDR for GAU is $.611 \pm 1.96 \times .112 = .611 \pm .219 = (.392, .830)$.

   $$\frac{\text{gau}}{\text{beta}} \doteq \frac{5}{1} \quad \text{at about } p = .92$$

4. Factor = beta (.5 | 19, 14)

   $E[\text{beta} (* | 19, 14)] = \frac{20}{35} = .572$

   $D^2[\text{beta} (* | 19, 14)] = .0068, D = .0824$

   ordinate $= \dfrac{1}{.0824}$ gau $(.5 - .572 | 0, 1) = \dfrac{.308}{.0824} = 3.7$

   The exact calculation, which is not difficult, gives 3.24.

5. For BETA we use the table and get $z = .177$. For GAU, left tail $= \frac{1}{101}$ at -2.330 σ's. Since

   $$E(p) = .292, \qquad D^2(p) = .00283, \qquad D(p) = .053,$$

   -2.330 σ's means $.292 - 2.33 \times .053 = .168$.

7. The exact value is .682. The approximate calculation goes

   $$E(s) = 25 \times .2 = 5, \qquad D^2(s) = 25 \times .2 \times .8 = 4, \qquad D(s) = 2,$$

   0   1   2 (3   4   5   6) 7   8   ...
              2.5   6.5

   $$\frac{6.5 - 5}{2} = .75, \qquad \frac{2.5 - 5}{2} = -1.25,$$

   left tail $(.75)$ − left tail $(-1.25) = .773 - .106 = .667$.

   This is not bad, for the distribution is certainly skew.

9. $s \sim$ BIN $(* | 100, .5); E(s) = 50, D^2(s) = 25, D(s) = 5$
   Approximate area $= 50 \pm 1.645 \times 5 = (41.78, 58.22)$. This means that $P(42 \leq s \leq 58)$ is close to, but a little more than, .90 (since it is the area from 41.5 to 58.5).

11. The exact sum from Table 5.4.3 is .310. Approximately, the area from 54.5 to 65.5 is
    GAU $(.534 | 0, 1) -$ GAU $(-.305 | 0, 1) = .325$.

12. Let's approximate by using *ordinate* $\times$ *width*. Here the width is 1, so all we need is the ordinate at $R = 40$:

    $$\frac{40 - 58.5}{13.1} = -1.41, \qquad \frac{1}{13.1} \text{ gau } (-1.41 | 0, 1) = \frac{.148}{13.1} = .0113.$$

    Then $1 \times .0113 = .0113$.

5.5 ENTROPY AND INFORMATION

The earlier sections of this chapter dealt with the distributions of numerical variables. When you work with numbers, there is a built-in order for the alternatives; you can point out the central region and the extremes; you can specify the distance of a point from the center. This leads to direct measures for the concentration of the probability, and the variance plays an important part. But when your alternatives are letters, fossils, electrical signals, or vegetables, there usually is no natural order for them. We need new kinds of measures of uncertainty. Instead of calculating expected values and variances, we can only determine how different one probability is from another and how *flat* or *rough* a set of probabilities is.

Example 5.5.1. We have a distribution with three alternatives. The set of probabilities

$$(.33, .34, .33)$$

is quite *flat*. We are very uncertain about what would happen if a random choice is made from this distribution. But the set of probabilities

$$(.90, .09, .01)$$

is quite *rough*. When we make a random choice from this distribution, it is quite likely that the first alternative will turn up. Our situation is not as uncertain as the previous one.

It would be helpful to have some way of characterizing the roughness or flatness of a distribution, to have some way of ordering distributions. There is no trouble, of course, with two-category distributions; ordering them by the size of the smaller probability suffices. But it's not clear what to do with distributions with three or more categories.

One possible approach is to see what a "typical" sample from a distribution looks like. Suppose we have a three-category distribution with the probabilities p, q, and r. When we make N independent trials, we expect to get about Np occurrences of the first alternative, Nq occurrences of the second alternative, and Nr occurrences of the third alternative. The *probability* of this result is

$$p^{Np} q^{Nq} r^{Nr}.$$

Now this is an unhandy expression. Let's take the logarithm:

$$Np \log p + Nq \log q + Nr \log r = N(p \log p + q \log q + r \log r).$$

Now we get rid of the N; that was arbitrary anyway. We're then effectively back to a single observation. Lastly, because $\log p$ is negative for $p < 1$, we change the sign. This gives

$$-p \log p - q \log q - r \log r.$$

That's for three categories. We can go through the same process for any distribution with probabilities p_i and get

$$H = \sum (-p_i \log p_i).$$

H is called the *entropy* of the distribution. [It is *not* the H that I used so often in the first chapters to mean the background information.]

This is not the way the expression for entropy was originally derived; you might like to read about that in Shannon and Weaver (1949). There the property that entropy is the negative logarithm of the probability of a typical sample is deduced from other properties. Entropy is the subject of *information theory* and is widely applied in the mathematical theory of communications. It has only recently appeared in statistics, and its potential is more suspected than realized. You may be among the first to exploit it!

Figures 5.5.1 and 5.5.2 show the entropy for distributions with two and three categories. In Fig. 5.5.1 you can read that $H(.4, .6) = .29$ and that $H(\frac{1}{2}, \frac{1}{2})$ is the largest. In Fig. 5.5.2 you can read that $H(.50, .28, .22) = .45$ and that $H(\frac{1}{3}, \frac{1}{3}, \frac{1}{3})$ is the largest. These charts are given only to show the general characteristics of entropy for simple distributions. There is a table of $-p \log p$ in Appendix B4. Each of the possible bases for the logarithm, 2, e, and 10, has some advantages, but I've chosen to work with logs to the base 10.

Example 5.5.2. What is $H(.90, .09, .01)$?

Probability	$-p \log p$ (from table)
.90	.0412
.09	.0941
.01	.0200
	$H = .1553 =$ entropy

The .0412 in the table was calculated this way:

$$\log_{10} .90 = .9542 - 1 = -.0458, \qquad -.90(-.0458) = .0412.$$

Example 5.5.3. What is $H(.90, .06, .03, .01)$?

Probability	$-p \log p$
.90	.0412
.06	.0733
.03	.0457
.01	.0200
	$H = .1792$

5.5 Entropy and Information

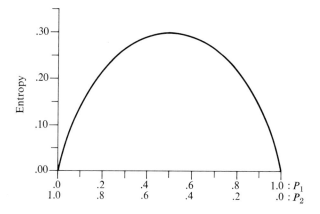

Fig. 5.5.1 Entropy for two categories.

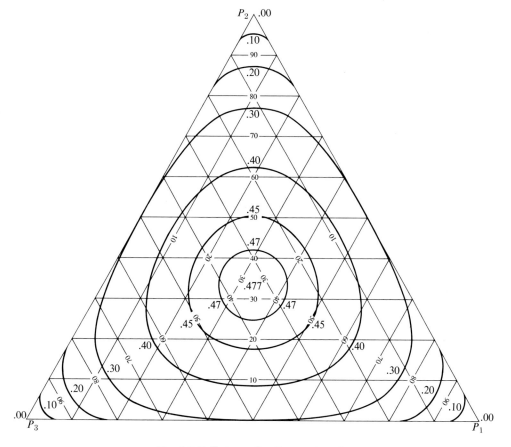

Fig. 5.5.2 Entropy for three categories.

Example 5.5.4. What is $H(.90, .10)$?

Probability	$-p \log p$
.90	.0412
.10	.1000
	.1412

You can see from Examples 5.5.2, 5.5.3, and 5.5.4 that

$$H(.90, .10) < H(.90, .09, .01) < H(.90, .06, .03, .01).$$

This has the interpretation that the grouped distribution is less uncertain than the more finely divided one. It can also be interpreted in another interesting way: we regard entropy as a measure of the average amount of information contained in an observation from a distribution. You get less information when categories are grouped, more when the subdivision is finer. It's easy to consider entropy as being an average or expected value of something because it can be written in the form $\sum (-\log p)p$.

Suppose that we have two variables:

$$X: \quad \frac{x_1 \quad x_2}{.6 \quad .4}, \quad H(X) = .2923,$$

$$Y: \quad \frac{y_1 \quad y_2}{.3 \quad .7}, \quad H(Y) = .2653.$$

What is the entropy of their joint distribution? It turns out that if the variables are independent, the entropy of the joint distribution is the sum of the two entropies; if they are dependent, the joint entropy is less than the sum of the individual entries (there is then not as much information contained in the joint distribution).

Example 5.5.5. X and Y are independent.

	.6	.4	
	.18	.12	.3
	.42	.28	.7

Probability	$-p \log p$
.18	.1341
.12	.1105
.42	.1582
.28	.1548
	.5576

and

$$.5576 = .2923 + .2653, \quad H(X, Y) = H(X) + H(Y).$$

Example 5.5.6. X and Y are dependent.

	.6	.4		Probability	$-p \log p$
	.24	.06	.3	.24	.1487
	.36	.34	.7	.06	.0733
				.36	.1597
				.34	.1593
					.5410

and
$$.5410 < .2923 + .2653, \quad H(X, Y) < H(X) + H(Y).$$

Logarithms are difficult to work with mathematically, and we like to avoid them if possible. When the probabilities of a c-category distribution are not too far from $1/c$—that is, when the probability distribution is almost flat—there is a convenient approximation to the entropy. We call

$$\rho = \sum p_i^2$$

the *repeat rate* of the distribution. [This is a Greek "rho."] By the law of compound probability, ρ is the probability that two independent observations from this population will be identical. The *bulge* of the repeat rate is

$$\beta = c\rho - 1 = c \sum p_i^2 - 1.$$

[This is a Greek "beta."] Our approximation is then

$$H \doteq \log_{10} c - .2172\beta.$$

Example 5.5.7. The probabilities are .27, .20, .25, .28:

p_i	p_i^2	$-p_i \log p_i$
.27	.0729	.1535
.20	.0400	.1398
.25	.0625	.1505
.28	.0784	.1548
1.00	.2538	.5986

Thus
$$H = .5986, \quad \rho = .2538, \quad \beta = 4(.2538) - 1 = .0152,$$
$$H \doteq \log_{10} 4 - .2172 \times .0152 = .6021 - .0033 = .5988$$

In other cases I like to work with the percentage deviations of the p_i's from $1/c$. I write

$$p_i = \frac{1}{c}(1 + \delta_i) \quad \text{(Greek "delta")}.$$

It turns out that

$$\beta = \frac{\sum \delta_i^2}{c};$$

we can stick this in the approximating formula, too.

Example 5.5.8 (*Continuation of Example 5.5.7*)

p_i	δ_i	δ_i^2
.27	.08	.0064
.20	−.20	.0400
.25	.00	.0000
.28	.12	.0144
	.00	.0608

You can see that

$$\beta = .0152 = \frac{.0608}{4}.$$

When the population is absolutely flat, all the δ's are 0; the entropy is then just $\log c$. You can see from Fig. 5.5.1 that

$$H(\tfrac{1}{2}, \tfrac{1}{2}) = .30 = \log_{10} 2,$$

and from Fig. 5.5.2 that

$$H(\tfrac{1}{3}, \tfrac{1}{3}, \tfrac{1}{3}) = .477 = \log_{10} 3.$$

This property is also easily proved from the definition. We sometimes consider 10^{entropy} as the *equivalent number of categories* in the distribution. When a distribution is very peaked, this approaches 1. I have seen this concept being used to specify when to stop making observations: you evaluate the entropy of your posterior distribution, and stop when the probabilities are peaked enough.

Example 5.5.9. I have two alternatives and want to sample until I get odds one way or the other of 999 to 1. What entropy is this equivalent to? Here

$$\text{one } p = \tfrac{1}{1000}, \quad \text{the other } p = \tfrac{999}{1000};$$
$$H = .0030 + .0004 = .0034;$$
$$10^{.0034} = 1.008.$$

What else is entropy being used for? As a measure of the relative roughness of a set of counts. You have to use f_i/N rather than p_i, but this works out all right. It has been applied to geologic studies of fossil mixtures in ancient bays and to studies of sediment size variations in sand dunes [see Miller and Kahn (1962), pages 425–439]. Then Rescigno and Maccacaro (1961) have used it to measure the effectiveness of various characteristics in taxonomy. It has even been used to measure the homogeneity of magazine subscriber lists when deciding what ploy will get the subscriber to renew his subscription!

Example 5.5.10. How do we rank these counts in roughness?

	f_1	f_2	f_3	f_4	N
A	7	10	6	2	25
B	8	8	1	8	25
C	6	7	6	6	25

We want to calculate

$$-\sum \frac{f_i}{N} \log \frac{f_i}{N} = -\frac{1}{N} \sum f_i \log f_i + \log N.$$

We get .5504 for A, .5309 for B, and .6010 for C. Thus C is the flattest and B is the roughest. This is only an empirical measure, of course; we have to use other methods if we want to ask whether B and C came from equally flat *populations*.

Jaynes (1958) and Tribus (1962) have suggested that entropy could help in selecting prior distributions to express certain types of information. The rule is to pick the distribution which accounts for the given information but which has the maximum entropy (uncertainty) under these conditions.

Example 5.5.11. Our possible alternatives are all the integers from 0 on up. We are told that the mean of the distribution is 5. What is the distribution with maximum entropy under this condition?

The answer turns out to be $p_k = \frac{1}{6}(\frac{5}{6})^k$ for $k = 0, 1, 2, \ldots$ Thus

$$P(0) = \tfrac{1}{6}, \qquad P(1) = \tfrac{5}{36}, \qquad P(2) = \tfrac{25}{216}, \quad \text{etc.}$$

Other applications are continually appearing. For example, when several experiments are available, you might choose to perform the one that is expected to reduce entropy the most. I have included this section because I believe that entropy is an important concept for statistics, even though its role is not clear-cut. A book cannot display only the cut-and-dried techniques; it must gamble a bit on new developments.

PROBLEM SET 5.5

1. Prove that
$$H\left(\frac{1}{c}, \frac{1}{c}, \ldots, \frac{1}{c}\right) = \log c.$$

2. What is $H(\frac{1}{5}, \frac{1}{5}, \frac{1}{5}, \frac{1}{5}, \frac{1}{5})$?

3. Show that $-.7 \log .7 = .1084$.

4. If you use logs to the base 2, what is $H(.5, .5)$?

5. Which has the larger entropy,

$$(.1, .2, .3, .4) \quad \text{or} \quad (.2, .2, .5, .1)?$$

6. Which has the larger entropy,

$$(.25, .73, .02) \quad \text{or} \quad (.45, .55)?$$

7. Discuss the entropy of the following joint distribution:

.6	.4	
.0	.4	.4
.6	.0	.6

8. Find the entropy of the joint distribution in Example 2.3.5. Compare it with the marginal entropies.

9. Find the entropy of the joint distribution in Example 2.3.7. Compare it with the marginal entropies.

10. What is the *bulge* of the distribution

$$(.21, .19, .23, .16, .21)?$$

11. Use the bulge from Problem 10 to estimate the entropy. Check against the exact calculation.

12. If I had two independent observations from the distribution of Problem 10, what is the probability that they would be the same?

13. What is the probability that two independent observations from a flat five-category population will be the same?

14. Which of these two sets of counts is roughest?

$$A: \quad 382 \quad 641 \quad 526 \quad 451$$
$$B: \quad 357 \quad 593 \quad 584 \quad 466$$

15. What entropy corresponds to 9-to-1 odds?

16. Show that the probabilities in Example 5.5.11 sum to 1 and that the expected value is 5.

17. In Table 4.1.1 compare the entropies of the posterior distributions for $n = 0, 10,$ and 100.

18. When I want to find out whether my observations come from the population with probabilities

$$(p_1, p_2, \ldots, p_c)$$

or from the flat random population with probabilities

$$\left(\frac{1}{c}, \frac{1}{c}, \ldots, \frac{1}{c}\right),$$

I use log-factors. Show that when the observations really come from the $(p_1, \ldots, p_c)$ population, the expected log-factor per observation is

$$\log c - H(p_1, p_2, \ldots, p_c).$$

ANSWERS TO PROBLEM SET 5.5

3. $\log .7 = .8451 - 1 = -.1549;\ .7 \times .1549 = .1084$
4. 1
6. The entropies are .2843 and .2989.
8. $H(X) = .1412,\ H(Y) = .2653,\ H(Z) = .2923$

    $H(X) + H(Y) + H(Z) = .6988$

    Using linear interpolation I get $H(X, Y, Z) = .6979$; this is close considering the crudity of the interpolation.
10. $\sum p_i^2 = .2028;$ bulge $= 5 \times .2028 - 1 = .0140$
11. $H \doteq \log 5 - .2172 \times .0140 = .6990 - .0030 = .6960.$ Exactly, $H = .6957.$
12. .2028
13. $\frac{1}{5} = .2$
14. $H(A) = .5941,\ H(B) = .5936.$ It's very close!
17. $\quad H_0 = \log 9 = .9542, \quad\quad 10^{.9542} = 9.0,$

    $\quad\ \ H_{10} = .7197, \quad\quad\quad\quad\ \ 10^{.7197} = 5.2,$

    $\quad\ \ H_{100} = .3044, \quad\quad\quad\quad 10^{.3044} = 2.0$

CHAPTER 6

**Inferences
about the
Gaussian
Distribution**

In this chapter we again take up GAU, *the Gaussian or normal distribution. Previously we studied its use in approximating other distributions; here we are concerned with the way it itself describes the errors in our measurements. Please be advised that <u>no physical process or collection of observations has ever followed or will ever follow the normal distribution exactly.</u> Although* GAU *provides a good approximation to some processes, no measuring device has unlimited resolution: we get basically a discrete set of readings. No process is unbounded: we never measure infinite values. Yet the Gaussian distribution rolls on continuously from* $-\infty$ *to* $+\infty$. *But it's so attractive mathematically! If a mathematical method works at all, it most likely works for* GAU; *ergo, assume that* GAU *applies. Well, it doesn't always apply, and you must learn the conditions under which it is both useful and reasonably safe. We don't make power tools illegal because some people lose fingers; we prescribe safeguards. These safeguards are discussed in Chapter 7.*

6.1 CONTINUOUS LIKELIHOODS AND PRIORS

Throughout this book we are using the tried-and-true Bayesian formula

$$posterior \propto prior \times likelihood$$

as the basis for our inferences. I've given you graphical problems, numerical problems, and problems in which formulas came out nicely. By now you should have no trouble tackling a new problem by yourself numerically. The remainder of the book presents many cases in which a neat formula can be obtained. I like neat formulas. Their main advantage lies not in computation—though they can be labor-savers there—but in their ability to exhibit the general characteristics of a solution. You can see how the solution changes as some of the circumstances of the problem change. It generally is necessary to solve quite a few numerical cases in order to get the same feel.

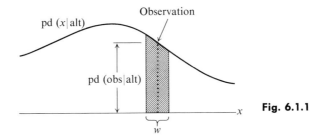

Fig. 6.1.1

A slightly new situation arises here. Our old likelihoods came from discrete distributions, but the Gaussian distribution prescribes a continuous distribution of errors. What do we use for the likelihood? I've mentioned that it doesn't make sense to talk about the probability of a point in a continuous distribution, since the area belonging to a point is the area of a line, 0. We therefore have to introduce the probability that the observation lies in a *small interval around the observed point* (Fig. 6.1.1).

Back in Example 4.2.6 I showed that, for narrow intervals, the probability of the interval is approximately

$$central\ ordinate \times width.$$

Therefore the likelihood we seek is approximately

$$pd\ (obs \mid alt) \times w.$$

Remember that the approximation gets better the smaller w becomes.

What happens during our analysis? We have the schema

Alternative	Prior	Likelihood	Joint
$\vdots$	$\vdots$	$\vdots$	$\vdots$
alt	$P_0(\text{alt})$	pd (obs | alt) × w	$P_0(\text{alt})$ × pd (obs | alt) × w
$\vdots$	$\vdots$	$\vdots$	$\vdots$

and naturally use the same w for all the alternatives. But that means that when we go from the joint to the posterior, the w gets absorbed in the normalization. It might just as well never have been there at all. So in talking about continuous likelihoods we ignore w; we just refer to the likelihood as

$$\text{pd (obs} \mid \text{alt)}.$$

When the distribution of the alternatives is also continuous, we have

$$\text{pd (alt} \mid \text{obs)} \propto \text{pd}_0 \text{ (alt)} \times \text{pd (obs} \mid \text{alt)}$$

as our rule of inference.

I now want to discuss the Gaussian density since its characteristics are important in making the mathematical analysis inviting. The definition was

$$\text{gau } (x \mid \mu, \sigma^2) = \frac{1}{\sigma\sqrt{2\pi}} \exp\left[-\frac{1}{2\sigma^2}(x-\mu)^2\right],$$

so the likelihood for *one* observation is

$$\text{gau } (x_1 \mid \mu, \sigma^2) = \frac{1}{\sigma\sqrt{2\pi}} \exp\left[-\frac{1}{2\sigma^2}(x_1-\mu)^2\right].$$

This gives the probability of x_1 in terms of the parameters μ and σ. Our task in this chapter will be to deduce information about μ or σ or both, depending on what is unknown. I am going to study σ even though σ^2 is officially the parameter. Usually, of course, we have several observations; if they are independent, the likelihoods multiply. The thing to note is that multiplication of Gaussian likelihoods leads to *addition* of the *exponents:*

$$\text{gau } (x_1 \mid \mu, \sigma^2) \times \text{gau } (x_2 \mid \mu, \sigma^2) \times \cdots \times \text{gau } (x_n \mid \mu, \sigma^2)$$

$$= \frac{1}{\sigma^n(\sqrt{2\pi})^n} \exp\left[-\frac{1}{2\sigma^2}\sum(x_i-\mu)^2\right].$$

This expression simplifies. We get rid of the $(\sqrt{2\pi})^n$ since we need retain only those terms which involve the parameters μ and σ. Everything else just gets canceled in the normalization anyway. More importantly, the *sum of squares* $\sum (x_i - \mu)^2$ can be rewritten

$$\sum (x_i - \mu)^2 = \sum (x_i - \bar{x})^2 + n(\mu - \bar{x})^2.$$

This is an identity, true for any μ and any set of x's and their average, $\bar{x}$. Gaussian analysis becomes quite easy if we can split up complicated sums of squares into the sum of simpler parts. Here the part $\sum (x_i - \bar{x})^2$ contains observational data only, while the other part contains μ and just the *average* of the x's.

Some of these terms appear over and over again, so I'll use a few special abbreviations:

$$S_x = \sum x \qquad (\bar{x} = S_x/n),$$
$$S_{xx} = \sum x^2,$$
$$S_{\bar{x}\bar{x}} = \sum (x - \bar{x})^2.$$

Thus

$$\sum (x_i - \mu)^2 = S_{\bar{x}\bar{x}} + n(\mu - \bar{x})^2.$$

At last we can write

$$\text{likelihood } (\mathbf{x}) = \prod \text{gau } (x_i \mid \mu, \sigma^2) \propto \frac{1}{\sigma^n} \exp \left\{ -\frac{1}{2\sigma^2} [S_{\bar{x}\bar{x}} + n(\mu - \bar{x})^2] \right\}.$$

This expression will crop up all over the place, and you should be able to recognize it.

Next, what about prior distributions? The Gaussian distribution and several others we will study have two parameters: μ, a *location* parameter, and σ, a *scale* parameter. If you are called upon to supply a prior distribution for one of them when it is unknown and under study, you must use all the information you have. There are more records available for this purpose than some people admit. It is only as a *last resort* that you try to express absolute ignorance about a variable. Sometimes, though, you have to.

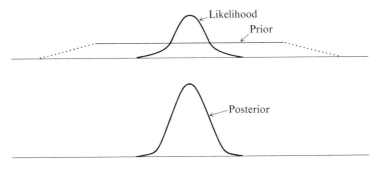

Fig. 6.1.2

Let's take μ first. If we have no knowledge of it, it can range widely in both directions, perhaps out to $-\infty$ and $+\infty$. A prior expressing ignorance will be flat in the region of interest. It can't be flat all the way to $-\infty$ and $+\infty$, because then the area under the curve would be infinite. So we say that the prior is flat in the region of interest and then falls off gradually to zero, so that the area stays finite. Ordinarily the likelihood decreases so rapidly far from the region of interest that the exact behavior of the prior way out there is not important. The height of the level prior is also not important because it gets normalized out (see Fig. 6.1.2). I call this prior FLAT.

The situation is different with the scale factor σ, since σ is essentially positive: $0 < \sigma < \infty$. Often we think about σ^2 or even $1/\sigma^2$ instead of σ. Compelling arguments have been made by Jeffreys (1961) to show that this implies that the *prior of the logarithm of σ should be flat*. We are led to the prior

$$\text{pd}_0(\sigma) \propto \frac{1}{\sigma}.$$

Again there are difficulties with infinite areas, so we have to say $\text{pd}_0(\sigma) \propto 1/\sigma$ in the region of interest and falls off fast enough outside it so that the total area is finite. I call this distribution LFLAT for *logarithmically flat*. The arguments about the form of this prior are certainly pertinent, but the best test is that in several important cases it gives results agreed upon by statisticians of all schools. This confirmation by experience gives us confidence in its use in other situations.

PROBLEM SET 6.1

1. Given $X \sim$ GAU $(* \mid 0, 1)$, show that $P(-.1 < X < .1) \doteq$ gau $(0 \mid 0, 1) \times (.2)$.
2. Given $X \sim$ GAU $(* \mid 0, 1)$, show that $P(1.9 < X < 2.0) \doteq$ gau $(1.95 \mid 0, 1) \times (.1)$.
3. Given $X \sim$ GAU $(* \mid 0, 1)$, show that gau $(3.5 \mid 0, 1)$ is not a very good approximation to $P(3 < X < 4)$.
4. Approximate $P(-3 < Y < -2.9)$, where $Y \sim$ GAU $(* \mid 1, 4)$.
5. What is $P(.48 < p < .52)$, approximately, if $p \sim$ BETA $(* \mid 10, 20)$?
6. Using the values
$$n = 4, \quad x = 6, 3, 4, 7, \quad \mu = 6,$$
verify the formula $\sum (x_i - \mu)^2 = S_{\bar{x}\bar{x}} + n(\mu - \bar{x})^2$.
7. Prove that $\sum (x_i - \mu)^2 = S_{\bar{x}\bar{x}} + n(\mu - \bar{x})^2$.
8. A continuous probability density is given by
$$\text{pd}(x) = \begin{cases} \dfrac{2x}{\theta} & \text{for } 0 \le x \le \theta, \\ \dfrac{2(1-x)}{(1-\theta)} & \text{for } \theta \le x \le 1 \end{cases}$$

(see Fig. 6.1.3). Assume that $pd_0(\theta) = 1$ from 0 to 1. [No worries about infinite areas.] We make three observations on X: .2, .4, and .6. Calculate the posterior distribution of θ at intervals of 0.1 from 0 to 1.
9. What is $E(\theta)$ in Problem 8?

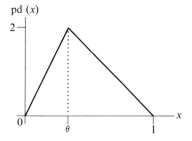

Fig. 6.1.3

ANSWERS TO PROBLEM SET 6.1

1. GAU $(.1 \mid 0, 1)$ − GAU $(-.1 \mid 0, 1) = .540 − .460 = .080$
   gau $(0 \mid 0, 1) = .3989$, $.3989 \times .2 = .07978$

3. GAU $(4 \mid 0, 1)$ − GAU $(3 \mid 0, 1) = .9999683 − .99865 = .00132$
   gau $(3.5 \mid 0, 1) = .0009$

5. beta $(.5 \mid 10, 20) \times .04 = \dfrac{31!}{10!\,20!} (.5)^{30} \times .04 = .035$

8.

θ	$pd_0(\theta)$	$pd(.2\mid\theta)$	$pd(.4\mid\theta)$	$pd(.6\mid\theta)$	Joint	Coeff	Posterior
.0	1	$\dfrac{1.6}{1}$	$\dfrac{1.2}{1}$	$\dfrac{.8}{1}$	1.54	1	.89
.1	1	$\dfrac{1.6}{.9}$	$\dfrac{1.2}{.9}$	$\dfrac{.8}{.9}$	2.11	4	1.22
.2	1	2	$\dfrac{1.2}{.8}$	$\dfrac{.8}{.8}$	3.00	2	1.73
.3	1	$\dfrac{.4}{.3}$	$\dfrac{1.2}{.7}$	$\dfrac{.8}{.7}$	2.61	4	1.51
.4	1	$\dfrac{.4}{.4}$	2	$\dfrac{.8}{.6}$	2.67	2	1.54
.5	1	$\dfrac{.4}{.5}$	$\dfrac{.8}{.5}$	$\dfrac{.8}{.5}$	2.05	4	1.18
.6	1	$\dfrac{.4}{.6}$	$\dfrac{.8}{.6}$	2	1.78	2	1.03
.7	1	$\dfrac{.4}{.7}$	$\dfrac{.8}{.7}$	$\dfrac{1.2}{.7}$	1.12	4	0.65
.8	1	$\dfrac{.4}{.8}$	$\dfrac{.8}{.8}$	$\dfrac{1.2}{.8}$	0.75	2	0.43
.9	1	$\dfrac{.4}{.9}$	$\dfrac{.8}{.9}$	$\dfrac{1.2}{.9}$	0.53	4	0.31
1.0	1	$\dfrac{.4}{1.0}$	$\dfrac{.8}{1.0}$	$\dfrac{1.2}{1.0}$	0.38	1	0.22

Area $= 52.00 \times \left(\dfrac{.1}{3}\right) = 1.733$

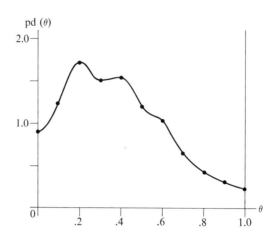

Fig. 6.1.4

"Coeff" gives the coefficients for the numerical integration used in normalizing the posterior distribution. A graph of the posterior density shows two humps (Fig. 6.1.4).
9. .40

6.2 INFERENCES ABOUT μ WITH KNOWN σ

In this section I assume that we have independent observations from a Gaussian population whose σ we know; we want to use them to get information about μ. This is not the most frequent situation, but it is sufficiently common to justify some study. I guess I could characterize it as the situation in which we have a measuring instrument whose precision is known and constant. This is not unusual in a laboratory, and there the measurement errors may be the main source of error. In biological experimentation, however, the error of measurement is often so much less than the variation due to natural causes that we usually have to consider σ as unknown also. That situation will be discussed in the next section.

All right, suppose we are given a set of independent measurements from the same GAU distribution. Our natural inclination, if we have no strong previous information, is to take the average of these values. We know from our discussion of expected value that ordinarily

$$\bar{x} = \text{Ave}(x) \to E(X) = \mu.$$

Thus $\bar{x}$ can be taken as an estimate of μ. But we have as yet not the slightest idea how close $\bar{x}$ might be to μ. We need to determine the *uncertainty* of the estimate.

It's back to the familiar path:

$$\text{posterior} \propto \text{prior} \times \text{likelihood}.$$

6.2 Inferences about μ with Known σ

I will take $\mathrm{pd}_0(\mu)$ = FLAT and use the likelihood we discussed at length in Section 6.1:

$$\mathrm{pd}(\mathbf{x} \mid \mu, \sigma^2) \propto \frac{1}{\sigma^n} \exp\left\{-\frac{1}{2\sigma^2}[S_{\bar{x}\bar{x}} + n(\mu - \bar{x})^2]\right\}.$$

Since σ is known, we eliminate all multipliers depending on σ but not on μ, our unknown. This means that we toss out

$$\frac{1}{\sigma^n} \quad \text{and} \quad \exp\left(-\frac{1}{2\sigma^2} S_{\bar{x}\bar{x}}\right)$$

and arrive at

$$\mathrm{pd}(\mu \mid \mathbf{x}) \propto \exp\left[-\frac{n}{2\sigma^2}(\mu - \bar{x})^2\right] = \exp\left[-\frac{1}{2\frac{\sigma^2}{n}}(\mu - \bar{x})^2\right].$$

If you now look back in the Table of Kernels, you can identify this posterior distribution of μ as GAU. [In that table x always represents the variable. We match the form $\exp[-a(x - b)^2]$ by matching $x = \mu$, $a = n/2\sigma^2$, $b = \bar{x}$.] When we consult the GAU page itself, we recognize that

$$\mathrm{pd}(\mu \mid \mathbf{x}) = \mathrm{gau}\left(\mu \mid \bar{x}, \frac{\sigma^2}{n}\right).$$

The mean of this posterior distribution of μ is $\bar{x}$, the sample average or sample mean; the variance of the distribution is σ^2/n. You will often hear the terminology *mean* for *mean of the posterior distribution of* μ, and *variance of the mean* and *s.d. of the mean* for the *variance* and *standard deviation of the posterior distribution of* μ.

Example 6.2.1. Suppose we are given the ten observations

$$\mathbf{x} = 10.84, 10.69, 10.19, 9.28, 8.97, 9.88, 10.24, 9.22, 9.45, 9.80$$

and are told that $\sigma = 1$. The formula immediately gives

$$[\mu \mid \mathbf{x}] \sim \mathrm{GAU}(* \mid 9.856, \tfrac{1}{10}).$$

The average of the ten observations was 9.856, and $\sigma^2/n = \tfrac{1}{10}$. And that's all there is to the problem. If we want to ask what the probability is that μ lies in a certain interval, we just look up the area in a table of GAU:

$$90\% \text{ HDR} = 9.856 \pm 1.645\sqrt{\tfrac{1}{10}}$$
$$= 9.856 \pm .52$$
$$= (9.34, 10.38).$$

Thus

$$P(9.7 < \mu < 10) = \text{GAU}(10 \mid 9.856, .1) - \text{GAU}(9.7 \mid 9.856, .1)$$
$$= \text{GAU}(.46 \mid 0, 1) - \text{GAU}(-.49 \mid 0, 1)$$
$$= .677 - .313 = .364,$$

$$\mathbb{O}\left(\frac{\mu < 10.0}{\mu > 10.0}\right) = \frac{\text{GAU}(.46 \mid 0, 1)}{1 - \text{GAU}(.46 \mid 0, 1)} = \frac{.677}{.323} = 2.1.$$

Let me review some of our assumptions.

1. *The observations were independent.* This may be justified, but you have to think about it. Sometimes instruments stick so that successive measurements are close together. Sometimes you actually observe a cumulative phenomenon where the *changes* are independent, but the successive totals are not.

2. *The observations were distributed according to* GAU. This may be a good approximation in the center, but it is not necessarily good in the tails. However, by the *Central Limit Theorem* the *average* of a group of observations is usually more nearly Gaussian than the individual observations. This should give you confidence, but don't press your luck.

3. *The observations had identical* GAU *distributions.* Unfortunately we sometimes run into populations which are mixtures of different distributions. I treat this in Chapter 8.

4. σ *was known.* In the following section I treat the case of unknown σ.

Note that:

5. Although a single observation may have a perceptible variance, you may drive down the variance of the mean as far as you please by taking more observations. But there's always the risk that the population may change with time, that you can't get an indefinite number of homogeneous observations.

6. As you take more and more observations, the *variance of the mean*, σ^2/n, goes down as $1/n$, while the *standard deviation of the mean*, $\sigma/\sqrt{n}$, goes down only as $1/\sqrt{n}$. Thus it takes *four times* as many observations to halve the uncertainty, *one hundred times* as many observations to cut the standard error of the mean by a factor of ten. It's usually better to look for new experimental methods than to plan on taking a very large number of observations.

7. $\bar{x}$ and n describe the sample completely when σ^2 is known. $\bar{x}$ and $\sigma/\sqrt{n}$ describe the inference about μ completely. [They are *sufficient.*] It is now customary in print to report observations as

$$\bar{x} \pm \frac{\sigma}{\sqrt{n}}$$

and to mention in the heading of the table that this is the

mean ± s.d. of the mean.

In the example above we would report

9.856 ± .316.

When data is plotted and the standard deviation of the mean is known, it is also customary to indicate it on the plot by some symbol like

which shows the mean and one standard deviation on either side.

This may all be customary, but unfortunate habits of interpretation may arise. You can check in your tables that mean ± 1 s.d. of the mean contains only 68% of the probability. Thus the chance is 1 in 3 that the true value lies outside the indicated interval. You have to keep your wits about you. It is gratifying, though, that reporting is getting more standardized. Not too long ago when you saw 12.34 ± .56 you didn't have the slightest idea of what was meant; sometimes it was one σ, sometimes two, and sometimes .6745 σ's, which gives the 50% HDR. In most advertisements it's still not clear. See Eisenhart (1968).

In all this discussion we have been analyzing only the *random* variation. Sometimes there is *systematic error* where a constant error is superimposed on the random error. Calibrations may be off, theory may be wrong, you may just measure one thing while believing you are measuring something else. Systematic errors cannot be detected in a single set of homogeneous observations. The variance and standard deviation of the mean refer only to the random error; it is up to the experimenter to prevent or compensate for systematic error. He can do this by making as many cross-checks as possible. *Comparison* is the watchword of statistics.

Example 6.2.2. $\sigma = 3$. How many observations are necessary so that the 95% HDR for μ is .50 wide?

95% HDR goes out to ±1.96 σ's, so

$$\frac{3.92\sigma}{\sqrt{n}} = .50, \qquad \sqrt{n} = \frac{11.76}{.50} = 23.52, \qquad n = 551.$$

We've derived the posterior distribution of μ. Occasionally we then ask: What is the probability distribution for the next observation if we make one? This is just an ordinary problem in compound probability. To continue the analogy, we make the values of μ be the urns in which there were different mixtures of colored balls, and the values of the next observation, those colored balls. The main change is that we must integrate rather than sum, for μ is a continuous variable:

$$\text{pd (next obs} \mid \mathbf{x}, \sigma^2) = \int \underbrace{\text{pd (next obs} \mid \mu, \sigma^2)}_{\substack{\text{pd of next obs} \\ \text{given definite } \mu \\ = \text{gau (next obs} \mid \mu, \sigma^2)}} \underbrace{\text{pd } (\mu \mid \mathbf{x}, \sigma^2)}_{\substack{\text{pd of } \mu \\ = \text{gau } (\mu \mid \bar{x}, \sigma^2/n)}} d\mu.$$

As usual, integrations involving the normal distribution work out quite easily. The result is

$$\text{next obs} \sim \text{GAU}\left(* \mid \bar{x}, \frac{n+1}{n}\sigma^2\right).$$

In other words, it's a normal distribution centered on $\bar{x}$ but with a little more variation allowed because of the uncertainty in μ. As n gets larger, our knowledge of μ gets better, and

$$\bar{x} \to \mu, \qquad \frac{n+1}{n}\sigma^2 \to \sigma^2.$$

For a sizable n, the extra factor is almost negligible. Moreover, many "known" variances are not known to such an accuracy anyway.

Example 6.2.3

$$\bar{x} = 6.3, \qquad n = 4, \qquad \sigma = .4.$$

What is $P(\text{next obs} < 6.0)$?

$$\text{next obs} \sim \text{GAU } (* \mid 6.3, \tfrac{5}{4}[.16]) = \text{GAU } (* \mid 6.3, .20),$$

$$P(\text{next obs} < 6.0) = \text{GAU}\left(\frac{6.0 - 6.3}{.45} \mid 0, 1\right) = .253.$$

If we knew that $\mu = 6.3$, then

$$P(\text{next obs} < 6.0) = \text{GAU}\left(\frac{6.0 - 6.3}{.4} \mid 0, 1\right) = .227.$$

Sometimes we are not entirely ignorant about μ before we start an experiment. We may have our own data from a previous experiment, or an experiment done by someone else may have been the inspiration for ours. If this other data is directly applicable, we want to combine all the information. Obviously we must describe this other information in our prior for μ. We can handle any prior numerically, but the only one that gives a concise formula at the end is a normal distribution, say,

$$\text{GAU}\left(* \mid \bar{x}_0, \frac{\sigma_0^2}{n_0}\right).$$

We use the familiar procedure. The likelihood of a normal sample was derived a short time ago; now we have to combine it with our Gaussian prior. After a little algebra in the exponent we find

$$\left[\mu \mid x, \bar{x}_0, \frac{\sigma_0^2}{n_0}\right] \sim \text{GAU}\left(* \left| \frac{\frac{1}{\left(\frac{\sigma^2}{n}\right)}\bar{x} + \frac{1}{\left(\frac{\sigma_0^2}{n_0}\right)}\bar{x}_0}{\frac{1}{\left(\frac{\sigma^2}{n}\right)} + \frac{1}{\left(\frac{\sigma_0^2}{n_0}\right)}}, \frac{1}{\frac{1}{\left(\frac{\sigma^2}{n}\right)} + \frac{1}{\left(\frac{\sigma_0^2}{n_0}\right)}}\right.\right).$$

This looks bad, but really isn't. The mean of this posterior distribution is the weighted average of the mean of the new sample ($\bar{x}$) and the mean of the prior distribution ($\bar{x}_0$). And what weights are used? The weight for $\bar{x}$ is proportional to the reciprocal of its variance: σ^2/n; the weight used for the mean of the prior distribution is proportional to the reciprocal of its variance: σ_0^2/n_0. In other words, the *reciprocal of the variance* is the measure of the importance of one observed mean relative to another. This is quite reasonable intuitively. If a mean has infinite variance, it should be ignored, for $1/\infty = 0$. If a mean has 0 variance, it is right on the head, and everything else should be disregarded. The thing that intuition can't decide, of course, is whether we want the reciprocal of the variance or the reciprocal of the standard deviation or perhaps even some other power of the standard deviation.

Example 6.2.4. $\text{pd}_0(\mu) = \text{FLAT}$.

Experiment	σ^2	n	$\bar{x}$
1	2	5	1.6
2	1	3	1.9

$$[\mu \mid \text{both experiments}] \sim \text{GAU}\left(* \left| \frac{\frac{1}{\left(\frac{2}{5}\right)} \times 1.6 + \frac{1}{\left(\frac{1}{3}\right)} \times 1.9}{\frac{1}{\left(\frac{2}{5}\right)} + \frac{1}{\left(\frac{1}{3}\right)}}, \frac{1}{\frac{1}{\left(\frac{2}{5}\right)} + \frac{1}{\left(\frac{1}{3}\right)}}\right.\right)$$

$$= \text{GAU}(* \mid 1.76, .182).$$

PROBLEM SET 6.2

1. $x = 6.02, 5.84, 5.96, 6.02, 5.95, 6.00, 5.90$. Find $\bar{x}$.
2. If $\sigma^2 = 4$, what is the variance of the mean when $n = 4$? when $n = 10$? when $n = 50$?
3. If $\sigma^2 = 4$, what is the standard deviation of the mean when $n = 4$? when $n = 10$? when $n = 50$?
4. If $\sigma = 10$, what must n be to make the standard error of the mean equal to .1?
5. If $\sigma^2 = 5$, what must n be to make the 68% HDR have a width of .2?

6. If $\sigma^2 = .07$, $n = 14$, and $\bar{x} = 3.01$, what is the posterior distribution of μ?
7. If $\bar{x} = 9.87$, $\sigma = .5$, and $n = 20$, what is the posterior distribution of μ?
8. $\sigma^2 = .16$ and $x = 1.66, 2.35, 1.72, 1.77, 2.77, 1.28$. What is pd $(\mu \mid x)$?
9. The standard deviation of one observation on a balance used in the XYZ laboratory is .01 gm (grams). The same object is weighed 11 times, and the average of the readings is .162 gm. What is a 95% HDR for the weight of the object?
10. An astronomer finds that under standard observing conditions his positional readings have individual standard errors of .17″. How many times must he observe one object to reduce the standard error of the mean to .02″?
11. Amino acid X is routinely assayed by chromatography plus spectrophotometry. The method typically has a standard deviation of one observation equal to .16. A set of 30 analyses of samples of the same substance gives $\bar{x} = 2.441$. What is the probability that the true value lies in the interval (2.403, 2.479)?
12. Draw a random sample of 25 from the population GAU $(* \mid 3, .64)$. After getting the sample, imagine that you know σ but have no information about μ. Give the 75% HDR for μ.
13. Compare $P(-2 < \text{next obs} < 2)$ when
    (a) $\sigma = 1$ and we have two observations with $\bar{x} = 0$,
    (b) $\sigma = 1$ and μ is known to be 0.
14. Dr. Q has just reported that the coefficient of linear expansion of a diamond is $1.203 \pm .071$ (in units of $10^{-6}/C°$). Since your laboratory has the necessary equipment and the experiment might get you some good publicity, you borrow your fiancée's engagement ring and set things up. The standard error of one determination is .29. You make 12 measurements and your $\bar{x} = 1.16$. What is the posterior distribution of this coefficient if all the information is combined?
15. You set out to measure a physical constant and have only the most general idea of its value. Two experiments are performed by different methods:

Experiment	σ^2	n	$\bar{x}$
1	.40	6	1.6
2	.20	4	1.8

The theoreticians seem pretty sure that the two methods actually measure the same thing, so you want to combine the results. What is the posterior distribution?

16. Suppose that $pd_0(\mu) = $ FLAT, and we have these three compatible experiments:

Experiment	Mean ± s.d. of the mean from experiment
1	11.62 ± .17
2	12.14 ± .39
3	11.73 ± .11

What is the combined information about μ?

17. Suppose that
$$\text{pd}_0(\mu) = \frac{\mu^3 e^{-\mu}}{6}, \quad \bar{x} = 3.4, \quad \frac{\sigma^2}{n} = .25.$$
What is the posterior distribution of μ?

18. Sixteen observations are made on a Gaussian variable. It is known beforehand that $\sigma = 2$ and that $1 < \mu < 3$. In the sample $\bar{x} = 0.8$. What are
$$\Theta\left(\frac{\mu < 2}{\mu > 2}\right)?$$

ANSWERS TO PROBLEM SET 6.2

2. 1; .4; .08

4. 10,000

6. GAU (* | 3.01, .005)

8. gau (μ | 1.925, .0267)

10. 72. Do you think that if he makes that many observations in one night, the readings are likely to be independent or the variance constant?

13. (a) .896   (b) .9545

14. GAU $\left(* \left| \dfrac{\dfrac{1}{.00503} \times 1.203 + \dfrac{1}{.00701} \times 1.16}{\dfrac{1}{.00503} + \dfrac{1}{.00701}}, \dfrac{1}{\dfrac{1}{.00503} + \dfrac{1}{.00701}} \right. \right)$

$= $ GAU (* | 1.185, .00294)

16. Exactly the same scheme can be extended to three or more experimental results. You continue to weight the individual results proportional to the reciprocals of their variances; the resulting variance is still the reciprocal of the sum of the reciprocals of the individual variances:

$$\text{posterior mean} = \frac{\dfrac{11.62}{.0289} + \dfrac{12.14}{.1521} + \dfrac{11.73}{.0121}}{\dfrac{1}{.0289} + \dfrac{1}{.1521} + \dfrac{1}{.0121}} = 11.69,$$

$$\text{posterior variance} = \frac{1}{\dfrac{1}{.0289} + \dfrac{1}{.1521} + \dfrac{1}{.0121}} = .00806.$$

You would report 11.69 ± .09.

17. Multiply the likelihood and the prior point by point. Figure 6.2.1 shows a rough sketch. Please note that the posterior density is very close to being normal. What is the reason? The prior curve varied quite slowly in the region where the likelihood was large. What the prior did the rest of the time was unimportant since the likelihood was essentially zero. [Of course, you can't allow sudden peaks of a million or so which would overcome the almost zero likelihood.] Thus it often makes little difference that our prior distribution is not exactly FLAT so long as it varies slowly.

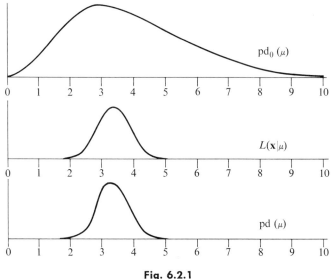

Fig. 6.2.1

6.3 INFERENCES ABOUT BOTH μ AND σ

In the last section we assumed that we knew σ and wanted information about μ. This may be realistic under laboratory conditions when the random variation is due mainly to measurement errors. But even in laboratories the precision of a measurement may depend on the care of the experimenter, the humidity of the room, or a host of other conditions. And when the error also depends on natural causes perhaps little understood and less controlled, you have to assess the variation anew in each experiment. It is true that you may have a little knowledge of the variation and the region in which the observations fall: you need a tinge of knowledge to choose the proper measuring instruments. You may also have developed some "feel" for the scheme of things when you work in an area for some time; one of the valued attributes of a good experimenter is the ability to predict which experiments will produce definitive information. But all this doesn't obviate the need to pin down the variation with information from the sample.

Thus we will be trying to make inferences about μ and σ simultaneously; a bivariate distribution will describe our knowledge of the two parameters. If we stick to geometric terminology, we need one more dimension. (See Fig. 6.3.1.) The two variables are plotted along the two axes in a plane; the probability density is represented by the height above the plane and forms a surface rather than a curve. Probability is now volume instead of area, but is still normalized to 1.

In this section I assume that the observations are independent and come from the same GAU distribution. I also assume that we know nothing very precise about

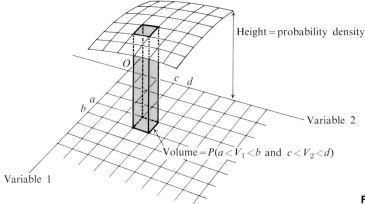

Fig. 6.3.1

μ and σ, and will use the FLAT and LFLAT prior distributions described in Section 6.1. Further, I will assume that our prior knowledge (or lack of it) about μ is independent of our prior knowledge (or lack of it) about σ. This isn't necessarily true sometimes (we know that big trees vary more than small trees), but I'm assuming no information at the moment. This is a last-resort procedure, as I've stated before. The prior distribution is then

$$\text{pd}_0(\mu, \sigma) = \text{FLAT} \times \text{LFLAT}(\sigma) \propto \frac{1}{\sigma}.$$

The posterior distribution is formed by multiplying this joint prior by the likelihood and normalizing:

$$\text{pd}(\mu, \sigma \mid \mathbf{x}) \propto \frac{1}{\sigma^{n+1}} \exp\left\{-\frac{1}{2\sigma^2}[S_{\bar{x}\bar{x}} + n(\mu - \bar{x})^2]\right\},$$

where

$$S_{\bar{x}\bar{x}} = \sum (x_i - \bar{x})^2.$$

In Fig. 6.3.2 I've drawn a picture of the joint posterior distribution of μ and σ for

$$n = 8, \quad \bar{x} = 1.459, \quad \text{and} \quad S_{\bar{x}\bar{x}} = .000120.$$

Figure 6.3.3 is a contour map of the same surface showing, as the picture cannot, where the probability density has diminished to $\frac{1}{10}$, $\frac{1}{100}$, $\frac{1}{1000}$, and $\frac{1}{10000}$ of the value at the maximum.

Because of the term

$$\exp\left[-\frac{1}{2\sigma^2}(\mu - \bar{x})^2\right]$$

the joint posterior density cannot be written as a product of two factors, one involving only μ and the other involving only σ. This means that the joint density is

192   Inferences about the Gaussian Distribution   6.3

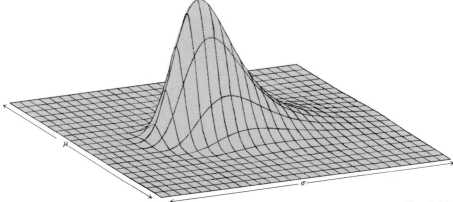

Fig. 6.3.2

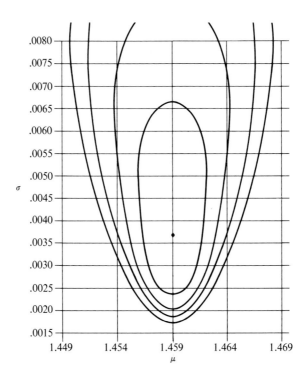

Fig. 6.3.3 pd $(\mu, \sigma \mid n = 8, \bar{x} = 1.459, S_{\bar{x}\bar{x}} = .000120)$.

not a product of marginal densities, and therefore our posterior knowledge of the two parameters is not independent. When we talk of something that depends on both μ and σ (like the probability distribution of the next observation) we have to use the joint distribution. But for many purposes we can form the marginal distributions of μ and σ separately and look at their individual behavior.

Before we do that, let's try to predict what we're going to find. How do the observations enter into the joint density? Only as $\bar{x}$ and $S_{\bar{x}\bar{x}}$. It's quite evident, then, that $\bar{x}$ will again give us an estimate of the parameter μ, while something involving $S_{\bar{x}\bar{x}}$ is going to give us an estimate of σ. After all, $S_{\bar{x}\bar{x}}/n$ is the sample variance; you'd expect it to be related to the population variance and the standard deviation. We can also expect that our marginal distributions will depend heavily on n, the number of observations. I'll tell you now that the degree of dependence between the two parameters decreases as n gets large.

All right, how do we get the marginal densities? When we had discrete variables we just added across rows or columns. [Review Section 2.1 if you're hazy on it.] We have to add here, too, but because the distribution is continuous the addition process becomes integration. Perhaps an analogy would help. Imagine that the drawing in Fig. 6.3.2 is a solid sausage—of unorthodox shape, of course. To get the marginal density of σ we put our sausage into a sausage-slicing machine which cuts lots of very thin vertical slices parallel to the μ-axis (Fig. 6.3.4). We keep all the slices in order, weigh each of them on a scale, and graph their weights. That graph shows the marginal density of σ. To get the marginal density of μ, we have to slice in the other direction.

Fortunately all this slicing and weighing can be done mathematically, thanks to the impeccable breeding of the Gaussian distribution. When we define

$$s^2 = \frac{S_{\bar{x}\bar{x}}}{n-1} \quad \text{and} \quad q^2 = \frac{S_{\bar{x}\bar{x}}}{n},$$

we get

$$\text{pd}(\mu \mid x) = \text{stu}\left(\mu \mid \bar{x}, \frac{s^2}{n}, n-1\right),$$

$$\text{pd}(\sigma \mid x) = \text{igam}(\sigma \mid q, n-1).$$

These are the *Student* and *Inverse Gamma* distributions, respectively. You'll find their formulas in the Gallery.

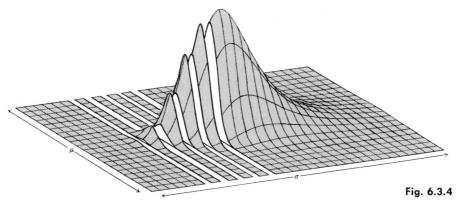

Fig. 6.3.4

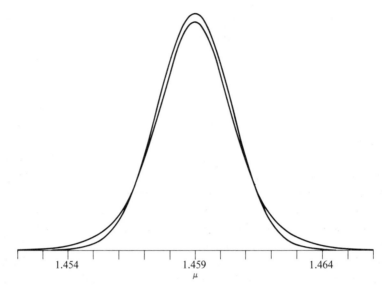

Fig. 6.3.5 Posterior density of μ.

The parameters in STU look much like those of the previous section where $\mu \sim \text{GAU}(* \mid \bar{x}, \sigma^2/n)$. The center of the distribution is $\bar{x}$, as before. Instead of σ^2/n in the second place we have s^2/n. As we anticipated, $s^2 = S_{\bar{x}\bar{x}}/(n-1)$ is playing a role much like the variance; of course it is only an observed quantity, not an exact one like the σ^2 of the preceding section. The third parameter is called the *number of degrees of freedom*. Roughly it means this: We had n independent pieces of information in our n observations. We needed to get a sum of squares to estimate our variation. We couldn't use $\sum (x - \mu)^2$ because μ wasn't known. We did the next best thing and took the sum of squares about $\bar{x}$, the only pinned-down location in sight. But $\bar{x}$ did not come free: it had to be calculated from our data. This took away one of our pieces of information; just $n - 1$ were left.

Figure 6.3.5 shows two curves. The shorter, broader one is the marginal posterior density of μ:

$$\text{stu}(\mu \mid \bar{x}, s^2/n, n-1),$$

in this case

$$\text{stu}(\mu \mid 1.459, .00000214, 7).$$

The slightly taller, narrower curve is

$$\text{gau}(\mu \mid \bar{x}, s^2/n),$$

in this case

$$\text{gau}(\mu \mid 1.459, .00000214).$$

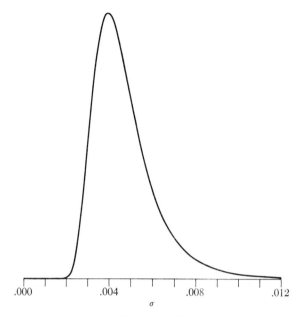

Fig. 6.3.6 Posterior density of σ.

I would have gotten the Gaussian curve if this s^2 had been the *known* precise value of σ^2 rather than an estimate. You can see that there is more uncertainty attached to the STU curve; in particular, the probability of extreme deviations is noticeably larger.

Figure 6.3.6 shows the posterior marginal density of σ:

$$\text{igam} (\sigma \mid q, n - 1),$$

in this case

$$\text{igam} (\sigma \mid .00388, 7).$$

Note how skew it is. Low values of σ are pretty well eliminated, but some astoundingly high values still have noticeable probability. The mode of this distribution is .00388, the mean .00467. Note that the mode is not at the same value of σ as the mode of the joint distribution; that was at .0037.

Example 6.3.1. I will go through one complete calculation in great detail. We have eight measurements of the index of refraction of frosted quartz for wavelength 5893 Å (angstroms):

1.466, 1.459, 1.461, 1.457, 1.464, 1.456, 1.457, 1.454.

Table 6.3.1

Observation = x	Residual = $x - \bar{x}$	Residual² = $(x - \bar{x})^2$
1.466	+.007	.000049
1.459	.000	.000000
1.461	+.002	.000004
1.457	−.002	.000004
1.464	+.005	.000025
1.456	−.003	.000009
1.457	−.002	.000004
1.454	−.005	.000025
S_x = 11.674	(+.002)	$S_{\bar{x}\bar{x}}$ = .000120
$\bar{x} = \dfrac{11.674}{8} = 1.459$	(=0 except for rounding)	

In Table 6.3.1 I calculated $\bar{x}$ and $S_{\bar{x}\bar{x}}$. The label *residual* is given to the difference between an observation and the *fitted* value, the one that comes from putting the observations together according to a particular model. Here I assumed that I was measuring the same thing all the time; the fitted value was $\bar{x}$. In a simple case like this the residuals exhibit little more than the observations themselves. In more complicated cases they may provide warnings that something is amiss. If you get your calculations done by machine, try to arrange to have the residuals printed out.
Next:

$$s^2 = \frac{S_{\bar{x}\bar{x}}}{n-1} = \frac{.000120}{7} = .0000171.$$

Many people quote s^2 as *the* estimate of the variance σ^2 because $E(s^2) = \sigma^2$. This expectation is true for any set of independent observations and any distribution, not just for GAU. I've already mentioned though that the consequences of an estimate must be taken into account, that we can't choose the best on purely mathematical grounds.
Next:

$$\frac{s^2}{n} = \frac{.0000171}{8} = .00000214.$$

Just as the variance of a mean is $(1/n)$th the variance of an observation, so is our *estimated* variance of the mean $(1/n)$th the *estimated* variance of an observation.
And:

$$\sqrt{.00000214} = .0015,$$

ν = number of degrees of freedom = 8 − 1 = 7.

[This is a lower-case Greek "nu."]

All this leads to the posterior distribution

$$\mu \sim \text{STU}(* \mid 1.459, .00000214, 7),$$

and we report

$$1.459 \pm .0015 \qquad (8),$$

giving the number of observations in parentheses. If you have only a small number of observations, it is vital to quote the number. STU with a small ν is very different from GAU. When ν is large, the practical difference is minor. We interpret this distribution for μ by using the table of STU $(* \mid 0, 1, \nu)$ in the Gallery much as we used the table of GAU $(* \mid 0, 1)$; the difference is that the additional parameter ν has to be used. To find the 90% HDR for μ, we find .9 in the row labeled "center" and move down that column to the row $\nu = 7$. There we find the number 1.90. Our 90% HDR is then

$$\bar{x} \pm 1.90 \times \text{(estimated s.d. of the mean)}$$
$$= 1.459 \pm 1.90 \times .0015 = (1.456, 1.462).$$

Note that because of the greater uncertainty, we have to go out *more* estimated s.d.'s of the mean than precise s.d.'s of mean. Had the distribution been GAU, we would have had to go only 1.645 instead of 1.90. If you look down the column of the STU table we were using, you will find that 1.64 in the row marked $\nu = \infty$.

You might be interested to know that the name *Student* was a pseudonym for W. S. Gosset, a statistician for an English brewery. He used t to denote the standardized Student variable:

$$t = \frac{\mu - \bar{x}}{\sqrt{s^2/n}},$$

so the distribution is often called the *t*-distribution.

Now for σ:

$$q^2 = \frac{S_{\bar{x}\bar{x}}}{n} = \frac{.000120}{8} = .000015,$$
$$q = \sqrt{.000015} = .00388.$$

So

$$\sigma \sim \text{IGAM}(* \mid .00388, 7).$$

Tables for IGAM are rather scarce; the one in the Gallery may not be too useful. To find an HDR for σ, you first find the interval for IGAM $(* \mid 1, 7)$ and then multiply the endpoints of that interval by q. I did a little graphical interpolation for the 90% value and found the values (.67, 1.72). When I multiplied by .388 I got (.26, .67); that is,

$$P(.26 < \sigma < .67) \doteq .90.$$

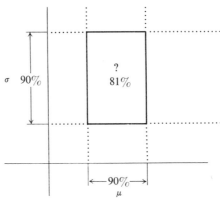

Fig. 6.3.7

Finally, the relation between the HDR's of the marginal distributions and the HDR's of the joint distribution is not very direct. In the first place, if you marked off 90% intervals in both variables, the intersection (a rectangle) would contain 81% of the joint probability only if the variables were independent (Fig. 6.3.7). In the second place, such a rectangle would not usually be an HDR; the ordinate has to be the same all along the boundary. Boundaries of joint HDR's usually look like the contour lines in Fig. 6.3.3.

PROBLEM SET 6.3

1. What is s^2 if $n = 5$ and $S_{\bar{x}\bar{x}} = 131.4$?
2. $t \sim$ STU $(*\,|\,0, 1, 5)$. What is $P(t < -2.02)$?
3. $t \sim$ STU $(*\,|\,0, 1, 30)$. What is $P(t > .26)$?
4. $t \sim$ STU $(*\,|\,0, 1, 8)$. What is a so that $P(-a < t < a) = .90$?
5. $t \sim$ STU $(*\,|\,0, 1, 10)$. What is $P(-1 < t < 1)$?
6. $\bar{x} = 10.0$, $n = 6$, $\sqrt{s^2/n} = .8$; the observations came from the same GAU distribution, and there was no prior information about μ and σ. What is $P(\mu > 11)$?
7. $\bar{x} = 16.7$, $n = 31$, $\sqrt{S_{\bar{x}\bar{x}}/n(n-1)} = 2.8$; the observations came from the same GAU distribution, and there was no prior information about μ and σ. What is $P(\mu < 10)$?
8. $\bar{x} = 100.13$, $n = 142$, $S_{\bar{x}\bar{x}} = 286.74$; the observations came from the same GAU distribution, and there was no prior information about μ and σ. What is the 99% HDR for μ?
9. Here are 15 observations on the melting point of villiaumite (data in °C):

$$991,\ 998,\ 959,\ 984,\ 1001,\ 997,\ 990,\ 976,$$
$$977,\ 975,\ 995,\ 981,\ 974,\ 995,\ 980.$$

Complete the analysis and discuss the accuracy of this result and of the method in general.

10. You make six measurements of the specific gravity of benzofluoride dichloride:

$$1.3171, \quad 1.3119, \quad 1.3142, \quad 1.3159, \quad 1.3161, \quad 1.3121.$$

How do you report your findings?

11. A pharmacologist is interested in the effect of increased amino acid intake on the chemical content of the brain. To establish a control for his future experiments, he feeds 16 rats a normal diet for a specified period. He then lops off their heads and quick-freezes them. He removes the brains, homogenizes them in a blender, and analyzes them chemically. The serotonin content (in millimicrograms per gram of brain weight) shows $\bar{x} = 475$, $S_{\bar{x}\bar{x}} = 54109$. What conclusions can he draw?

12. Twelve measurements of the viscosity of dimethylaniline at 20°C give (in centipoises):

$$146, \quad 154, \quad 141, \quad 140, \quad 136, \quad 132, \quad 147, \quad 140, \quad 147, \quad 139, \quad 140, \quad 140.$$

What is $P(\text{viscosity} > 140 \text{ cP})$?

13. A fertilizer was developed and applied to a strip of ground. A test crop was planted, and the yield recorded for each of seven successive subplots of equal size (see Table 6.3.2). Analyze the data.

Table 6.3.2

Plot	Yield
1	5.44
2	6.84
3	8.91
4	8.33
5	9.86
6	10.07
7	10.78

14. One dark night two physics students, A and B, decide to repeat a famous experiment. Armed with flashlights, they take up positions exactly 1000 ft apart. A switches on his flashlight and starts his stopwatch simultaneously. B switches on his flashlight as soon as he sees A's light. A stops his watch as soon as he sees B's light. They do the experiment 100 times and report the time taken by light to travel 2000 ft (there and back) as

$$1.50 \pm .025 \text{ sec} \quad (100).$$

Then they estimate the velocity of light as

$$\frac{2000 \text{ ft}}{1.50 \text{ sec}} = 1333 \text{ ft/sec}.$$

Comment.

15. Show that $\sum (x_i - a)^2$ is least when $a = \bar{x}$.
16. Show that $E(s^2) = \sigma^2$.
17. Suppose someone made ten observations on a Gaussian variable with unknown mean and variance, but reported only the largest and smallest of the observations. How can you use this information to get estimates of μ and σ?

ANSWERS TO PROBLEM SET 6.3

2. .05

4. 1.86

6. $\nu = 6 - 1 = 5$; when $\mu = 11$, $t = \dfrac{11 - 10}{.8} = 1.25$
$P(t > 1.25) \doteq .13$

8. $S_{\bar{x}\bar{x}}/(141 \times 142) = .0143$; $\sqrt{.0143} = .12$. $\nu = 141$ is so large we use the ($\nu = \infty$)-line. The limits are then $100.13 \pm 2.33 \times .12 = (99.85, 100.41)$.

9. We would report that

$$\text{melting point} = 985 \pm 3 \quad (15)$$

and mention somewhere that this meant

$$\text{mean} \pm \text{estimated s.d. of the mean} \quad \text{(no. obs).}$$

A 90% HDR would be $985 \pm 1.76 \times 3 = (980, 990)$.

When we have this many observations we can try to do a little checking for anomalies. There seems to be no particular trend in magnitude or sign. The estimated σ is $\sqrt{1914/14} = 11.7$. As a *very rough* check we see that

$$11 \text{ residuals out of 15 are within } 1 \times 11.7,$$
$$14 \text{ residuals out of 15 are within } 2 \times 11.7,$$
$$15 \text{ residuals out of 15 are within } 3 \times 11.7.$$

We don't know μ and σ exactly, so a Gaussian evaluation is not quite cricket; but if it were, the expected numbers would be 10, 14, 15 (68%, 95%, 99.7%). There's no striking reason to disbelieve the model; even so, no one believes a model completely. It's a tool to work with, something to hang your ideas on, not one of those ultimate realities. Here we've performed a trivial experiment; we haven't carried out a systematic investigation. The statistical conclusion we reach is always conditional, conditional on the assumptions being satisfied. Did we check to see whether the samples were contaminated with other minerals? Did we calibrate our equipment recently against a standard? The situation might be that shown in Fig. 6.3.8. A single set of homogeneous samples like this measures only the random variation, not the systematic error. It's comparison and cross-checking in a planned series of experiments that locates the systematic errors.

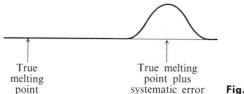

Fig. 6.3.8

Here the 90% HDR for the σ of our method is

(8.5, 16.2).

This describes the *internal consistency* of the measurements.

11. Serotonin content = 475 ± 15   (16)
12. About .86 under the usual assumptions.
13. A cursory look at this data should convince you to put away the adding machine. As in many agricultural experiments, there is a noticeable trend in the fertility of the land. This may be due to drainage, variation in soil type, or fertilization from past experiments. When you want to compare two varieties of grain, you do not plant one in one field and the other in another field. Variations in the fertility of the fields would ruin the comparison. Instead you divide up the available area into smaller *blocks*, hopefully each homogeneous within itself, and plant some of each type of grain in each block (Fig. 6.3.9). Then you analyze the *differences within the blocks*. The assignment of varieties to plots is usually made randomly to avoid any possible biases on the part of the experimenter; hopefully, unknown factors may also average out. We wouldn't want grain A always to be north of B, always upwind, or always at a higher elevation.

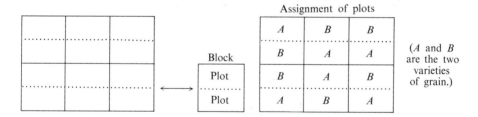

Fig. 6.3.9

14. Our experimenters ignored other knowledge about the velocity of light, but we'll overlook that. And their stopwatch readings and computations are probably quite accurate. The trouble is that statistical conclusions depend on the assumptions, and here the assumptions are not true. The students think they measured the travel time of light. Actually they measured

travel time of light + reaction times

+ switch-movement times + filament-heating times,

and the latter three factors completely obscure the first. The variation they see is due mainly to the reaction times and the switch-movement times. How do they disentangle the situation? By varying the circumstances of the experiment. When the students stand 2000 or 4000 ft apart and find that they get approximately the same results, they will begin to sense their error. In performing experiments, we have to go beyond *internal consistency* and make sure that we achieve consistency with the real world.

16. $\sum (x - \mu)^2 = \sum (x - \bar{x} + \bar{x} - \mu)^2 = S_{\bar{x}\bar{x}} + n(\bar{x} - \mu)^2$. But $E[(x - \mu)^2] = \sigma^2$ and $E[(\bar{x} - \mu)^2] = \sigma^2/n$, since the variance of a mean is $(1/n)$th the original variance. Therefore

$$E(S_{\bar{x}\bar{x}}) = n\sigma^2 - n\frac{\sigma^2}{n} = (n-1)\sigma^2 \quad \text{and} \quad E\left(\frac{S_{\bar{x}\bar{x}}}{n-1}\right) = \sigma^2.$$

17. The likelihood using only the largest and smallest of ten is (calling the largest $x_{(10)}$ and the smallest $x_{(1)}$)

$$\text{pd}(x_{(10)}, x_{(1)} \mid \mu, \sigma^2) \propto \text{gau}(x_{(1)} \mid \mu, \sigma^2)[\text{GAU}(x_{(10)} \mid \mu, \sigma^2)$$
$$- \text{GAU}(x_{(1)} \mid \mu, \sigma^2)]^8 \text{gau}(x_{(10)} \mid \mu, \sigma^2)$$

and

$$\text{pd}(\mu, \sigma \mid x_{(10)}, x_{(1)}) \propto \frac{1}{\sigma} \text{pd}(x_{(10)}, x_{(1)} \mid \mu, \sigma^2).$$

You can find your posterior marginal distributions numerically. To show you what happens, I took ten random observations from GAU $(* \mid 0, 1)$,

.289, .716, $-.436$, -1.931, .437, .686, .381, .550, $-.182$, 1.198,

and analyzed σ both ways. Figure 6.3.10 shows the comparison.

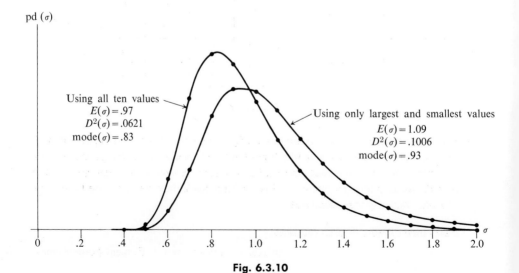

Fig. 6.3.10

CHAPTER 7

Nonnormality *Most of the work of the last two chapters was based on the normal, or Gaussian, distribution. In many cases* GAU *approximates well the distribution of errors we encounter; in other cases it definitely does not. Since a great variety of distributions occur in nature, we should not be surprised that all cannot be governed with the same reins. When we meet a nonnormal symmetric one-humped distribution, we have the alternatives of going to a different mathematical form or trying to modify our Gaussian approach to meet the challenges presented. In this chapter I will discuss (1) some other symmetric distributions and their relationship to* GAU, *(2) the problem of guarding against the contamination of a* GAU *distribution by observations from a longer-tailed distribution, and (3) ways in which some nonnormal distributions can be made more normal.*

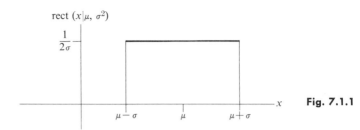

Fig. 7.1.1

7.1 TWO COUSINS OF GAU

Here I want to talk about two distributions which depart from GAU in opposite directions. The first is

$$\text{RECT}(* \mid \mu, \sigma^2),$$

the *rectangular distribution* (see Fig. 7.1.1). RECT has parameters called μ and σ, as does GAU, but they have the general meaning

$$\mu = \text{location parameter},$$
$$\sigma = \text{scale parameter}.$$

[In the Gallery you can see that $D[\text{rect}(* \mid \mu, \sigma^2)] = \sigma/\sqrt{3}$.] When μ changes, the distribution shifts. When σ changes, the distribution expands or contracts.

How does RECT contrast with GAU? The main difference is that it is truncated, while GAU's tails taper off gradually all the way to $\pm \infty$. RECT is a transition between the one-humped *bell-shaped* distributions and the *U-shaped* ones. We analyze it in the usual way. Without prior information, I set

$$\text{pd}_0(\mu, \sigma) = \text{FLAT} \times \text{LFLAT}(\sigma) \propto \frac{1}{\sigma}$$

and

$$\text{pd}(\mathbf{x} \mid \mu, \sigma^2) = \frac{1}{(2\sigma)^n}.$$

Then

$$\text{pd}(\mu, \sigma \mid \mathbf{x}) \propto \frac{1}{\sigma} \times \frac{1}{(2\sigma)^n} \propto \frac{1}{\sigma^{n+1}} \quad \text{for} \quad \mu - \sigma < x_{(1)} < x_{(n)} < \mu + \sigma.$$

[When we take a sample of n observations and sort them into ascending order, we label the smallest $x_{(1)}$, the second smallest $x_{(2)}$, and so on up to the largest observation, $x_{(n)}$.]

This joint posterior density is pictured in Fig. 7.1.2. Note that it is actually zero in a large region, a phenomenon we haven't encountered before. The contour lines are straight horizontal lines. The whole distribution depends on only the number

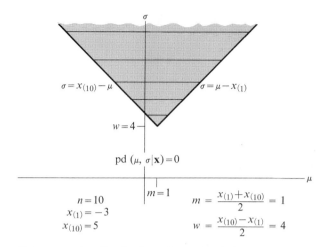

Fig. 7.1.2 Joint posterior distribution of μ and σ derived from a sample of 10.

of observations and the largest and smallest of them; the precise values of the interior observations are unimportant. Contrast this with GAU where, in estimating μ and σ, we give each observation equal weight. In RECT the outer ones are given a weight of 1; the inner ones are given a weight of 0. The two outer observations and the total number of observations constitute sufficient statistics for RECT.

The posterior marginal density of μ is easily found by calculus. Call

$$m = \frac{x_{(1)} + x_{(n)}}{2} \quad \text{and} \quad w = \frac{x_{(n)} - x_{(1)}}{2}.$$

Then

$$\text{pd}(\mu \mid x) = \begin{cases} \dfrac{(n-1)}{2} \dfrac{w^{n-1}}{(w+m-\mu)^n} & \text{for } \mu < m, \\ \dfrac{(n-1)}{2} \dfrac{w^{n-1}}{(w+\mu-m)^n} & \text{for } \mu > m. \end{cases}$$

This density is pictured in Fig. 7.1.3. We calculate

$$E(\mu) = m,$$

where m is the *midrange*, the average of the two extreme observations.

An error distribution that is extreme in the other direction is

$$\text{STU}(* \mid \mu, \sigma^2, 1),$$

also called the *Cauchy distribution*. This one has exceedingly long tails; so long, in

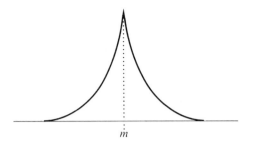

Fig. 7.1.3

fact, that the average of a thousand observations has the same sampling properties as a single observation! We will never meet anything quite this bad in practice, but it makes a splendid example. How do we estimate the parameters? Same old method:

$$\text{pd}_0\,(\mu, \sigma) \propto \frac{1}{\sigma},$$

$$\text{stu}\,(x \mid \mu, \sigma^2, 1) = \frac{1}{\pi\sigma} \frac{1}{1 + \left(\frac{x - \mu}{\sigma}\right)^2},$$

so

$$\text{pd}\,(\mu, \sigma \mid x) \propto \frac{1}{\sigma^{n+1}} \prod \frac{1}{1 + \left(\frac{x_i - \mu}{\sigma}\right)^2}.$$

Unfortunately, no easy formula arises here; we have to multiply it out. In Problem 4.2.3 you did the calculation for three samples from STU $(* \mid \mu, 1, 1)$, a simpler case in which σ was known.

Doing a calculation like this is no great trouble if you have a computer. It turns out, however, that the *sample median* provides a good approximation to the center of the posterior distribution of μ. Table 7.1.1 shows some sampling results that I ran off. I tested three possible estimators for μ,

>sample mean,
>sample median,
>mode of posterior distribution of μ,

and looked to see which was closer to the actual center in 1900 cases. [I really should have used the expected value of the posterior distribution, but I didn't have the time.]

The table reads this way: I took a sample of size five 200 times in all. In 23 cases the *sample mean* was the closest of the three estimates to the true value; in 31 cases the *sample median* was the closest; in 146 cases the *mode of the posterior distribution*

Table 7.1.1

Number of Times Estimate Was Closest to Center

Sample size, n	Number of repetitions, R	Sample mean	Sample median	Mode of posterior distribution of μ
5	200	23	31	146
10	200	18	41	141
15	400	24	64	312
20	200	12	42	146
25	200	9	29	162
35	100	3	19	78
50	200	11	48	141
75	200	2	48	150
100	200	8	48	144

of μ was the closest. The sample median outshines the sample mean, and the calculation using the posterior distribution greatly outshines both. From this data you can infer only *average* behavior. In any particular case the sample mean might just by chance be the best, but the odds are against it.

When we have several possible estimators, we want to have a measure of their relative values. It is common to quote the *efficiency* of one estimator compared with another. If $\hat{\mu}_1$ and $\hat{\mu}_2$ are two estimators of μ, then we define

$$\text{efficiency of } \hat{\mu}_1 \text{ versus } \hat{\mu}_2 = \frac{\text{variance }(\hat{\mu}_2)}{\text{variance }(\hat{\mu}_1)}.$$

That is, if the variance of $\hat{\mu}_1$ is the larger, then $\hat{\mu}_2$ is the more efficient. The efficiency has a heuristic meaning: it indicates the relative number of observations required by each estimator to achieve the same average precision of estimation. [Precision as measured by the variance.] If estimator $\hat{\mu}_1$ is 50% as efficient as estimator $\hat{\mu}_2$, on the average $\hat{\mu}_1$ will require $1/.50 = 2$ times as many observations to achieve the same precision. Again I emphasize that this is only an average value, a preexperiment planning value. Efficiency is not the sole criterion in choosing an estimator. If observations are cheap but computation is dear, an inefficient but easy-to-evaluate estimator may prove best. If you are striving to get every last bit of information out of a sample, you vote for computation.

Let's compare m and $\bar{x}$ in the RECT case:

$$\frac{D^2(m)}{D^2(\bar{x})} = \frac{\frac{2w^2}{(n-2)(n-3)}}{\frac{\sigma^2}{3n}} = \frac{6w^2}{\sigma^2} \frac{n}{(n-2)(n-3)}.$$

You can see that the efficiency of $\bar{x}$ relative to m actually goes to 0 as n gets large; the midrange becomes infinitely more efficient. Of course, that's when the sample is from RECT. If it were from GAU, $\bar{x}$ would be vastly more efficient. [In the above formula $w \leq \sigma$, so

$$\frac{D^2(m)}{D^2(\bar{x})} \leq 6 \frac{n}{(n-2)(n-3)} \doteq \frac{6}{n}.\text{]}$$

In the case of STU ($* \mid \mu$, 1, 1), I have only experimental mean-squared errors to present. They are

$$\frac{2.25}{n} \quad \text{for the mode of the posterior distribution,}$$

$$\frac{2.72}{n} \quad \text{for the sample median,}$$

out of this world for the sample mean.

This means that the median is about 83% as efficient as the mode of the posterior distribution; you would need about 20% more observations (on the average) to bring it up to the same precision. If you had to analyze STU ($* \mid \mu$, 1, 1) observations without computing facilities, the median would be a good bet. The mean would be disastrous.

In Problem 6.3.17 I displayed the posterior distribution of σ from a GAU distribution. I calculated it from a sample of 10 in two ways: by using all the observations, and by using only the largest and smallest. The relative efficiency in this case was about 62%. If you have plotted the curves but haven't calculated the variances, you can get a quick value for the efficiency by taking the square of the ratio of the maximum values or the square of the half-widths. Using the maxima, we get

$$\left(\frac{1.57}{1.97}\right)^2 = .63.$$

Using the half-widths (the distance between the ordinates which are each half the maximum value), we have

$$\left(\frac{1.80}{2.25}\right)^2 = .64.$$

You might look back to Problems 4.1.2 and 4.1.3 to see why these relations are true.

I hope you realize that what you should do with your observations depends largely on what the underlying distribution is. You don't always take sample means; you don't always use medians or midranges. A large number of descriptive statistics have been produced by statisticians, and each has some use somewhere. Don't think that all have to be calculated for each distribution; you use the ones that are pertinent to *your* distribution. If you don't know what it is, you may have to improvise until you find out. General experience indicates, for example, that more distributions are long-tailed than truncated. It might be wise to use techniques appropriate to longer tails if you're really in doubt. The rest of the chapter is devoted to studying these long-tailed beasts.

PROBLEM SET 7.1

1. Plot the posterior distribution of μ when you have a sample of 10 from RECT $(* \mid \mu, \sigma^2)$ with
$$x_{(1)} = -3, \quad x_{(10)} = 5.$$

2. This is a sample of 15 from RECT $(* \mid 6, 25)$:

    4.371, 1.147, 8.645, 4.842, 1.364, 7.499, 10.294, 10.022, 1.139, 8.920, 2.580, 7.214, 10.789, 5.735, 1.958.

    If you didn't know μ and σ, how would you estimate μ?

3. Generate a sample of 25 from RECT $(* \mid 0, 4)$. Compare m and $\bar{x}$. Find $\sum m^2$ and $\sum \bar{x}^2$ for the class.

4. Generate a sample of 25 from STU $(* \mid 0, 1, 1)$. Compare medians and means. [STU $(x \mid 0, 1, 1) = \frac{1}{2} + (1/\pi) \tan^{-1} x$.]

ANSWERS TO PROBLEM SET 7.1

1. $\text{pd}(\mu \mid \mathbf{x}) = \begin{cases} \dfrac{9}{2} \dfrac{4^9}{(5-\mu)^{10}} & \text{for } \mu < 1, \\ \dfrac{9}{2} \dfrac{4^9}{(\mu+3)^{10}} & \text{for } \mu > 1 \end{cases}$

    Some values are:

μ	pd (μ)
1	1.250
$1 \pm .2$	.692
$1 \pm .5$	.348
1 ± 1.0	.121
1 ± 1.5	.047
1 ± 2.0	.020
1 ± 3.0	.004

2. $\hat{\mu} = m = \dfrac{10.789 + 1.139}{2} = 5.964$

7.2 PROBABILITY PAPER

I want to show you a simple device that aids in exploring the shapes of distributions. To pin down shapes exactly, you have to keep careful records and do well-planned experiments. It's really quite hard to get information about the tails of distributions. Suppose you ask, "What is $P(X < 3)$?" and the probability is really .01. In, say, 300 times you'll observe something in that region just three times on the average; the variation will often make it, say, 1 to 6. We went through all this with the BETA distribution. To pin down a frequency you need many observations.

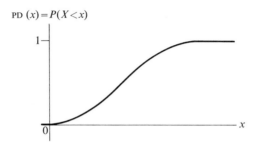

Fig. 7.2.1

But sometimes you can get an idea of the overall shape by plotting the *cumulative distribution*. That was the function that went from 0 to 1 as the variable went from $-\infty$ to $+\infty$. For GAU and many other curves it looks generally like the curve in Fig. 7.2.1, an *S*-curve. We used it for drawing samples from a continuous distribution.

Here we use the curve in straightened-out form. You can always achieve this by changing the vertical scale in a systematic manner. Suppose we want to straighten out GAU $(* \mid 0, 1)$. We mark .001 at -3.09 on the vertical axis because

$$\text{GAU}\,(-3.09 \mid 0, 1) = P(X < -3.09) = .001;$$

then we mark .01 at -2.33, .05 at $-1.645, \ldots,$ .50 at 0.0, $\ldots,$ .999 at $+3.09$. All we have to do is use the table for the left tail of GAU $(* \mid 0, 1)$. Paper with these graduations on it is sold as *arithmetic probability paper;* it looks like Fig. 7.2.2. What good is it? Now GAU $(x \mid 0, 1)$ plots as a straight line (Fig. 7.2.3). It hits .001 at -3.09, .01 at -2.33, .50 at 0, .999 at $+3.09$. You can see that any GAU $(x \mid \mu, 1)$ will also plot as a straight line shifted to the left or right of this one. In fact, *any* GAU cumulative curve will plot as a straight line because

$$\text{GAU}\,(x \mid \mu, \sigma^2) = \text{GAU}\,\left(\frac{x - \mu}{\sigma} \,\middle|\, 0, 1\right).$$

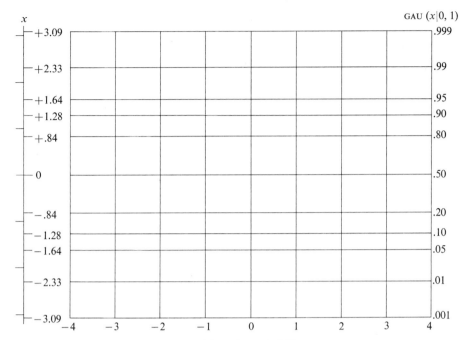

Fig. 7.2.2 Arithmetic probability paper.

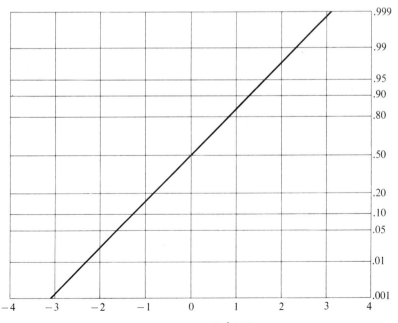

Fig. 7.2.3 GAU $(x \mid 0, 1)$.

The value of μ controls the left-right location, and the value of σ controls the slope. [Slope = $1/\sigma$.] A plot of GAU $(x \mid -1, 2)$ is shown in Fig. 7.2.4.

I have been talking about arithmetic (normal) probability paper because GAU is everyone's favorite distribution. You can also make probability paper for any other distribution which depends on a location parameter and a scale parameter. This includes RECT, STU, and WEIB. If, like STU and WEIB, the distribution has a third or fourth parameter too, you need a separate probability paper for each value of these other parameters. All STU $(x \mid \mu, \sigma^2, 1)$'s plot on one paper; all STU $(x \mid \mu, \sigma^2, 2)$'s plot on another, etc. The WEIB I mentioned above is the *Weibull* distribution; it is used in industry to model failure rates.

Although it's soul-satisfying to see these perfect straight lines, the purpose of this whole exercise is to plot *samples*. Samples plot only as *approximate* straight lines; but by looking at a number of related plots you may be able to tell how your samples jibe with the distribution which determined the paper. It's an exploratory procedure, not one leading to neat numbers.

There are a few problems in plotting samples. First, you have to arrange the observations in order of size from smallest to largest. Second, you should not follow your natural inclinations and plot $x_{(1)}$ at $1/n$, $x_{(2)}$ at $2/n$, and $x_{(n)}$ at $n/n = 1$. The largest member of your sample is seldom the largest allowable value

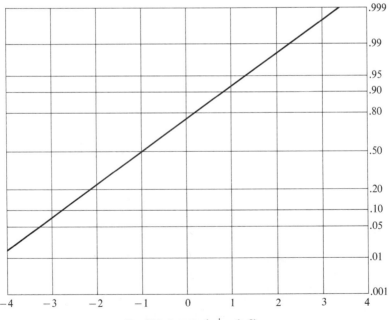

Fig. 7.2.4 GAU $(x \mid -1, 2)$.

from your population; you don't want any point at cumulative probability 1. Besides, your population may go to $\pm\infty$, and these values are off the paper, anyway. In addition, when you have an odd number of observations, you want the middle one to fall at the 50% point. The solution is to put

$$x_{(1)} \text{ at } \frac{1}{n+1},$$

$$x_{(2)} \text{ at } \frac{2}{n+1},$$

$$x_{(3)} \text{ at } \frac{3}{n+1},$$

$$\vdots$$

$$x_{(n)} \text{ at } \frac{n}{n+1}.$$

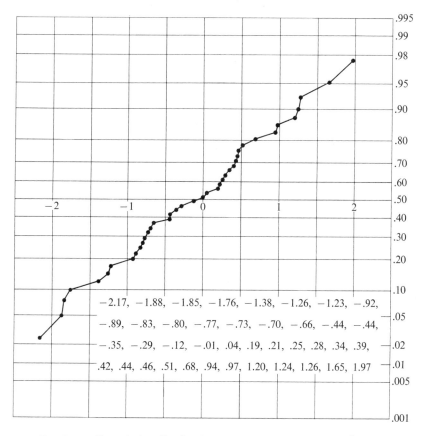

Fig. 7.2.5 Cumulative distribution of 40 values from GAU $(* \mid 0, 1)$.

Now there's symmetry, and nothing falls at 0 or 1. In case you've forgotten,

$x_{(1)}$ means the smallest observation,
$x_{(2)}$ means the next-to-smallest observation,
$x_{(n)}$ means the largest observation

in a sample of n. They are called the *order statistics*.

In Fig. 7.2.5 I've plotted a random sample of 40 from GAU ($* \mid 0, 1$) on arithmetic probability paper. Here

$x_{(1)} = -2.17$ is put at $\frac{1}{41} = .024,$
$x_{(2)} = -1.88$ is put at $\frac{2}{41} = .049,$
$\vdots$
$x_{(40)} = 1.97$ is put at $\frac{40}{41} = .976.$

The plot looks like a drunken column of stragglers, but the ghost of a straight line is there. The ghost is often fainter for smaller samples. If you'd like to see a plot of 1000 observations which fall neatly on a line, see Gabbe, Wilk, and Brown (1967).

What do we see on such a plot when the distribution of the samples is different from the distribution of the paper? If the sample distribution has longer tails, the line will tend to look like Fig. 7.2.6. If the sample distribution has shorter tails, the line will tend to look like Fig. 7.2.7. Whatever the case, it takes quite a few observations to develop much confidence in a particular shape.

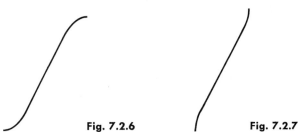

Fig. 7.2.6   Fig. 7.2.7

PROBLEM SET 7.2

1. Plot GAU ($x \mid 1, 4$) on arithmetic probability paper.
2. Take four different samples of 10 from GAU ($* \mid 1, 4$) and plot them separately on arithmetic probability paper. Then mix them together to form one sample of 40 and plot it on arithmetic probability paper.
3. Construct probability paper for RECT.
4. Draw a sample of 15 from RECT ($* \mid 1, 4$) and plot it on the paper you made in Problem 3.

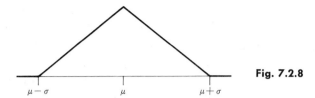

Fig. 7.2.8

5. Plot that sample of 40 Gaussian observations from Problem 2 on the paper you made in Problem 3.
6. Construct probability paper for the distribution whose density is shown in Fig. 7.2.8.
7. Plot the sample of 25 from STU (∗ | 0, 1) that you got in Problem 7.1.4 on arithmetic probability paper.
8. Construct probability paper for STU (∗ | ∗, ∗, 1).

$$[\text{STU}\,(x \mid 0, 1, 1) = \tfrac{1}{2} + (1/\pi)\tan^{-1} x.]$$

9. Plot the sample from Problem 7.1.4 on the paper you made in Problem 8.
10. Plot the residuals of Problem 6.3.9 on arithmetic probability paper.

7.3 LONG-TAILED DISTRIBUTIONS

I mentioned before that when a symmetrical law of error deviates from GAU, it's more likely to go in the direction of long tails than toward truncation. That is, the probability of getting observations far from the mean is greater than in the Gaussian distribution. This is just a general statement; *your* case needs personal attention. But let me discuss why long tails arise.

I first want to rule out of the discussion the extreme observations that are suspect for obvious external reasons. You rightly reject observations if you've bumped the instrument or dropped the chemical, or if the lights dimmed peculiarly when you turned on the switch. You reject data on forms so rain-soaked you cannot read them clearly; you reject punched cards which have too many holes or holes in the wrong columns. [You may be able to reconstruct some of this data with care, but that's beside the point.] And if 99 of your observations lie between 5 and 10, and the 100th is out at 269—Well, it must be a special case, and no one would dream of throwing it in with the others. Such deviants indicate either gross blunders or new phenomena. You study the former to improve your experimental technique, the latter to learn new things about nature.

The observations that people worry about are the ones that do *not* appear to be way off to one side by themselves. They worry about distributions which may be superficially all right, but which may have some ringers in there somewhere. They wonder whether the underlying distribution is really close to GAU or has more probability far from the center. The first thing to ask is "Do I have any good evidence that my distribution is not Gaussian?" Not nearly enough postmortems are made to see whether experiments were well behaved. Not nearly enough people realize what strange phenomena can arise from the purely random variations in a GAU distribution. All our principles say that the questions to be raised are "Is the phenomenon that I regard as suspicious much more likely on the possible alternative hypotheses than on the hypothesis that variation is purely random?" and "How likely are these alternatives in the first place?" Progress in science has been steady because simple hypotheses have been preferred to more complicated *ad hoc* ones.

I don't mean to minimize the difficulties that some people do have with outliers, these more extreme observations. Undeniably, they are sometimes present. Astronomers can usually find them; it's claimed that they also appear frequently in engineering. I just want to point out that you may be chasing will-o'-the-wisps. It's difficult to do much on a casual basis or with a small sample. But when you are involved in one kind of experimentation for a long period, it is your duty to carefully investigate the random behavior of the processes you study. Furthermore, you must not regard this as being done forever when you have done it once. Sudden changes can be quite startling, but the gradual change is the more insidious. Unoiled machinery or deteriorating chemicals can cause great unsuspected trouble.

On the other hand, there is such a thing as a safety play. We were willing to sacrifice some efficiency in an estimate in return for ease of computation. We may just as readily sacrifice some efficiency in order to safeguard against unlikely but possible dangers. Much work in present-day statistics concerns *robustness*. A robust procedure gives good results in a range of situations even though it is not the best possible procedure in any one case. We use a robust procedure because we are really quite uncertain about which distribution is applicable. We could do better if we knew the likelihood function exactly, but might do worse if we chose the wrong one. Strictly, of course, we should do the calculations for each possible likelihood function and weight the results according to our estimate of the probability that it applies.

All right, suppose we have a distribution that may be subject to more extreme deviations than GAU. There are several problems:

1. How do we describe the distribution?

2. How do we estimate the location of the center without being deflected by the extreme observations?

3. How do we get a measure of the shape of the central part of the distribution?

4. How do we estimate what fraction of the population belongs to the wilder element and how much to the central core?

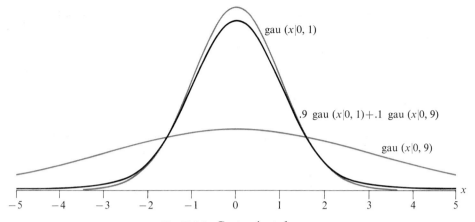

Fig. 7.3.1 Contaminated GAU.

Much active research is going on, and I feel competent only to make a few remarks about problems 1 and 2. How interested in the subject you are depends on how nonnormal *your* distributions are.

First, descriptions. You may be able to use one of the standard forms. We saw how the STU densities had longer tails than GAU; perhaps one of them will work. Jeffreys (1961) believes that many observed sets of observations can be satisfactorily described by STU (∗ | ∗, ∗, 7), and goes through the usual mechanics of finding the posterior distributions of the parameters. Other values of ν give other shapes. [See Fig. 6.3.5 for a comparison of gau $(x \mid 0, 1)$ and stu $(x \mid 0, 1, 7)$.] Box and Tiao (1962) suggest densities of the form

$$\frac{\text{constant}}{\sigma} \times \exp\left(-\frac{1}{2}\left|\frac{x-\mu}{\sigma}\right|^k\right).$$

By adjusting k (which is 2 for GAU), they achieve a wide variety of shapes and still have something that can be handled reasonably well by formal mathematics.

Possibly more closely parallel to many physical situations is the mixing of several Gaussian distributions with the same mean but different variances. That is, the observations you see are not homogeneous, but have a variety of variances. This kind of distribution is harder to handle mathematically, but has great conceptual advantages. Newcomb (1886) gives an early and excellent treatment. Tukey (1962 and elsewhere) is the modern protagonist of this approach, and supplies the term *contaminated distribution*. Figure 7.3.1 shows

$$\text{pd}(x) = .9 \text{ gau}(x \mid 0, 1) + .1 \text{ gau}(x \mid 0, 9).$$

Here gau (∗ | 0, 1) is the basic density and gau (∗ | 0, 9) is the contamination. The contamination appears only one-ninth as often as the basic density, but has a standard deviation three times as large. [You calculate pd (x) by weighting the two densities and adding them point by point.] It turns out that

$$D^2(X) = .9 \times 1 + .1 \times 9 = 1.8.$$

The contaminating part contributes as much to the overall variance as the great mass in the center!

There are several ways to locate the center. One is to say, "I see the end observations; I won't note their exact values for fear they're too large; I'll just record that they're more extreme than the next ones in toward the center." This loses information if everything is really Gaussian, but protects us to some extent from outliers. The Bayesian analysis goes as before. If we treat one observation at each end in this fashion, the likelihood is

$$\text{pd } (x \mid \mu, \sigma) = \prod_{i=2}^{n-1} [\text{gau } (x_{(i)} \mid \mu, \sigma^2)][\text{GAU } (x_{(2)} \mid \mu, \sigma^2)][1 - \text{GAU } (x_{(n-1)} \mid \mu, \sigma^2)].$$

The product corresponds to observations $x_{(2)}$ to $x_{(n-1)}$; the next term says something is smaller than $x_{(2)}$; the last term says something is larger than $x_{(n-1)}$. The formulas are a little nasty, but a first approximation to the center of the posterior distribution of μ is described as *winsorizing*. You replace $x_{(1)}$ by $x_{(2)}$, replace $x_{(n)}$ by $x_{(n-1)}$, sum, and take the average as usual.

Example 7.3.1. Here are 19 observations which may be a little nonnormal:

.65, −.99, −2.83, −.46, −.38, .23, .59, −1.22, 1.43,
.64, 2.52, .42, .16, .92, .26, .25, −1.90, −.17, −1.13.

Arranged in order they are

−2.83, −1.90, −1.22, −1.13, −.99, −.46, −.38, −.17,
.16, .23, .25, .26, .42, .59, .64, .65, .92, 1.43, 2.52.

To find the *winsorized mean*, I replace the −2.83 by −1.90, replace the 2.52 by 1.43, and add:

$$-1.90 - 1.90 - 1.22 - \cdots + .92 + 1.43 + 1.43 = -1.27.$$

Then $-1.27/19 = .067$. The ordinary sample mean is .058. You can't prove a thing by one case, of course. The *double winsorized mean*, .049, is found by replacing both $x_{(1)}$ and $x_{(2)}$ by $x_{(3)}$, and replacing both $x_{(19)}$ and $x_{(18)}$ by $x_{(17)}$. The sample, by the way, is straight out of the table of normal deviates in Appendix C2. The cumulative plot on probability paper in Fig. 7.3.2 does look a little suspicious, though.

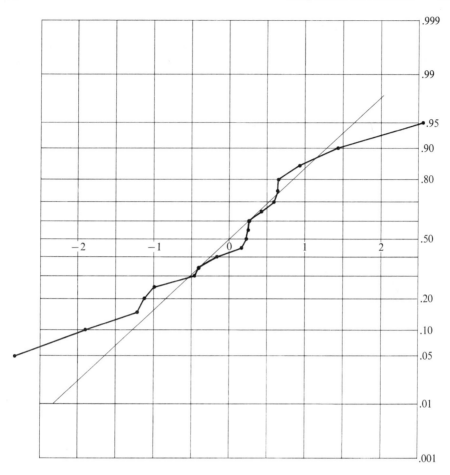

Fig. 7.3.2

Other procedures basically amount to weighting the observations and taking the weighted average. In treating pure GAU we weighted all observations equally: we just added them up as they stood. The weighting scheme varies greatly depending on your ideas of the shape of the distribution. When Jeffreys fits STU (* | *, *, 7), the weights range from 1.0 at the center down to 0.5 at about 2.3 σ's from the center and down to 0.1 at about 6.5 σ's from the center. The more you believe there really are outliers, the less weight you want to give the extreme observations. You can, in fact, reject some entirely; this gives them a weight of 0. Such weighting procedures do not brand with a scarlet letter those observations they weight low or 0; there is no way at all to tell what is a legitimate observation. These are just ways to safeguard your conclusions.

220 Nonnormality

The ultimate in weighting schemes is to use the *sample median*. This weights the center observation 1 and all the rest 0. When the distribution actually is GAU, the efficiency of the median relative to the sample average is only 64%. That is, it takes $1/.64 = 1.6$ times as many observations using the median to get the precision obtained with the sample average. But as the probability in the extreme tails increases, the median quickly becomes more efficient than the sample average. It doesn't take much contamination at all to reach this point.

It is possible to derive a posterior distribution for the *median* of a distribution by assuming only that the underlying distribution is symmetric and has a location and scale parameter. We get

$$P(x_{(r)} < \text{median} < x_{(r+1)}) = \binom{n}{r}\left(\frac{1}{2}\right)^n.$$

Example 7.3.2. For six observations, Table 7.3.1 gives the probabilities that the median of the underlying population lies in the intervals between the observations.

Table 7.3.1

$x_{(1)}$

$$\binom{6}{0}\left(\frac{1}{2}\right)^6 = \frac{1}{64} = .016$$

$x_{(2)}$

$$\binom{6}{1}\left(\frac{1}{2}\right)^6 = \frac{6}{64} = .094$$

$x_{(3)}$

$$\binom{6}{2}\left(\frac{1}{2}\right)^6 = \frac{15}{64} = .234$$

$x_{(4)}$

$$\binom{6}{3}\left(\frac{1}{2}\right)^6 = \frac{20}{64} = .312$$

$x_{(5)}$

$$\binom{6}{4}\left(\frac{1}{2}\right)^6 = \frac{15}{64} = .234$$

$x_{(6)}$

$$\binom{6}{5}\left(\frac{1}{2}\right)^6 = \frac{6}{64} = .094$$

$$\binom{6}{6}\left(\frac{1}{2}\right)^6 = \frac{1}{64} = .016$$

From this table we see that the probability that the median of the underlying population lies between $x_{(1)}$ and $x_{(2)}$—whatever they are—is .094.

Jeffreys has produced a beautiful approximation to this distribution when n is large. Those probabilities that the median lies between the order statistics are probabilities from BIN, the *binomial distribution*. Binomial probabilities are approximated well by GAU, and you can derive HDR's for the median by *counting observations* left and right of the sample median instead of by *measuring distances*. Our report of the uncertainty in the median would look like

$$\text{median} = x_{\left(\frac{n+1}{2} \pm \frac{\sqrt{n}}{2}\right)},$$

which means that the 68% HDR runs roughly from

$$x_{\left(\frac{n+1}{2} - \frac{\sqrt{n}}{2}\right)} \quad \text{to} \quad x_{\left(\frac{n+1}{2} + \frac{\sqrt{n}}{2}\right)}.$$

Since these order subscripts rarely turn out to be whole numbers, you have to interpolate. The easiest way to do this is to use the cumulative plot on probability paper.

Example 7.3.3. Consider those 19 observations given in Example 7.3.1. The 68% HDR runs from

$$x_{\left(\frac{19+1}{2} - \frac{\sqrt{19}}{2}\right)} = x_{(7.8)} \quad \text{to} \quad x_{\left(\frac{19+1}{2} + \frac{\sqrt{19}}{2}\right)} = x_{(12.2)}.$$

You remember that we plotted the sample cumulative distribution by putting $x_{(1)}$ at $\frac{1}{20}$, $x_{(2)}$ at $\frac{2}{20}$, etc. Therefore we interpolate by finding the values on the cumulative plot corresponding to cumulative probabilities of $7.8/20 = .39$ and $12.2/20 = .61$. I get

$$(-.20, +.28).$$

You might be amused that the usual STU analysis gives a 68% HDR for the mean as

$$(-.22, +.34).$$

This one case proves nothing except that using the median analysis does not give way-out-of-line answers.

A last result. We can also calculate the probability that a *new observation* from the underlying population will fall between our order statistics. The simple result is that the probability is $1/(n + 1)$ that a new observation will fall between any adjacent pair of order statistics or off either end. This gives additional justification for the way we plotted cumulative samples on probability paper. We started at $1/(n + 1)$ and separated each adjacent pair by $1/(n + 1)$.

Example 7.3.4. The probability is $\frac{1}{20} = .05$ that a new observation from the population giving rise to those 19 observations will be less than $x_{(1)} = -2.83$. That, of course, is an estimate based upon the present information. In our omniscience we

know that
$$\text{GAU}(-2.83 \mid 0, 1) = .0023.$$

The order statistics provide an example of *nonparametric* or *distribution-free* statistics, measures which apply with a minimum of assumptions. They are often attractive in the early groping phases of a problem when you have a minimal amount of information. But certainly they decrease in importance as you uncover more of the structure of your problem. The literature in this field is vast, confusing, and ill-organized. It constitutes more a collection of ideas than a theory.

PROBLEM SET 7.3

1. If you make two observations, what is the probability that the true median lies between them?
2. If you make two observations, what is the probability that a future observation from the same population will fall between them?
3. If you have a sample of 24 from a symmetric population, what is the probability that the median of the population lies between $x_{(11)}$ and $x_{(14)}$?
4. What is an approximate 90% HDR for the median of the population underlying the 40 observations plotted in Fig. 7.2.5?
5. You weigh an object three times and get 1.26, 1.27, and 1.35 gm. It is your experience that
$$\text{pd}(w) = .8 \text{ gau}(w \mid \mu, .0004) + .2 \text{ gau}(w \mid \mu, .0036)$$
describes such weighings. What is the estimate for the true weight?
6. Generate a random sample of 100 from
$$.9 \text{ gau}(x \mid 0, 1) + .1 \text{ gau}(x \mid 0, 9)$$
and plot it on arithmetic probability paper.
7. Take the sample you procured in Problem 6 and calculate the winsorized mean obtained by replacing five observations at the top and five at the bottom.
8. To reduce the effects of extreme observations in estimating σ, we sometimes use the *mean deviation*
$$\frac{1}{n} \sum |x_i - \text{sample median}|.$$

The usual estimate of σ involves *squares* of the deviations from the sample *mean*. This statistic involves the *first* power of the deviations from the sample *median*.

Take samples of 25, 50, 100, and 200 from GAU ($* \mid 0, 1$) and find out which of these relations is true:

(a) $\hat{\sigma} \doteq \sqrt{\dfrac{\pi}{2}}$ (mean deviation)

(b) $\hat{\sigma} \doteq \dfrac{1}{\sqrt{2\pi}}$ (mean deviation)^2

(c) $\hat{\sigma} \doteq \dfrac{1}{\sqrt{2\pi n}}$ (mean deviation)

9. Draw a sample of 50: 47 from GAU ($* \mid 30, 3$) and 3 from GAU ($* \mid 30, 9$).

   (a) What is
   $$\frac{\text{maximum deviation from } \bar{x}}{\hat{\sigma}}?$$

   (b) Find a rough 90% HDR for the center of the distribution using Jeffrey's method.

10. Table 7.3.2 gives empirical samples of
    $$\frac{\text{largest deviation from } \bar{x}}{\sqrt{S_{\bar{x}\bar{x}}/(n-1)}}$$

    for some GAU and contaminated GAU populations. The first figure in the parentheses at the top of each column gives the size of the sample—25, 50, or 100—and the second figure gives the number of contaminated observations present (from a population with a standard deviation three times as large). The first 4 in column (25, 0) says that in 4 samples out of 150 of this composition, the statistic specified above was between 1.6 and 1.7. Smooth these empirical distributions, normalize them, and plot them on the same axes. This should give you a little feeling for how large a deviation you can get by chance.

11. Use Table 7.3.2. Suppose you know that a particular sample of 25 either is pure GAU or has two contaminated observations of the type described. Discuss the effect of establishing some criterion for totally rejecting an observation.

12. Explain why the variance of the distribution

    $$\text{pd}(x) = .9 \text{ gau}(x \mid 0, 1) + .1 \text{ gau}(x \mid 0, 9)$$

    is *not*

    $$D^2(X) = (.9)^2 \times 1 + (.1)^2 \times 9 = .90.$$

13. You have a record of the largest spring flood on a river for each year of the past 30 years. What is the probability that the largest spring flood next year will exceed all these 30?

Table 7.3.2

Contaminated GAU (a little $3 \times \sigma$)

	(25, 0)	(25, 2)	(50, 0)	(50, 1)	(50, 4)	(100, 0)	(100, 5)
1.6–<1.7	4						
1.7–<1.8	5	1					
1.8–<1.9	9	2	2				
1.9–<2.0	12	2	5				
2.0–<2.1	16	2	12	2	1	4	
2.1–<2.2	21	8	20	4	–	10	
2.2–<2.3	24	4	42	6	1	19	2
2.3–<2.4	16	4	41	6	–	50	–
2.4–<2.5	12	1	39	3	4	61	1
2.5–<2.6	10	3	38	5	–	69	1
2.6–<2.7	9	7	37	5	2	91	2
2.7–<2.8	2	3	30	2	2	80	5
2.8–<2.9	4	1	22	2	1	79	–
2.9–<3.0	3	1	16	2	1	75	2
3.0–<3.1	1	3	19	–	5	67	3
3.1–<3.2	1	2	10	2	4	31	–
3.2–<3.3	1	1	5	1	1	35	4
3.3–<3.4		–	3	–	2	29	2
3.4–<3.5		–	3	–	2	12	2
3.5–<3.6		2	1	2	1	13	3
3.6–<3.7		–	4	1	3	12	1
3.7–<3.8		2	1	–	–	4	1
3.8–<3.9		–	–	–	2	6	2
3.9–<4.0		–	–	–	3	2	1
4.0–<4.1			–		4	–	1
4.1–<4.2		1		–	2	1	2
4.2–<4.3				2	4		2
4.3–<4.4				–	2		1
4.4–<4.5				–	1		2
4.5–<4.6				1	1		1
4.6–<4.7				1	–		–
4.7–<4.8				1	–		2
4.8–<4.9				1	–		1
4.9–<5.0				–	–		2
5.0–<5.1				1	–		–
5.1–<5.2					1		–
5.2–<5.3							2
5.3–<5.4							–
5.4–<5.5							–
5.5–<5.6							1
5.6–<5.7							1
Total	150	50	350	50	50	750	50

ANSWERS TO PROBLEM SET 7.3

1. $\frac{1}{2}$

3. $\left[\binom{24}{11} + \binom{24}{12} + \binom{24}{13}\right]\left(\frac{1}{2}\right)^{24} = .46$

4. $x_{\left(\frac{41}{2} \pm 1.64 \frac{\sqrt{40}}{2}\right)} = (x_{(15.3)}, x_{(25.7)}) = (-.66, +.28)$

5. pd $(\mu \mid w) \propto \prod [.8 \text{ gau } (w_i \mid \mu, .0004) + .2 \text{ gau } (w_i \mid \mu, .0036)]$. By numerical integration, $E(\mu) = 1.275$, while $\bar{x} = 1.293$.

7. This means that you are to replace each of

$$x_{(1)}, x_{(2)}, x_{(3)}, x_{(4)}, \text{ and } x_{(5)} \quad \text{by} \quad x_{(6)}$$

and each of

$$x_{(100)}, x_{(99)}, x_{(98)}, x_{(97)}, \text{ and } x_{(96)} \quad \text{by} \quad x_{(95)}.$$

8. (a)

11. Example of possible rule: If

$$\frac{\text{maximum deviation}}{\hat{\sigma}} > 3,$$

reject that observation. When the sample is actually pure, about 2% of the time you will throw away a good observation.

7.4 TRANSFORMATIONS

Suppose you have a lot of nonnormal observations to organize. If you know their distributional form, you can always use the Bayesian formula

posterior $\propto$ *prior* $\times$ *likelihood*

to study them. But it is sometimes advantageous to make the distribution more normal, or at least more civilized, by transforming the variable. That is, you take the square root of the variable, or its logarithm, or its square. At the start of your investigation you may have had a choice of several variables to work with, and by ill fate you happened to pick an awkward one. Instead of working with velocity, say, you would have been better off working with the kinetic energy, which is proportional to the velocity squared. Or the other way around. When you have a choice of variables and get into distributional complexities, always look around to see whether a different formulation of the problem would straighten you out.

There are also times when the transformation of a variable simplifies the analysis even though the transformed variable has no evident natural meaning. Again, you can always go to the Bayesian formula, but again, for practical considerations, you may simplify your overall task by transforming to another variable. There are, for example, a tremendous number of variables whose *logarithms* seem to be approximately normally distributed:

- size of oil deposits,
- score in sports car rallies,
- sedimentary bed thickness,
- size of small particles in sediments,
- size of incomes and bank deposits,
- radar cross section of birds.

The analysis becomes more convenient when we work with the logarithm of the variable—which has no intuitive meaning—rather than with the original variable itself. There is always a little trouble, though, in transferring prior information from one system to the other.

In biology there are cases in which the *probit* (*probability integral*) transformation seems helpful. For an original *percentage p*, you substitute the value of x for which GAU $(x \mid 0, 1) = p$. Thus for $p = .001$, you get $x = -3.09$. This is just the reverse of the transformation we used in making normal (arithmetic) probability paper. Usually 5 is added to x to make all the values positive. This works because GAU $(-5 \mid 0, 1)$ is very small. That is, you rarely encounter x's less than -5. The advantage here may seem strange: from the limits 0 to 1 you go to the limits $\pm \infty$. But this latter freedom helps in drawing straight lines.

A transformation of a different type is called for when you have just a *ranking* of quantities instead of scaled values. It may be profitable to transform these ranks into scaled values so you can do arithmetic on them. There is no guarantee that it'll work, but a transformation to try is one that produces *normal scores*. In Owen (1962) see Table 7.1: Expected values of order statistics from a normal distribution. In Fisher and Yates (1953) see Table XX: Scores for Ordinal (or Ranked) Data.

Lastly, I want to talk about transformations made on empirical data when you don't know the underlying distribution. If you can find a functional relationship between the standard error and the mean in samples of your data, you might try the transformations given below. They may simultaneously make your variance almost constant and bring the distribution near normality. Ordinarily you will have only estimated standard errors and means to put into the formula, so the procedure is a little precarious.

$$\text{If } \hat{\sigma} \propto \hat{\mu}^2, \quad \text{try } x = \frac{1}{v};$$

$$\text{if } \hat{\sigma} \propto \hat{\mu}, \quad \text{try } x = \log v;$$

$$\text{if } \hat{\sigma} \propto \sqrt{\hat{\mu}}, \quad \text{try } x = \sqrt{v};$$

$$\text{if } \hat{\sigma} \propto f(\hat{\mu}), \quad \text{try } x = \int \frac{dv}{f(v)}.$$

7.4 Transformations

The last is the general formula. You naturally don't worry about multiplicative constants, but it sometimes helps to add or subtract a constant before transforming.

On the next few pages I will demonstrate these facts in reverse. I will take an honest GAU variable, make the *inverse* transformation, which skews things, and show you that the predicted behavior between $\hat{\sigma}$ and $\hat{\mu}$ holds—sort of. It takes more material than I want to display here to let you actually recognize and untangle what is happening; I just want to suggest that the theory is reasonable.

In Table 7.4.1 you see four random samples of 10 observations each; they are labeled I, II, III, IV. The column "(0, 1)" gives random values from GAU $(* \mid 0, 1)$. The column "$(\mu, 1)$" adds 5 to Sample I, 7 to Sample II, 9 to Sample III, and 10 to Sample IV in order to make their means different. Call an entry in this second column "x". Columns "$10/x$", "e^{x-7}", and "$x^2/10$" apply these three transformations to each x.

Example

.42 is a random value from GAU $(* \mid 0, 1)$,
5.42 is thus a random value from GAU $(* \mid 5, 1)$,
$1.84 = 10/5.42$,
$.21 = e^{5.42-7}$,
$2.94 = 5.42^2/10$.

At the bottom of the table are the estimates

$$\hat{\mu} = \bar{x}, \qquad \hat{\sigma} = \sqrt{S_{\bar{x}\bar{x}}/(n-1)}$$

for each of the samples as originally given and for each of the three transformations.

Example

$\left.\begin{array}{l}\hat{\mu} = 4.73 \\ \hat{\sigma} = 1.05\end{array}\right\}$ come from the values 5.42, 4.56, ..., 4.99;

$\left.\begin{array}{l}\hat{\mu} = 10.5 \\ \hat{\sigma} = 2.29\end{array}\right\}$ come from the values 10.6, 12.5, ..., 8.7.

The entries in the second column show that the random generator was not behaving too badly. The $\hat{\mu}$'s are near 5, 7, 9, and 10; the $\hat{\sigma}$'s are not too far from 1.

In the column labeled "$10/x$" I took my legitimate GAU variable x and converted it into a non-GAU variable v by calculating $v = 10/x$. Imagine that we are presented with v; its distribution is skew and non-GAU. How do we transform it into a more well-behaved variable? By using the formula $x = 10/v$, which is the inverse of $v = 10/x$. Our list of transformations tells us that this is the thing to try if $D(v) \propto [E(v)]^2$. We have some estimates of the relation: the four points

Table 7.4.1

		(0, 1)	(μ, 1)	$10/x$	e^{x-7}	$x^2/10$
I		.42	5.42	1.84	.21	2.94
		− .44	4.56	2.19	.09	2.08
		− .83	4.17	2.40	.06	1.74
		−1.85	3.15	3.18	.02	.99
		−1.26	3.74	2.68	.04	1.40
		− .29	4.71	2.12	.10	2.22
		.34	5.34	1.87	.19	2.85
		1.97	6.97	1.43	.97	4.86
		− .73	4.27	2.34	.07	1.82
		− .01	4.99	2.00	.13	2.49
II		−1.88	5.12	1.95	.15	2.62
		.44	7.44	1.34	1.55	5.53
		− .36	6.64	1.51	.70	4.41
		−1.38	5.62	1.78	.25	3.16
		.46	7.46	1.34	1.58	5.58
		.97	7.97	1.25	2.64	6.37
		−1.23	5.77	1.73	.29	3.33
		− .66	6.34	1.58	.52	4.02
		.19	7.19	1.39	1.21	5.18
		.39	7.39	1.35	1.48	5.48
III		.94	9.94	1.01	19.0	9.9
		− .92	8.08	1.24	2.9	6.5
		− .44	8.56	1.17	4.8	7.3
		.04	9.04	1.11	7.7	8.2
		− .12	8.88	1.13	6.5	7.9
		1.24	10.24	.98	25.5	10.5
		− .80	8.20	1.22	3.3	6.7
		.25	9.25	1.08	9.5	8.6
		−1.76	7.24	1.38	1.3	5.2
		− .77	8.23	1.21	3.4	6.8
IV		.28	10.28	.97	27	10.6
		1.20	11.20	.89	57	12.5
		− .89	9.11	1.10	8	8.3
		−2.17	7.83	1.28	2	6.1
		1.26	11.26	.89	71	12.7
		1.65	11.65	.86	105	13.6
		.21	10.21	.98	25	10.4
		.51	10.51	.95	33	11.1
		.68	10.68	.94	40	11.4
		− .70	9.30	1.07	10	8.7
I	$\hat{\mu}$		4.73	2.21	.19	2.34
	$\hat{\sigma}$		1.05	.48	.28	1.07
II	$\hat{\mu}$		6.69	1.52	1.04	4.57
	$\hat{\sigma}$		.95	.27	.80	1.25
III	$\hat{\mu}$		8.77	1.15	8.4	7.8
	$\hat{\sigma}$		.90	.12	7.8	1.61
IV	$\hat{\mu}$		10.20	.99	38	10.5
	$\hat{\sigma}$		1.16	.13	32	2.29

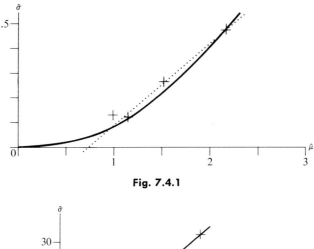

Fig. 7.4.1

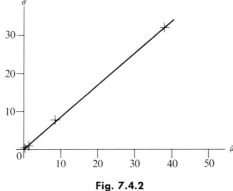

Fig. 7.4.2

(2.21, .48), (1.52, .27), (1.15, .12), and (.99, .13). [See the bottom of the third column.] Do they indicate the relation $\hat{\sigma} \propto \hat{\mu}^2$? They're only estimates, of course. Figure 7.4.1 shows that they fall near the correct parabola, but it's not a relation you would have guessed without a clue. The material here is just too scanty. In fact, if you had come in without our secret information, the dotted line might have been your guess, indicating a transformation $x = \log v$. [Since the line doesn't go through the origin, the transformation would actually have been $x = \log(v - .7)$.]

In the column labeled "e^{x-7}" I computed $v = e^{x-7} = e^x/1100$. How do I get from my skew variable v back to x? By the inverse, $x = \ln(1100v) = 7 + \ln v$. According to our list of transformations, this would be justified if the estimates of $D(v)$ and $E(v)$ lay on a straight line. Figure 7.4.2 shows that they are reasonably close, though a curve is by no means ruled out.

Similarly, in the column labeled "$x^2/10$" I created a skew variable v by $v = x^2/10$. If I want to go back to the docile variable x, I have to use the transformation $x = \sqrt{10v}$. If I didn't have all this knowledge, this would be the

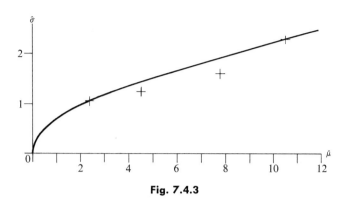

Fig. 7.4.3

thing to try if $D(v) \propto \sqrt{E(v)}$. Figure 7.4.3 shows a rather lousy fit. In all of these attempts, four points are not enough to let you progress. The random variation is just too large.

PROBLEM SET 7.4

1. A temperature T_C on the centigrade scale is converted to a fahrenheit reading T_F by the formula $T_F = \frac{9}{5}T_C + 32$. Given a measurement with mean 38.3 and standard deviation 2.7 on the centigrade scale, what are the corresponding mean and standard deviation on the fahrenheit scale?
2. $X \sim$ RECT $(* | 1.5, .25)$; that is, it's flat from 1 to 2. Find, roughly, the distribution of $Y = X^2$.
3. What transformation is suggested by the following collection?

$\hat{\mu}$	4.73	6.69	8.77	10.20
$\hat{\sigma}$	1.05	.95	.90	1.16

4. If I have five ranked values, what normal scores do I substitute for them?
5. What is the probit corresponding to $p = .95$?

ANSWERS TO PROBLEM SET 7.4

2. However you get it, check that you've approximated

$$\text{pd}(y) = \frac{1}{2\sqrt{y}} \quad \text{for } 1 < y < 4.$$

3. It appears that $\hat{\sigma}$ is not dependent on $\hat{\mu}$. That is, $f(\hat{\mu}) = 1$. This says that you shouldn't transform. These values are actually those of the untransformed variable in Table 7.4.1.
4. $-1.163, -0.495, 0, 0.495, 1.163$

CHAPTER **8**

**Making
Decisions**

Throughout this book I have been showing you how to accumulate evidence about the alternatives under consideration. We kept this information in the form of probability distributions and modified these distributions on the basis of observation or experiment. Occasionally, the matter may end there. More often, the information that we have must be used to help us decide on a course of action. We all work under the pressures of time and money: Does this line of inquiry look promising enough to pursue for three months? Is that equipment versatile enough to justify its cost and the space it requires? How should production be allocated this week? The evidence we have must be brought to bear or we will have gathered it in vain. In this chapter we will study criteria for making decisions, what to do when the alternatives are poorly specified, and deciding when to stop taking observations. In decision-making, as in the stating of conclusions, there is ample opportunity for error: no one need be without.

8.1 A CRITERION FOR DECISION-MAKING

What kinds of decisions will we consider? Let me rule out a few. At one extreme we have a chess master who, by his extensive knowledge of the game, sees a line of play by which he can win in six moves regardless of his opponent's efforts. This situation has no statistical interest for us save, perhaps, in our finding out how early in a game and how often such a forced win arises. Next we have the general who is planning a campaign. Many military doctrines emphasize the philosophy of evaluating the enemy's *capabilities* rather than his *intentions*. Thus our general may try to find a course of action which offers good chances against any possible enemy countermove; it may not be the one which would be optimal if the enemy's intentions were known, but the general will sacrifice an uncertain best for a certain good. This also is not our realm, but the subject of the *theory of games*. It is not especially interesting to us because it makes no use of the *probabilities* that the enemy will follow various plans.

What then can I present? Problems that demand use of uncertain information but in which no opponent is deliberately trying to counter our efforts. This may appear to be a limited field, but actually it has enormous possibilities. We will explore all those elements which are involved in decision-making, elements which often are handled intuitively, but superficially and inconsistently. The greatest recommendation one could ask for is that both businessmen and engineers are finding that the theory pays off! It is not just an academic procedure, even though it was developed by an abstract mathematician [Wald (1950)].

What are the elements involved in making a decision?

1. There are a number of possible actions. [The list may include doing nothing at all.] We have to choose one of them.

2. The situation at hand—a *state of the world*—is one of a list of known possibilities. But we do not know which one it is.

3. For every combination of action and state of the world there is a consequence, often, though not necessarily, measured in money. These consequences are known or estimated.

4. We have some estimates of the probabilities of the various states of the world.

If we are to make decisions which utilize all our information and satisfy certain consistency standards, we want, according to the most appealing criterion, to maximize our *expected income*. This is the same expected value we have talked about so much. The very presence of the word *expected* shows that we are trying to take our uncertainties into account.

Example 8.1.1. I am offered the chance to participate in a rather speculative business deal.

1. *Possible actions:* I participate.
   I don't participate.
2. *States of the world:* The deal succeeds.
   The deal does not succeed.
3. *Consequences*

	Deal succeeds	Deal does not succeed
I participate	Make much money	Lose stake
I don't participate	No gain	No loss

4. *Information:* The probability of success is p.

That's the formal structure; now let's put a little meat on the bare bones:

1. *Possible actions:* I invest $5000.
   I don't invest.
2. *States of the world:* Success with return of $40,000.
   Failure with loss of $5000.
3. *Consequences*

Income	Deal succeeds	Deal does not succeed
I participate	$35,000	−$5000
I don't participate	$0	$0

4. *Information:* The probability of success is .2.
   The probability of failure is .8.

Now I calculate the expected income:

$$E(\text{income} \mid \text{participate}) = (\$35{,}000)(.2) + (-\$5000)(.8)$$
$$= \$3000,$$
$$E(\text{income} \mid \text{don't participate}) = (\$0)(.2) + (\$0)(.8) = \$0.$$

Since $3000 > $0, I choose to participate in the deal.

Whoa, there are a number of hidden assumptions in all this. I guess the two listed actions might exhaust the possibilities. But are the two states of the world the only ones? There might be intermediate states between complete success and complete failure. And is the profit actually known so exactly? Again, I have given the consequences in dollars; this is all right if the amounts of money involved are small compared with the wealth of the participants. If my total wealth is only $5000, or if $40,000 seems a fortune, I can't count every dollar equally. Also, if I don't use the money in business and this venture takes quite a while, I can get interest on the money in a bank. Then there are taxes, too. On the other hand, if I have a reputation for being a "dealer," my not participating might cause me to lose face. Lastly, the evaluation of the probability of success can be mighty subjective.

You may complain that I gave a simple example and then tried to complicate it unnecessarily or denigrate it. I just want you to see both sides: the simplicity of the theory and the possible complexity of the analysis of the subject matter. The example serves as both a challenge and a warning. But you don't necessarily have to know all the details as exactly as I've given them. Suppose we leave the consequences as they are and look at the probability of success:

$$E(\text{income} \mid \text{participate}) = (\$40{,}000)p - \$5000.$$

When is this better than \$0? Whenever $p > .125$. So, if you are relatively sure of the consequences, all you really have to do is decide whether $p \lessgtr .125$, not determine its exact value.

Or, suppose $p = .2$ seems a fair estimate of the chance of success, but you're not too sure of the size of the return (if successful):

$$E(\text{income} \mid \text{participate}) = R(.2) - \$5000.$$

When is this better than \$0? Whenever $R > \$25{,}000$. So, if you are fairly sure of the probability of success, all you really have to do is decide whether $R \lessgtr \$25{,}000$, not determine its exact value.

There are some technical terms in statistical decision theory which are useful to know. The word *loss* is used in a special way. *Loss* means the *distance below the largest expected income*.

Example 8.1.2. In Example 8.1.1, the largest expected income is $3000:

Choosing *participate* meant a loss of $3000 - \$3000 = \$0$.
Choosing *not participate* meant a loss of $3000 - \$0 = \$3000$.

Just remember that we are *not* using "loss" in the sense of the opposite of "profit." However, we can restate our criterion for decision-making as: *Minimize the expected loss.*

In some problems the consequences are naturally expressed in terms of income; in others, in terms of losses. You can see at once an ideal opportunity for confusion. I shall do my part in preventing it by writing *Income* or *Losses* in the upper left corner of every consequence table. Just for the record, let me display the loss table for Example 8.1.1. According to the definition, I must arrange it so that in *each column* (state) the best possible result has loss = 0. This causes no trouble: you can add or subtract the same amount from each element in a column of either an income or a loss table without affecting your decision. Try it.

Example 8.1.3. The loss table for Example 8.1.1 is:

Losses	Deal succeeds	Deal does not succeed
I participate	$0	$5000
I don't participate	$35,000	$0

$$E(\text{loss} \mid \text{participate}) = (\$0)(.2) + (\$5000)(.8) = \$4000,$$
$$E(\text{loss} \mid \text{don't participate}) = (\$35{,}000)(.2) + (\$0)(.8) = \$7000.$$

To minimize the expected loss we choose participation. The advantage is $3000, as we calculated before.

There are two other terms you should learn; they have to do with the cost incurred because of incomplete knowledge. What is absolutely the largest expected value we could conceive of? Why, it's what we would get if we knew the true state of nature before we made our choice, rather than being bound to make a choice, once and for all, ahead of time. This value is called the *expected income under certainty*. The difference between this and the *expected income under uncertainty* is then called the *loss due to uncertainty*.

Example 8.1.4 (*Continuation of Example 8.1.1*). Our expected income under uncertainty was the $3000. Now if the deal will succeed, we choose to participate:

$$\text{income} = \$35{,}000, \quad \text{probability} = .2;$$

if the deal will not succeed, we choose not to participate:

$$\text{income} = \$0, \quad \text{probability} = .8.$$

Therefore

$$\text{expected income under certainty} = (\$35{,}000)(.2) + (\$0)(.8) = \$7000,$$
$$\text{loss due to uncertainty} = \$7000 - \$3000 = \$4000.$$

Lastly, the word *utility* is used as a general term for value; we can measure in *utiles* instead of dollars. How dollars, hours, or lives are converted to utiles is a problem of subject matter. I will continue to use dollars and assume that all the amounts involved truly reflect the values to the participants. Although this simple statement covers up a multitude of difficulties which are discussed at great length in economics books, I don't feel inclined to pursue the subject here. There is some discussion of it in Schlaifer (1959). Grayson (1961) considers decisions about drilling oil wells. Lisco (1967) looks at commuters' evaluations of their traveling times. Cook and Garner (1966) look at baseball. Porter (1967) discusses extra-point strategy in football. Market research people are hot on the subject of utility; they want to plan advertising strategy and set selling prices for products.

It's also possible that your unit of utility may vary during the course of a problem because you can't see far enough ahead to evaluate the ultimate consequences. When you measure the speed of light, you concentrate on, say, the mean-square error rather than the possible uses of the value.

PROBLEM SET 8.1

1. *Extension of Example 8.1.1.* Using p (the probability of the deal's success) as the x-variable and R (the return if the deal succeeds) as the y-variable, plot the combinations of p and R which recommend participation.

2. Suppose we have the opportunity to enter a business venture. The situation is summarized by this table:

	Income	\multicolumn{5}{c	}{States = degrees of success}
		1	2	3	4	5
Action	Enter	$10,000	$5000	$0	−$5000	−$10,000
	Don't enter	$100	$100	$100	$100	$100
Probability		.2	.3	.2	.2	.1

Assume that the data and probabilities are reasonably correct. What is $E(\text{income} \mid \text{enter})$?

3. Use the data of Problem 2. What is $E(\text{income} \mid \text{don't enter})$?

4. Use the data of Problem 2. What is the decision?

5. I probably have been too bullish. Exhibit a set of probabilities for the states in Problem 2 so that $E(\text{income} \mid \text{enter}) = -\$100$.

6. Exhibit the loss table for Problem 2.

7. What is the expected loss due to uncertainty in Problem 2?

8. What is the connection between the expected loss due to uncertainty and the $E(\text{loss} \mid \text{best decision})$? Demonstrate in Example 8.1.1 and Problem 2.

9. Here is a miniature inventory problem. Every day a newsdealer has to order copies of turf magazine XYZ. They cost him $0.20 each, and he sells them for $0.50 each. He is unable to return the unsold copies, so he doesn't want to order too many. The dealer has been studying statistical decision theory in night school and decides to apply it. To estimate his probabilities he makes a record of the demand for this magazine over a reasonably long period and finds:

Demand	0	1	2	3	4	5	6	≥ 7
Probability	.15	.19	.25	.21	.13	.06	.01	.00

To simplify the problem, assume that if the dealer is out of the magazine when it is requested, he loses neither goodwill nor future patronage. Also assume that we are talking about run-of-the-mill days; special events would create a larger demand. How many copies should he order?

10. When we were talking about continuous posterior distributions, I said that if we regard the loss as proportional to the square of the error, we should choose the mean as our estimate. Prove this.

11. Suppose that you have to choose one of two alternatives, one of which is correct and the other wrong, and that the consequences of one of the two possible errors is twice as serious as the other. How do you choose? Relate your answer to the odds of the alternatives.

12. Suppose that a rationalistic jury believed that it was 100 times as serious to convict an innocent man as to set a guilty man free. When would they convict?

13. In modern digital communications the length of a symbol is the number of *bits* ("0"s and "1"s) in it; spaces are not counted. [The word *bit* comes from *bi*nary dig*it*.] You have to adopt one of these two codings for an eight-letter alphabet:

Letter	Probability	Code I	Code II
A	.27	000	00
B	.23	001	01
C	.16	010	100
D	.10	011	101
E	.08	100	1100
F	.07	101	1101
G	.05	110	1110
H	.04	111	1111
	1.00		

Assume that the transmission cost is proportional to the total number of bits; which code should be adopted?

14. The data of Problem 2.1.10 show that when two evenly matched chess masters play the Sicilian Defense, the player with the white pieces wins about 60% of the time. Set up a scoring system for this opening (ignoring draws) so that
    (a) one point is awarded per game,
    (b) white and black both expect half a point.

    You will have to make some other reasonable assumption to get a unique solution.

15. The probability that a hurricane will strike a particular Florida coastal town tonight is .90. The owner of a store there must decide whether to board up his large display window or not. If the hurricane strikes, the probability that the window will be broken is .10 if the window is not boarded up, and .01 if it is boarded up. The cost of replacing the window plus possible water damage amounts to $1000; boarding up the window costs $100. What should the store owner do? [Assume that $1000 is not an overwhelming loss for the store.]

16. A statistician does not know which estimator to use because he is unsure of the type of underlying distribution. The expected squared errors are:

Losses	Really distribution A	Really distribution B
Use sample mean	.26	.61
Use sample median	.45	.42
	.7	.3
	Estimated probabilities	

    Which estimator should he pick?

ANSWERS TO PROBLEM SET 8.1

1. The dividing curve is $Rp = \$5000$, a hyperbola. The decision problem reduces to deciding which side of the curve you're on.

3. $100

5. There are many solutions. One is (.1, .2, .38, .22, .1).

6.
Losses	1	2	3	4	5
Enter	$0	$0	$100	$5100	$10,100
Don't enter	$9900	$4900	$0	$0	$0

7. E(income under certainty) $= (\$10{,}000)(.2) + (\$5000)(.3)$
   $+ (\$100)(.2) + (\$100)(.2) + (\$100)(.1) = \$3550$

   E(loss due to uncertainty) $= \$3550 - \$1500 = \$2050$

9.

	Income	0	1	2	3	4	5	6
	0	$0.00	$0.00	$0.00	$0.00	$0.00	$0.00	$0.00
	1	− 0.20	0.30	0.30	0.30	0.30	0.30	0.30
	2	− 0.40	0.10	0.60	0.60	0.60	0.60	0.60
Dealer's order	3	− 0.60	− 0.10	0.40	0.90	0.90	0.90	0.90
	4	− 0.80	− 0.30	0.20	0.70	1.20	1.20	1.20
	5	− 1.00	− 0.50	0.00	0.50	1.00	1.50	1.50
	6	− 1.20	− 0.70	− 0.20	0.30	0.80	1.30	1.80

(Customer's demand as column header)

The possible actions are the number of copies the dealer orders; the possible states of the world are the number demanded by his customers. If, for example, he orders 4 copies and 3 are demanded, his cost is $0.80, his gross is $1.50, and his net is $0.70.

$E(\text{income} \mid \text{order } 0) = \$0.00$,

$E(\text{income} \mid \text{order } 1) = (-\$0.20)(.15) + (\$0.30)(.19) + (\$0.30)(.25)$
$+ (\$0.30)(.21) + (\$0.30)(.13) + (\$0.30)(.06)$
$+ (\$0.30)(.01) = \$0.225 = 22\frac{1}{2}$ cents.

The others go the same way, and the complete list is:

Copies ordered	Expected income
0	$0.000
1	0.225
2	0.355
3	0.360
4	0.260
5	0.095
6	− 0.100

The conclusion is that he should order 3 copies. Of course, the expected profit for 2 is so similar that he can't take the distinction seriously. But he should order a moderate number and be willing to let some of the demand go unsatisfied.

11. Our situation is this: $S1$ and $S2$ are the alternatives, v and $2v$ the consequences (losses):

	Losses	Truth	
		S1	S2
Choice	S1	0	v
	S2	$2v$	0
		p_1	p_2 $[p_1 + p_2 = 1]$

Therefore
$$E(\text{loss} \mid \text{choose } S1) = (0)(p_1) + (v)(p_2) = p_2 v,$$
$$E(\text{loss} \mid \text{choose } S2) = (2v)(p_1) + (0)(p_2) = 2p_1 v.$$

To minimize the expected loss we

choose $S1$ when $p_2 v < 2 p_1 v$ or $\dfrac{p_1}{p_2} > \dfrac{1}{2}$;

choose $S2$ when $p_2 v > 2 p_1 v$ or $\dfrac{p_1}{p_2} < \dfrac{1}{2}$.

Since
$$\frac{p_1}{p_2} = O\left(\frac{S1}{S2}\right),$$

you shouldn't pick the alternative with the more serious consequence of error unless the odds for that alternative are at least 2.

14. One solution. This minimizes the size of the off-diagonal elements:

	White wins	Black wins
Score for white	$\tfrac{5}{6}$	0
Score for black	$\tfrac{1}{6}$	1

16. $E(\text{loss} \mid \text{mean}) = .7 \times .26 + .3 \times .61 = .345 \ \Longleftarrow$
    $E(\text{loss} \mid \text{median}) = .7 \times .45 + .3 \times .42 = .441$

8.2 DISCRIMINATION

Here we use a *continuous* measurement to choose among a *discrete* set of alternatives. A criterion for making the choice must also be supplied. This is an increasingly important problem in modern science and engineering, since the discrimination often has to be done extremely frequently, rapidly, and automatically. Consider, for instance, reading a digital magnetic tape. One senses the magnetic flux and tries to decide whether it represents a "0" or a "1"; there are two different mean values for the two types of signal, and the observations vary around them. [Note that we write the 0-bit and the 1-bit in quotes to distinguish them from the other numerals scattered across these pages.] In digital communications and in computer technology, the electronic circuits represent the "0"s and "1"s by two levels of voltage; the observed voltages show variations around these mean levels. In both these cases, thousands, or even millions, of decisions may have to be made each second. Biologists define species by the values of one or more physical measurements, and use the defining measurements to identify a new specimen when it is found, to classify it as one species or the other. These defining measurements show variations about some mean.

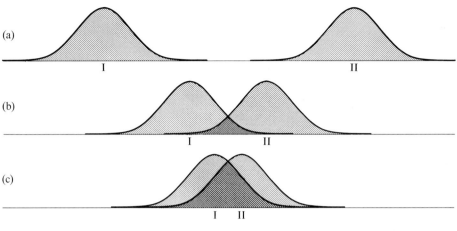

Fig. 8.2.1

Where does this variation come from? A rough answer might be: From inadequately controlled or inadequately understood factors. Time and money always determine how far you push the development of knowledge or of instrumentation. In most cases you just have to learn to live with some uncertainty. The well-publicized indeterminacies of quantum theory seldom introduce uncertainty comparable to other sources of variation. I apologize for glossing over a lot of details here, but every subject has its own peculiarities; I am endeavoring to present general considerations for a general audience. When you get involved in a particular field, you must learn the relevant details.

The important thing right now is that the variation can often be reasonably well represented by a GAU distribution. Figure 8.2.1 might clarify the situation.

In each part of this figure the two humps are the probability densities representing the variation within the two populations we are trying to distinguish. Hump I might represent the variation in the voltage for a "0", while hump II represents the variation in the voltage for a "1". Or, hump I might represent the variation in the length of the hinge line for one species of shellfish, while hump II represents the variation in the length for another species. You can see that the problem in discriminating between the two populations may range from trivial to severe or almost impossible. It obviously depends on how far apart the populations are compared with their within-population variations.

We will start out with two strong assumptions, and later see what happens when they are relaxed:

1. The probability densities of the two populations are known; that is, we are not trying to identify new observations and determine the populations at the same time.
2. The populations are Gaussian with *equal variances*.

Ultimately we will want to establish a dividing line between the populations for decision purposes. [When we make millions of decisions a second, the line must be calculated in advance.] But before we can consider any decision-making, we have to measure the uncertainty for every possible observation. Back again to

$$\text{posterior prob} \propto \text{prior prob} \times \text{likelihood}.$$

Again the prior probability represents our knowledge or ignorance before the observation. In the technological areas mentioned here, there often is excellent historical information available to guide us. Calculating the likelihood again boils down to calculating the ordinate of the density at the point of observation.

Example 8.2.1. Suppose 3 V is the mean value for "0"s, 5 V is the mean value for "1"s, and both populations have $\sigma^2 = .25$. What is the factor that an observation of 3.6 V came for a "0" rather than from a "1"?

$$\text{Population I:} \quad \text{GAU}(*\,|\,3, .25) \quad [\text{the ``0''s}],$$
$$\text{Population II:} \quad \text{GAU}(*\,|\,5, .25) \quad [\text{the ``1''s}],$$

This situation looks much like Fig. 8.2.1(b). There is a small overlapping region in the center, but the populations are reasonably well separated. [The means are 4 σ's apart.] The factor is the ratio of the ordinates:

$$\mathcal{F}\left(\frac{\text{I}}{\text{II}}\,\bigg|\,3.6\right) = \frac{\text{gau}(3.6\,|\,3, .25)}{\text{gau}(3.6\,|\,5, .25)}$$

$$= \frac{\dfrac{1}{.5\sqrt{2\pi}}\exp\left[-\dfrac{1}{2}\left(\dfrac{3.6-3}{.5}\right)^2\right]}{\dfrac{1}{.5\sqrt{2\pi}}\exp\left[-\dfrac{1}{2}\left(\dfrac{3.6-5}{.5}\right)^2\right]}$$

$$= \exp\left[-\frac{1}{2}\left(\frac{.36}{.25}\right) + \frac{1}{2}\left(\frac{1.96}{.25}\right)\right]$$

$$= \exp(3.2) = e^{3.2} = 24.5.$$

If the prior odds were 1 : 1—and this is often the case in an application like this—then the posterior odds that this observation comes from a "0" are also 24.5.

We can get a general formula that throws more light on the subject. Suppose our two populations are

$$\text{I:} \quad \text{GAU}(*\,|\,\mu_1, \sigma^2),$$
$$\text{II:} \quad \text{GAU}(*\,|\,\mu_2, \sigma^2),$$

and the measurement on a new individual is w. Then

$$\mathcal{F}\left(\frac{\mathrm{I}}{\mathrm{II}}\bigg|w\right) = \frac{\text{gau}(w\mid\mu_1, \sigma^2)}{\text{gau}(w\mid\mu_2, \sigma^2)}$$

$$= \frac{\dfrac{1}{\sigma\sqrt{2\pi}}\exp\left[-\dfrac{1}{2}\left(\dfrac{w-\mu_1}{\sigma}\right)^2\right]}{\dfrac{1}{\sigma\sqrt{2\pi}}\exp\left[-\dfrac{1}{2}\left(\dfrac{w-\mu_2}{\sigma}\right)^2\right]}$$

$$= \exp\left[-\frac{1}{2}\left(\frac{w-\mu_1}{\sigma}\right)^2 + \frac{1}{2}\left(\frac{w-\mu_2}{\sigma}\right)^2\right]$$

$$= \exp\left[\left(\frac{\mu_1-\mu_2}{\sigma}\right)\left(\frac{w-\dfrac{\mu_1+\mu_2}{2}}{\sigma}\right)\right].$$

That simplifies nicely, doesn't it? [Another dividend for assuming normality.] You can see that $(\mu_1 + \mu_2)/2$ is the point midway between the two means, so

$$\frac{w - \dfrac{\mu_1+\mu_2}{2}}{\sigma}$$

is the distance the observation is from the midpoint, measured in standard deviations. The other term, $(\mu_1 - \mu_2)/\sigma$, is the distance between the population means, also measured in standard deviations. There is no particular reason to remember this formula since you can always do the calculation directly; it does show, however, that all the pertinent elements have been taken into account.

If w lies on the μ_1-side of the midpoint, the exponent is greater than 0 and

$$\mathcal{F}\left(\frac{\mathrm{I}}{\mathrm{II}}\bigg|w\right) > 1;$$

if w lies on the μ_2-side of the midpoint, the factor is less than 1. This is as we would expect: an observation closer to μ_1 is evidence for I; an observation closer to μ_2 is evidence for II.

Example 8.2.2. Let's go back to Example 8.2.1 and check that the general formula gives the same result that we derived directly:

$$\mathcal{F}\left(\frac{\mathrm{I}}{\mathrm{II}}\bigg|3.6\right) = \exp\left[\left(\frac{3-5}{.5}\right)\left(\frac{3.6 - \dfrac{3+5}{2}}{.5}\right)\right]$$

$$= \exp\left[\left(\frac{-2}{.5}\right)\left(\frac{-.4}{.5}\right)\right] = \exp(3.2).$$

We also have

$$\log \mathcal{F}\left(\frac{\mathrm{I}}{\mathrm{II}} \middle| w\right) = \left(\frac{-2}{.5}\right)\left(\frac{w-4}{.5}\right)(.4343) = -3.4744(w-4).$$

If $w = 3 = \mu_1$, then

$$\log \mathcal{F}\left(\frac{\mathrm{I}}{\mathrm{II}} \middle| 3\right) = 3.4744, \qquad \mathcal{F} = 2981.$$

If $w = 5 = \mu_2$, then

$$\log \mathcal{F}\left(\frac{\mathrm{I}}{\mathrm{II}} \middle| 5\right) = -3.4744, \qquad \mathcal{F} = \tfrac{1}{2981} = 2981 \text{ against}.$$

When is $\mathcal{F} = 5$?
 When $\mathcal{F} = 5$,

$$\log \mathcal{F} = \log_{10} 5 = .6990 = -3.4744(w-4),$$

so

$$w - 4 = -.2, \qquad w = 3.8.$$

What is the probability of error in all of this?

Example 8.2.3. In Example 8.2.1, the probability that the observation representing a "0" is on the wrong side of the midpoint is

$$1 - \text{GAU}(4 \mid 3, .25) = 1 - \text{GAU}(2 \mid 0, 1) = .0228 = \tfrac{1}{44}.$$

Thus one time in 44 an observation representing a "0" will produce a factor pointing to a "1". The problem is quite symmetrical, so one time in 44 an observation representing a "1" produces evidence pointing to a "0".

Obviously this is not good enough if millions of trials are to be made.

Example 8.2.4. What would σ have to be for the probability of error to be only one in two million? Well,

$$\text{right tail} = .0000005 \quad \text{at} \quad 4.892 \ \sigma\text{'s},$$

so

$$\frac{4-3}{\sigma} = 4.892, \qquad \sigma = .205, \qquad \sigma^2 = .042.$$

This is rather academic, though, since hardly anyone believes GAU four or five σ's away from the mean. [I start to get dubious at two or three σ's.] Your population densities may have been obtained from extensive previous experimentation, but it would take *millions* of previous observations to establish whether GAU was realistic in this region. Fortunately, it's now possible to do millions of simple experiments under computer control, so there is a chance here of getting a good estimate of the actual probability.

The strength of your information depends greatly on σ. If you can reduce it in any way, you will improve your analysis. When σ describes the variation of shell or skull measurements, you can hardly breed the population again to make your discrimination better. But if σ is a characteristic of equipment you put together yourself, you may be able to reduce it by improving the gear. When the measurements are made on electrical signals you generate, you also may have the option of generating the signal a couple of times to get independent samples of the same thing. Then σ goes down to $\sigma/\sqrt{n}$.

Example 8.2.5. In Example 8.2.1, if we make four independent determinations on a signal, σ becomes $.5/2 = .25$. If 3.6 is the mean of the four observations, then

$$\mathfrak{F}\left(\frac{\mathrm{I}}{\mathrm{II}} \middle| 3.6\right) = \exp\left[\left(\frac{-2}{.25}\right)\left(\frac{3.6-4}{.25}\right)\right] = \exp(12.8) = 360{,}000.$$

The situation becomes much more interesting when we can regenerate signals on command. If our first observation is determinative, there will be no need for further observations. We make it a policy to obtain a second, third, or fourth observation only when the interpretation of the previous ones is dubious. In modern digital communications the portions of a transmission that contain dubious or unclear bits are automatically retransmitted. This procedure falls in the category of *sequential decisions*, which will be discussed in Section 8.4.

Example 8.2.6. For Example 8.2.1 we follow this procedure: Let w_1 be the first observation. Then

$$\text{if } w_1 < 3.8, \quad \text{call it a "0";}$$
$$\text{if } w_1 > 4.2, \quad \text{call it a "1".}$$

Otherwise take a second observation, w_2. Then

$$\text{if } w_2 < 4, \quad \text{call it a "0";}$$
$$\text{if } w_2 > 4, \quad \text{call it a "1".}$$

Consider the probability of error. Assume we really have a "0":

$$P(w_1 < 3.8 \mid \mathrm{I}) = \mathrm{GAU}\,(3.8 \mid 3, .25) = .9452,$$
$$P(4.2 < w_1 \mid \mathrm{I}) = 1 - \mathrm{GAU}\,(4.2 \mid 3, .25) = .0082,$$
$$P(3.8 < w_1 < 4.2) = 1 - .9452 - .0082 = .0466,$$
$$P(w_2 < 4 \mid \mathrm{I}) = \mathrm{GAU}\,(4 \mid 3, .25) = .9772,$$
$$P(w_2 > 4 \mid \mathrm{I}) = 1 - .9772 = .0228.$$

Then

$$P(\text{error} \mid \text{I}) = .0082 + .0466 \times .0228 = .0093 = P(\text{error} \mid \text{II}),$$
$$P(\text{error}) = .5 \times .0093 + .5 \times .0093 = .0093.$$

With only one measurement and a dividing point of 4,

$$P(\text{error}) = .0228, \quad \text{some } 2\tfrac{1}{2} \text{ times as large.}$$

But we have done only 1.0466 times as much work to achieve the smaller error rate! [This isn't stated quite fairly: extra equipment is required to control the repetition, and more time is used.]

If individuals have several distinguishing characteristics *and the random variation of one characteristic is independent of the variation of another*, you can multiply the factors obtained from analyzing each characteristic separately. Unfortunately, many physical measurements are quite entangled. Height and weight, for example, certainly do not vary independently; you could not multiply the factors obtained from them. But eye color and weight seem an innocuous combination.

Example 8.2.7. Suppose we have these two populations:

I: weight ~ GAU (* | 180, 100); $P(\text{brown}) = .25$, $P(\text{blue}) = .75$.
II: weight ~ GAU (* | 160, 100); $P(\text{brown}) = .55$, $P(\text{blue}) = .45$.

Then a blue-eyed individual of 180 lbs gives a factor of

$$\mathcal{F}\left(\frac{\text{I}}{\text{II}} \,\middle|\, \text{blue}, 180\right) = \frac{\text{gau }(180 \mid 180, 100)}{\text{gau }(180 \mid 160, 100)} \times \frac{.75}{.45}$$
$$= 7.40 \times 1.67 = 12.3$$

When the characteristics are correlated, you must use the multivariate distribution which describes the relationships between them.

Example 8.2.8. BIGAU $(*, * \mid \mu_x, \mu_y, \sigma_x^2, \sigma_y^2, \rho)$ is a bivariate continuous distribution which describes the random variation of two correlated GAU variables; it is given in detail in the Gallery. Here, as in other continuous distributions, the likelihood is effectively the ordinate of the density. Two shell populations are distinguishable by their length and their width, which are correlated:

I: (length, width) ~ BIGAU (*, * | 16, 24, 4, 9, .2),
II: (length, width) ~ BIGAU (*, * | 20, 26, 4, 9, .3).

We are presented with a shell and asked to identify it as one type or the other.

We calculate the factor

$$\frac{\text{bigau }(19, 28 \mid 16, 24, 4, 9, .2)}{\text{bigau }(19, 28 \mid 20, 26, 4, 9, .3)} = \frac{1}{5},$$

which we get when our unknown shell has length 19 and width 28. [We just substituted into the formula.] If we have no other information and if the two possible errors are about equally serious, we call this shell type II.

When we know the characteristics of the population only from a sample, we calculate the probability distribution of a *new* observation and use that for the likelihood. When we had GAU observations with known σ, that probability density was

$$\text{pd (new)} = \text{gau}\left(\text{new} \mid \bar{x}, \frac{n+1}{n}\sigma^2\right).$$

When σ is unknown, the probability density becomes

$$\text{pd (new)} = \text{stu}\left(\text{new} \mid \bar{x}, \frac{n+1}{n}s^2, n-1\right).$$

Example 8.2.9. Our underlying population is GAU. Five observations give $\bar{x} = 7.1$, $s^2 = 10$. The likelihood to be used for a new specimen with $x = 5$ is

$$\text{stu }(5 \mid 7.1, 12, 4) = .087.$$

I've been talking about identifying a new individual when we already know the parent population. Just as often, we first have to *define* those parent populations, to classify the mass of individuals we see into two or more groups.

Example 8.2.10. Variable stars come in a great variety of shapes and styles. Figure 8.2.2 shows an attempt to classify them into families by studying the shape of the overall frequency curve. Other stellar characteristics have to be used later to make the classification more meaningful.

This classification problem is often quite severe. With the good computing facilities now available, a large variety of techniques are being tried out. *Factor analysis, principal components, stepwise multiple regression, clustering, numerical taxonomy,* and *pattern recognition* are some of the terms associated with this problem. Ordinarily there are a good many simultaneously observed characteristics to play with, and this helps. The biggest trouble is that you seldom know what progress you've made until you can get some outside independent confirmation.

I hate not being able to show you a big problem, but I'll show you how you can get into trouble in a very simple case. Suppose there are two underlying

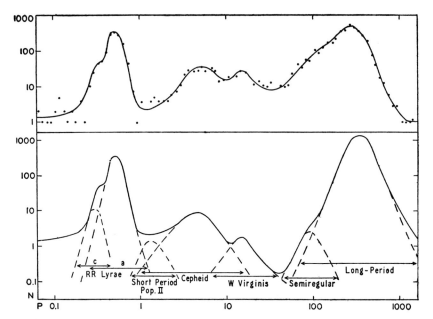

Fig. 8.2.2 Numbers of known variable stars (1950) within intervals in log P of 0.05. The upper part of the figure shows the observed numbers; the lower part is approximately corrected for volume surveyed. Both horizontal and vertical spacings are logarithmic. Broken curves in the lower figure show the domains of variable stars of several types. The curves are drawn to follow a Gaussian distribution, but the material is too incomplete for a decision as to whether such a distribution actually represents the data. Reproduced with permission from C. Payne-Gaposchkin (1954), *Variable Stars and Galactic Structure* (London: The Athlone Press), Figure 2.1, page 17.

subpopulations, each occurring about half the time. The probability density describing the mixture is then $\frac{1}{2} \text{pd}_1 + \frac{1}{2} \text{pd}_2$. If both subpopulations are GAU with the same variance, the population density itself is

$$\tfrac{1}{2} \, \text{gau}\,(x \mid \mu_1, \sigma^2) + \tfrac{1}{2} \, \text{gau}\,(x \mid \mu_2, \sigma^2).$$

The shape of the compound curve varies greatly depending on the distance between the means of the two component distributions. This distance is measured in standard deviations, as usual. See Fig. 8.2.3.

In each part of this figure the light lines show the pd's of the two component distributions and the heavy line shows $\frac{1}{2} \text{pd}_1 + \frac{1}{2} \text{pd}_2$. At 7σ the separation is essentially complete. As the means get closer together, the middle area fills in until, at 2σ-separation, the composite curve is no longer *bimodal* (having two peaks or modes) but is unimodal. It seems likely that many broad-tailed distri-

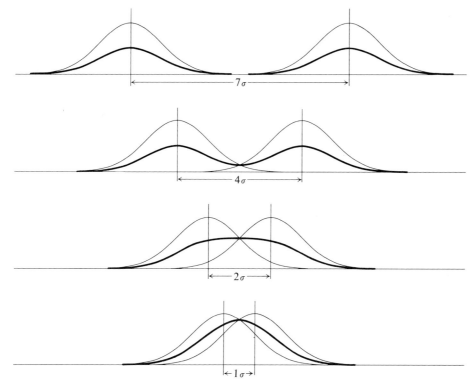

Fig. 8.2.3

butions are actually mixtures of distributions, but unfortunately the two peaks do not necessarily show up clearly in samples. Furthermore, you can't just draw a line down the middle to get the component distributions; some of the probability belonging to the left-hand distribution is on the right of the line, and vice versa. And if you calculate sample means of the two halves, they will be farther apart than the true means.

Example 8.2.11. You can, however, do the estimation properly. If we call the two means

$$m - \Delta \quad \text{and} \quad m + \Delta,$$

the likelihood of a sample of n is proportional to

$$\frac{1}{\sigma^n} \prod \left\{ \exp\left[-\frac{1}{2}\left(\frac{x_i - m + \Delta}{\sigma}\right)^2\right] + \exp\left[-\frac{1}{2}\left(\frac{x_i - m - \Delta}{\sigma}\right)^2\right] \right\}.$$

A computer can get successive approximations to m, Δ, and σ by hunting for the mode of the posterior density. Here are records of the progress of the approximations in some experimental sampling:

1. $n = 50$, $m = 0.0$, $\Delta = .50$, $\sigma = 1.0$. (1σ-separation)

Step	Estimate of m	Estimate of Δ	Estimate of σ
1	.1594	.7274	.7202
2	.1464	.6770	.7669
3	.1467	.6391	.7981
4	.1501	.6089	.8209

2. $n = 50$, $m = 0.0$, $\Delta = 2.00$, $\sigma = 1.0$. (4σ-separation)

Step	Estimate of m	Estimate of Δ	Estimate of σ
1	−.3968	1.6753	1.6588
2	−.3040	1.8137	1.5094
3	−.2314	1.9436	1.3416
4	−.1644	2.0388	1.1951
5	−.1121	2.0904	1.1042
6	−.0831	2.1118	1.0635

PROBLEM SET 8.2

In Problems 1 through 4, the two populations are

$$\text{I: } \text{GAU}(*\,|\,0, 1),$$
$$\text{II: } \text{GAU}(*\,|\,1, 1),$$

and

$$P_0(\text{I}) = P_0(\text{II}).$$

1. If an item of unknown type is measured to be .2, what are the odds that it belongs to population I rather than to population II?
2. At what point are the odds even?
3. If we set the criterion at .5, what is the probability of error?
4. Where should we set the criterion if it is three times as bad to misclassify a member of I as to misclassify a member of II?
5. If

$$\text{I: } \text{GAU}(*\,|\,0, 1),$$
$$\text{II: } \text{GAU}(*\,|\,1, 1),$$

and

$$P_0(\text{I}) = 5P_0(\text{II}),$$

at what point are the posterior odds even?

6. In Problem 5, what is the overall probability of error if the criterion is set at 2.11?
7. In Problem 5, where should we set the criterion if it is twice as bad to misclassify a II as to misclassify a I?
8. The sepal length is different in two species of violet:

$$\text{Species I: length} \sim \text{GAU} (* \mid 5, .64),$$
$$\text{Species II: length} \sim \text{GAU} (* \mid 6.8, .64).$$

If you knew that species II occurred twice as often as species I, how would you evaluate a flower with sepal length = 6.6 if it had to be one or the other?

9. The only obvious characteristic that distinguishes the two species of *Pseudomusca*, the giant Tibetan tsetse fly, is the length of the thorax:

Species	Distribution of length in millimeters
P. alba	GAU (* | 24, 4)
P. rosa	GAU (* | 28, 4)

If the two species are about equally abundant, how would you classify a quivering specimen with thoracial length of 27 mm, and how strong is the evidence?

10. If you make one million decisions per second, how many decisions do you make in a 24-hour day?

11. Suppose a bit is received satisfactorily 99.99% of the time. What is the probability that a block of 1024 bits will be received satisfactorily? [Assume independence.]

12. A notorious criminal has four favorite haunts: grid locations (2, 2), (3, 0), (1, −1), and (3, −3). The police have cars at positions (−3, 2) and (−3, −3) waiting for him to radio his henchmen; they then will take a bearing on him and close in quickly. Their direction finders have standard errors of 10° (independent of each other). This large error is due to their mobile mounts and the reflections off buildings. Let north be 0°, east 90°, etc. A transmission is heard; car A fixes the direction at 112°, car B at 59°. If the four sites were equiprobable to begin with, what is their posterior distribution on this information?

13. Draw a sample of 100 from the composite distribution whose means are 4 σ's apart. Make a histogram of the result using about 20 equal intervals. Is there any sign of bimodality?

14. Suppose that two variables are distributed according to BIGAU, and that observed values of the two variables are plotted in the plane. The two BIGAU populations may have different means, variances, and correlation coefficients (ρ). Show that the loci of observations having equal factors are conic sections. When are these loci straight lines?

15. You have five sample values:

$$88, 98, 91, 99, \text{off scale} > 100,$$

252   Making Decisions   8.2

which came from one of these two equally likely populations:

$$\text{I:}\quad \text{GAU}\,(* \mid 90, 25),$$
$$\text{II:}\quad \text{GAU}\,(* \mid 94, 4).$$

What are the posterior odds on the two populations?

ANSWERS TO PROBLEM SET 8.2

1. $e^{.3} = 1.35$ 4. At $+1.6$, where $\mathrm{O}\!\left(\dfrac{\mathrm{I}}{\mathrm{II}}\right) = \dfrac{1}{3}$ 5. 2.11

6. $P(\text{error} \mid \text{I}) = .0175$
   $P(\text{error} \mid \text{II}) = .866$
   $P(\text{error}) = \tfrac{5}{6} \times .0175 + \tfrac{1}{6} \times .866 = .159$

   If we had made no observation at all and had called everything "I", the error rate would have been .167.

8. $\mathrm{O}\!\left(\dfrac{\mathrm{II}}{\mathrm{I}} \,\Big|\, 6.6\right) = \dfrac{2}{1} \dfrac{\text{gau}\,(6.6 \mid 6.8, .64)}{\text{gau}\,(6.6 \mid 5.0, .64)}$

   $= 2 \dfrac{\text{gau}\,(-.25 \mid 0, 1)}{\text{gau}\,(2.0 \mid 0, 1)}$

   $= 2 \dfrac{.3864}{.0540} = 14.3$

   Strong evidence for II.

11. $(.9999)^{1024} = .903$

12. Make a careful plot to find the angles between the bearings and the sites (Fig. 8.2.4). I get:

Site	Deviation$_1$	Deviation$_2$	$\propto$ Likelihood	Posterior Probability
(2, 2)	22°	14°	exp (-3.900)	.017
(3, 0)	4°	4°	exp (-0.160)	.726
(1, −1)	15°	4°	exp (-1.205)	.256
(3, −3)	18°	31°	exp (-6.425)	.001

The first likelihood above is

$$\propto \exp\!\left[-\dfrac{1}{2}\!\left(\dfrac{22^2 + 14^2}{100}\right)\right].$$

These posterior probabilities aren't worth three decimal places, considering the crudeness of the deviations. Also, a GAU distribution might be a good approximation for small angular deviations, but it goes on to $\pm\infty$ while the deviations are limited

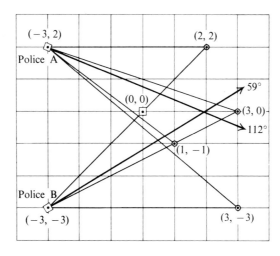

Fig. 8.2.4

to $\pm 180°$. In Section 10.2 we will study CIRG, a distribution which applies to some angular deviations.

15. Each likelihood calculation requires the multiplication of four Gaussian ordinates and one Gaussian tail area. The odds are $\frac{40000}{1}$.

8.3 TESTS OF SIGNIFICANCE AND GOODNESS OF FIT

I want to review the whole two-alternative decision structure in order to introduce some widely used terminology. Consider Fig. 8.3.1. We have two populations, I and II, and are going to make an observation to help decide between them. If it is a times as bad to misclassify a member of population I as it is to misclassify a member of population II, the decision threshold is set where

$$\mathcal{O}\left(\frac{\mathrm{I}}{\mathrm{II}}\right) = \frac{1}{a}.$$

When the odds are larger than $1/a$, we choose I; when the odds are smaller than $1/a$, we choose II. [See Problem 8.1.11 for a discussion.]

Until now we have been treating our hypotheses rather impartially; here a bit of favoritism enters. Hypothesis I will be regarded as an old friend, held in special esteem, and given the name *null hypothesis*. [Originally it was the hypothesis that something was zero.] The other, perhaps a young upstart, will be called merely the *alternative hypothesis*.

Suppose a sample is really from I (*null hypothesis true*), but the odds turn out to be so low that we call it a II (*reject null hypothesis when true*). We have made an *error of the first kind*. The probability of this error is called α ("alpha"). When a sample comes from II (*alternative hypothesis true*) but is classified as I (*accept*

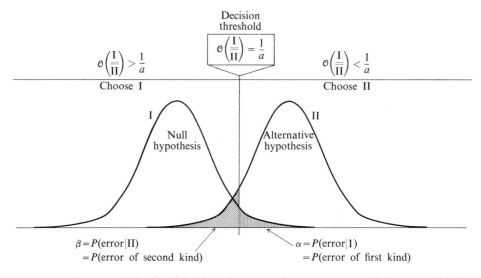

Fig. 8.3.1 An error of the first kind is *a* times as serious as an error of the second kind.

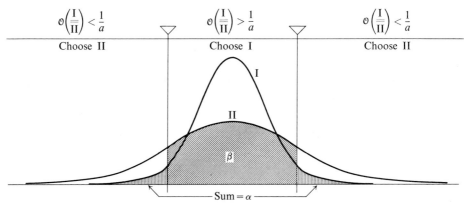

Fig. 8.3.2

null hypothesis when false), we have made an *error of the second kind*. The probability of this error is called β ("beta"). [An *error of the third kind* is giving the right answer to the wrong question!]

The same terminology and concepts apply in Fig. 8.3.2, which shows a case I neglected in the last section: the means of the two populations are the same, but their variances differ. We now choose I (accept null hypothesis) when the sample is close to the common mean. We choose II (reject null hypothesis) when the

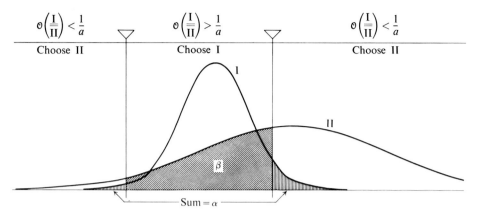

Fig. 8.3.3

sample is far from the mean on either side. Again

α = P(reject null hypothesis when true) = P(error of first kind),

β = P(accept null hypothesis when false) = P(error of second kind).

A more lopsided case is shown in Fig. 8.3.3, where both the means and the variances differ. I hesitated to bring these up before, because they run a little contrary to intuition. The *regions of rejection* of the null hypothesis in Fig. 8.3.2 may not seem too unnatural, but Fig. 8.3.3 gives everyone a start. Why, you think, should I accept the population with the larger mean when the observation turns out very small? You have to search your soul in any particular case to see if you really know the tail behavior of your distributions that far from the mean. Don't let formality overcome your common sense.

This whole decision structure is quite uncontroversial when our hypotheses are well specified. Trouble arises when the alternative is vague. We may have some idea or theory (the null hypothesis) which has stood the test of time or general acceptance. How do we decide whether we should continue to act as if it were true? [This is what accepting the null hypothesis means.] Many times there's no specific challenge, just the nagging worry that something might be wrong. You wonder, "Does my old theory really fit all this data?" This is the problem of *significance tests* and *tests of goodness of fit*.

A lot of people talk in language like this when they're really just interested in estimating a value. They talk about testing whether this fertilizer is more powerful than that one. Is there anybody who would believe that the two values are exactly the same? The problem is to get a reliable estimate for the difference. You want not statistical significance but practical significance. Or, is that value exactly 2.1468? Very likely not exactly; take more observations and get a more precise estimate. Once in a while, though, we feel we have a legitimate precise hypothesis to discuss. The problem then is to define the alternative hypothesis. [Saying "the opposite" is not good enough.] I've gone by this quickly at least twice. I asked, "Is $p = .5$?" and used a flat distribution for all the values from 0 to 1 for the alternative. I asked, "Are the two variables in this contingency table independent?" and assigned a distribution to the nonindependent cases.

Sometimes people refuse to assign a prior distribution to the alternatives or claim that it is impossible. What can they do? Well, the first step in the decision process is still plain. Even with vague alternative hypotheses, we know roughly in which regions we would accept the null hypothesis (where the odds for it are high) and in which we would reject the null hypothesis (where the odds for it are low).

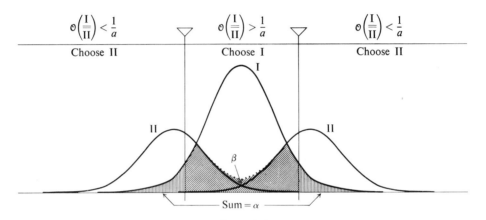

Fig. 8.3.4

Example 8.3.1. Suppose we wanted to decide whether some constant was very close to 0 (give or take a little). The alternative hypothesis would be made up of subalternatives on either side of 0. Some interval bracketing 0 would be the region of acceptance; intervals farther away on both sides would compose the region of rejection. This is shown in Fig. 8.3.4.

Example 8.3.2. Suppose we wanted to test whether a theory fit a set of data and we had some measure of the discordance. High values of that measure would invite rejection; low values would invite acceptance. There's no trouble in laying out the general scheme; the problem again is setting a criterion.

Sometimes the effect is so great that no precise criterion is needed. John Arbuthnot (1710) studied the records of births in London for 82 years. [The *records* covered 82 years, that is.] He found that in each of those years more males were born than females. Many people immediately say, "$(\frac{1}{2})^{82}$ is a terribly small number; this is obviously a very significant result." I agree it's significant, but such people miss the point. Any possible sequence of "male years" and "female years" has the same probability in a series of Bernoulli trials with $p = \frac{1}{2}$. The reason we regard this result as outstanding is that a simple alternative explanation has a reasonable prior probability and a much greater likelihood—the alternative that the birth rate for males is slightly different from that for females. However, Arbuthnot went a little further, the way I hope most of us would go. He said to himself something like: "This result is quite startling; I have this new tentative idea that more boys are born than girls and also an estimate of the ratio. Let's look at records in other countries and in other years to see if the same thing happens there too." Sure enough, it does, and this *confirmation* far outweighs any calculation, precise or imprecise, on the original data.

But what if I can't supply data so startling that rough calculation takes care of it? What if I can't think of my alternatives with any precision? First, many statistics are calculated to give a rough indication of the divergence between expectation and observation. Usually there's a whole mess of such results lying around. We can't look at everything, but when we are prospecting for ideas, such statistics provide a general ranking of phenomena; they help tell us in what order to look at things. We have to temper their message with an evaluation of the possible benefits of the investigation, but we should take such hints gratefully; they help us do our job more efficiently. But they can't give us the *probabilities* that new phenomena exist. Our attitude toward exactitude varies according to the stage of our investigation. At the beginning, to get oriented, we do things we would scorn later on when the situation is much cleaner.

However, people usually go farther than this; they calculate *significance levels*. Since the statistical and applied literature is full of significance tests, it is important for you to realize what they are and what they are not. What is left when the specific alternative hypothesis is gone? Only the probability distribution of the null hypothesis, and rough knowledge of the regions of acceptance and regions of rejection of the null hypothesis. The procedure people follow to calculate the significance level goes like this: They let the *observation* itself define the boundary of the rejection region, and then calculate the corresponding α. It may be either a one-tailed or a two-tailed region. If alpha is quite small, they say that they reject or doubt the null hypothesis. Note that they *cannot* get the probability from this. Certainly α is not that probability, and $1/\alpha$ is not the factor against the null hypothesis. There just is no way to set a true decision criterion without looking at the alternatives in more detail. Moreover, although you may have α, you do not have β. It is β that tells you the probability of missing a new phenomenon.

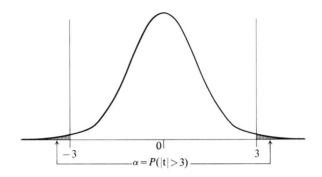

Fig. 8.3.5

Example 8.3.3. A physicist wants to test whether a certain constant is precisely 0. He makes some observations and calculates

$$t = \frac{\bar{x} - 0}{\sqrt{s^2/n}} = 3$$

with 10 degrees of freedom. He has in mind the vague alternative possibility that the value is slightly different from 0. If he allows both positive and negative alternatives, his conception would look like Fig. 8.3.4. To calculate the significance level he sets the pseudo decision line at the value of the observation: 3. The other line comes from symmetry. See Fig. 8.3.5. This is a *two-tailed t-test*. For α he gets

$$\alpha = P(|t| > 3 \mid t \sim \text{STU}(* \mid 0, 1, 10)) = .014.$$

Our physicist says,

"the significance level is .014"

or

"the result is significant at the 2% (.02) level"

or

"the result is not significant at the 1% (.01) level."

This does *not* mean that P(null hypothesis) = .014. It does *not* mean that the factor against the null hypothesis is 1/.014 = 70, because the alternatives are imprecisely specified. *The same significance level can correspond to different P(null hypothesis)'s when the alternatives are different.*

It would seem much more informative to try to determine

1. how far this value has to be from 0 to have any practical significance;
2. what the relative importance is of
   i) saying that the value is not 0 when it really is 0,
   ii) missing a new effect;
3. what the previous evidence is on the subject—can a small experiment tell much?

You must remember that by definition you will obtain a significance level of .014 once in 70 times when there is no new effect. In fact, by continuing a sequence of observations you can attain *any* significance level you want, even though the null hypothesis is true. Contrast this with the nonzero probability of never being able to get an arbitrarily large *factor* against a true hypothesis. Also, how many nonsignificant results have you discarded lately?

Jeffreys (1961) has analyzed the problem in Example 8.3.3 and many others to show how uninformative prior distributions for alternatives can be developed. They don't automatically take the place of careful consideration of the peculiarities of each problem, but they offer valuable hints. In this case he obtained

$$\mathcal{F}\left(\frac{\text{null}}{\text{alternative}}\right) \div \sqrt{\frac{\pi\nu}{2}} \frac{1}{\left(1+\frac{t^2}{\nu}\right)^{(\nu-1)/2}} = \frac{1}{4.5}.$$

It has long been Good's contention (1950, 1965, and elsewhere) that a significance level α often corresponds to a factor of something like $30/\alpha$ to $3/\alpha$, here $\frac{1}{2}$ to $\frac{1}{21}$. The conclusion from all of this is that this experiment gives a little weight against the hypothesis of 0, but nothing like a factor of $\frac{1}{70}$.

You probably are wondering how significance levels work in practice, since up to now they've been the standard approach. It's quite difficult to say. Certainly the right decision is often made, though the evidence may not be as strong as advertised. Wrong decisions are also made, but this may not be attributed to the statistics. I know of no study which tries to compare scientific results achieved with the significance level of the experiments. Remember that statistical significance has little to do with practical significance. A small effect can produce a high significance level if many observations are taken; a large effect may go unnoticed in a small experiment. My opinion, from limited observation, is that many people use significance levels as an ordering of future experimental possibilities. I also think many actually believe that the significance level is the probability of the null hypothesis—which it isn't—but have developed an automatic downgrading mechanism like Good's so they don't take it so seriously. Unfortunately, this mechanism also makes it hard for them to believe the actual probabilities of hypotheses coming out of Bayes' Theorem.

PROBLEM SET 8.3

1. Given

    null hypothesis: variance = 1,
    alternative hypothesis: variance > 1,

    what is the rejection region for the null hypothesis?

260   *Making Decisions*   8.4

2. The chi-square statistic measures discordance between data and theory. It is 0 when data exactly agree with expectation (a rather unlikely happening), greater than 0 otherwise. What is the rejection region for the null hypothesis when the alternative is that the data do not agree with the theory?

3. See Problem 2. What is the rejection region for the null hypothesis when the alternative is that the data have been forced to agree with the theory?

4. Given
$$\text{null hypothesis: value} = 0,$$
$$\text{alternative hypothesis: value} > 0,$$

what is the rejection region for the null hypothesis?

5. What is the ratio of the ordinates
$$\frac{\text{stu }(3 \mid 0, 1, 10)}{\text{stu }(3 \mid 3, 1, 10)}?$$

6. What is the ratio of the ordinates
$$\frac{\text{stu }(3 \mid 0, 1, 10)}{\text{stu }(3 \mid 6, 1, 10)}?$$

ANSWERS TO PROBLEM SET 8.3

2. The right tail   4. The right tail   5. $\frac{1}{34}$

8.4 SEQUENTIAL DECISIONS

The decisions we have been making so far have been *terminal decisions*. We had the data, and we made our decision on that basis. This is realistic in many circumstances: all available data have been acquired; or information naturally comes in big chunks. On the other hand, information may come only now and then and in small driblets; or the cost of acquiring information may not be so small that we can immediately ask for the maximum amount. Then we want to analyze our information as it comes in, and make our decisions when we have acquired the proper amount of information—no sooner, no later. There are also intermediate cases in which several observations fall together in a block, and we analyze block by block. All this is handled straightforwardly by Bayesian methods. We just keep using

$$\text{posterior prob} \propto \text{prior prob} \times \text{likelihood},$$

where the prior probabilities represent our knowledge before the current observation.

For example, take the problem of comparing the effectiveness of two drugs in actual clinical trials. At stake are not the lives of laboratory rats, but the lives of human beings. Compassion—if not the law—requires that decisions be reached as soon as possible to avoid any taint of experimentation for experimentation's sake. Or consider the problem of inspecting the finished product in a factory. If the product is produced in large lots, ordinarily only a sample of the lot is completely inspected. The size of the sample depends on the quality that the manufacturer desires and how the inspection proceeds. And in Examples 8.2.5 and 8.2.6 I mentioned the selective retransmission of radio signals if the quality is bad the first time.

Here we must explicitly include *cost* in the planning of the experiment. In the drug testing the cost was the effects of an inferior drug. In the factory inspection the cost was the cost of inspection and the ill will a bad product engenders. In the radio transmission the cost was the time taken for retransmission and the cost of the equipment used to store the data for possible retransmission. We always have to ask what is profitable in the light of the expected return. This sequential attitude requires careful specification of procedures, since much of the analysis or direction is in practice carried out by statistically unskilled persons.

I will present two important examples in which we try to decide between two hypotheses. The data in both is a sequence of Bernoulli trials. In the first example the cost is measured by the probabilities of the two types of decision errors. The cost of experimentation is regarded as relatively small, although we wish to finish as soon as possible while still controlling the errors. In the second example the cost of experimentation is major, and the analysis is more complicated. Both problems are phrased in terms of industrial inspection, but the methods illustrated are directly transferable to other fields. Only the names need be changed.

Example 8.4.1. We observe a sequence of Bernoulli trials. The two hypotheses are

$$\text{I:} \quad p = .7,$$
$$\text{II:} \quad p = .4.$$

Because of the consequences of the two types of error in this situation, we want their probabilities to be

$$\alpha = .15 = P(\text{choose II} \mid \text{really I}),$$
$$\beta = .20 = P(\text{choose I} \mid \text{really II}).$$

[The ratio of α to β is determined by the relative seriousness of the errors; their absolute size, by the total allowable cost.] It turns out that this procedure does the job: Look at the factor

$$\mathcal{F}\left(\frac{\mathrm{I}}{\mathrm{II}}\bigg|\,\mathrm{seq}\right) = \frac{P(\mathrm{seq}\,|\,\mathrm{I})}{P(\mathrm{seq}\,|\,\mathrm{II})}$$

that is building up. As long as

$$\frac{1}{5.33} < \mathcal{F} < 4.25,$$

take another observation. When

$$\mathcal{F} \leq \frac{1}{5.33},$$

decide on II. When

$$\mathcal{F} \geq 4.25,$$

decide on I. This secures the desired error rates while minimizing the amount of experimentation.

This is known in the statistical literature as the *sequential probability ratio test*. The proof that it works was given by Wald (1947); it is so simple and instructive that I have to show it to you. I will only ask you to assume that the process will eventually end. To start with, we know that the probability of an observed sequence of Bernoulli trials with s successes and f failures is

$$p^s(1-p)^f,$$

so

$$\mathcal{F}\left(\frac{\mathrm{I}}{\mathrm{II}}\bigg|\,s,f\right) = \left(\frac{.7}{.4}\right)^s \left(\frac{.3}{.6}\right)^f \quad \left[\left(\frac{p_1}{p_2}\right)^s \left(\frac{1-p_1}{1-p_2}\right)^f,\ \text{in general}\right].$$

We are going to sample as long as

$$\frac{1}{5.33} < \mathcal{F} < 4.25 \qquad [B < \mathcal{F} < A,\ \text{in general}].$$

Choose II when

$$\mathcal{F} \leq \frac{1}{5.33} \qquad [\mathcal{F} \leq B,\ \text{in general}].$$

Choose I when

$$\mathcal{F} \geq 4.25 \qquad [\mathcal{F} \geq A,\ \text{in general}].$$

Now imagine that you can list all those sequences of results that lead to the decision *choose* I. Call this set X_I. Since we assume that the experiment will eventually end,

$$P(X_\mathrm{I}\,|\,\mathrm{I}) = 1 - \alpha, \qquad P(X_\mathrm{I}\,|\,\mathrm{II}) = \beta,$$

because we err if a sample from II leads us to X_I; we do the right thing if a sample from I leads us there. In the same way imagine that you can list all those sequences of results that lead to the decision *choose* II. Call this set X_{II}. We have

$$P(X_{II} \mid I) = \alpha, \qquad P(X_{II} \mid II) = 1 - \beta,$$

because we err if a sample from I leads us to X_{II}; we do the right thing if a sample from II leads us there.

Now look at any one of the sequences of observations gathered together in X_I. Call it x_1. It is in X_I because it led to the decision *choose* I. This means that

$$A \leq \mathcal{F}\left(\frac{I}{II} \mid x_1\right) = \frac{P(x_1 \mid I)}{P(x_1 \mid II)},$$

so

$$A \times P(x_1 \mid II) \leq P(x_1 \mid I).$$

But this is true of every sequence $x_1, x_2, x_3, \ldots$ from X_I:

$$A \times P(x_1 \mid II) \leq P(x_1 \mid I),$$
$$A \times P(x_2 \mid II) \leq P(x_2 \mid I),$$
$$A \times P(x_3 \mid II) \leq P(x_3 \mid I),$$
$$\vdots \qquad\qquad \vdots$$

Now add. Since $x_1, x_2, x_3, \ldots$ exhaust the set X_I,

$$P(x_1 \mid II) + P(x_2 \mid II) + P(x_3 \mid II) + \cdots$$

is just the probability of ever reaching X_I when the sample really comes from II. This is the probability of an error of the second kind, that is, β. And

$$P(x_1 \mid I) + P(x_2 \mid I) + P(x_3 \mid I) + \cdots$$

is the probability of ever reaching X_I when the sample is really from I. This is $1 - \alpha$. Therefore

$$A\beta \leq 1 - \alpha.$$

By similar reasoning, we find that

$$\alpha \leq B(1 - \beta).$$

When the situation is investigated numerically, it turns out that most of the time these inequalities are pretty close to being equalities, and that we can set

$$A\beta \doteq 1 - \alpha, \qquad \alpha \doteq B(1 - \beta).$$

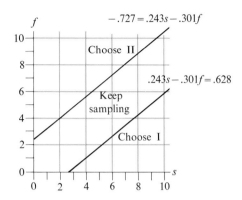

Fig. 8.4.1

For Example 8.4.1 we can then calculate

$$A \doteq \frac{1-\alpha}{\beta} = \frac{.85}{.20} = 4.25 \quad \text{and} \quad B \doteq \frac{\alpha}{1-\beta} = \frac{.15}{.80} = \frac{1}{5.33}.$$

The item-by-item decision calculation can be done by a simple chart. The requirement for continuing to sample is

$$\frac{1}{5.33} < \left(\frac{.7}{.4}\right)^s \left(\frac{.3}{.6}\right)^f < 4.25.$$

We can take logarithms:

$$\log \frac{1}{5.33} < s \log \left(\frac{.7}{.4}\right) + f \log \left(\frac{.3}{.6}\right) < \log 4.25,$$

or

$$-.727 < .243s - .301f < .628.$$

The equations giving the *boundaries* of this *keep-sampling* region are

$$-.727 = .243s - .301f \quad \text{and} \quad .243s - .301f = .628.$$

If we plot s and f on two axes, we see that these two equations are the equations of two parallel lines (Fig. 8.4.1). To evaluate any particular sequence of observations, we start at the origin (0, 0), move one step to the right for every success and one step up for every failure. Whenever we hit or jump over one of the boundaries, we make that decision. The sequence of observations

$$s, s, f, s, s$$

is shown in Fig. 8.4.2. It leads to the decision *choose* I.

The decisions made with charts like this are so precise and convenient that the scheme was widely used in factories during World War II. It even got a security

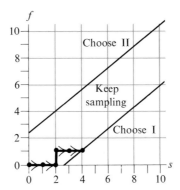

Fig. 8.4.2

classification! There are variants to handle other types of sequential problems, but this simple example gives the basic flavor. By the way, if you work with odds, you sample as long as

$$B\Theta_0 < \Theta < A\Theta_0.$$

Example 8.4.2. Back in Section 3.5 I talked about the ultimate probability of error, about the probability of being able to drive down the probability of a true hypothesis. The answer I gave then was that the probability that I could ever get a factor of $1/a$ or less for the true hypothesis was at most $1/a$. This follows directly from the formulas of Example 8.4.1:

$$\alpha = \frac{1}{a}, \quad \beta = 0, \quad B \geq \frac{\alpha}{1-\beta} = \frac{1}{a},$$

$$P(\mathcal{F} \leq B) = \alpha, \quad P\left(\mathcal{F} \leq \frac{1}{a}\right) \leq \frac{1}{a}.$$

Now I take up the sequential decision problem when the cost of sampling cannot be neglected. We will extend Example 3.2.1.

Example 8.4.3. An automatic machine produces 24 large parts per day. On 90% of all days it is in good working condition, and an average of 95% of the parts are good. On 10% of the days it is in a poor condition, and only 70% of the parts are good. The cause of the machine's getting into the bad state are not understood, but it has something to do with temperature changes during the night when the machine is not operating.

Now if you are willing to sacrifice the production time for one part, a mechanic can inspect the machine and put it into the good state if it is not already there. Of course, if it is already good, you've wasted good production time, and can turn out only 23 parts. Each good part is worth $200, a bad part is worth $0; and the mechanic's time for the adjustment is worth $15. Should you plan to have the mechanic adjust the machine the first thing each day? [I will assume that bad parts, if any, are produced at random times during the day.]

Now we set up our table of consequences. When we do not adjust, we have 24 opportunities to produce parts at the appropriate rate. When we do adjust, we have only 23. The expected income is

$$\$200 \times (\text{number of periods}) \times (\text{prob of good part}).$$

Therefore the table of consequences is:

Income	Good state	Bad state
Adjust	$200 × 23 × .95 − $15 = $4355	$200 × 23 × .95 − $15 = $4355
Don't adjust	$200 × 24 × .95 = $4560	$200 × 24 × .70 = $3360
	.9	.1

Then

$$E(\text{income} \mid \text{adjust}) = (\$4355)(.9) + (\$4355)(.1) = \$4355,$$
$$E(\text{income} \mid \text{don't adjust}) = (\$4560)(.9) + (\$3360)(.1) = \$4440. \Leftarrow$$

With this information we maximize our expected income by not adjusting. We expect $85 more per day this way. We can also write the loss table as:

Losses	Good state	Bad state
Adjust	$205	$0
Don't adjust	$0	$995

Then

$$E(\text{loss} \mid \text{adjust}) = .9(\$205) = \$184.50,$$
$$E(\text{loss} \mid \text{don't adjust}) = .1(\$995) = \$99.50. \Leftarrow$$

The same difference of $85 appears. We also see that the loss due to uncertainty is $99.50.

All right, it was judged not worth while to adjust the machine at the beginning of the day even though it might have been (unknowingly) in a bad state. Suppose that the mechanic actually had business in the neighborhood and could perform the adjustment, if desired, *after* the first part has been produced and quickly tested. Are there any situations now in which it would pay to adjust the machine? Here our sampling is free: we would produce the part anyway, and the mechanic is around. [You know that the answer to the question is *yes* or I wouldn't have brought up the matter; but please bear with me!]

If the first part produced is good, the posterior distribution for the state of the machine is calculated as before:

Alternative	Prior	Likelihood	Joint	Posterior
Good	.9 ×	.95	= .855 →	.9243
Bad	.1 ×	.70	= .070 →	.0757
	1.0		.925	1.0000

If we adjust, there will be 22 production cycles left; if not, then 23. We must consider the expected income *after* this first part:

Income	Good	Bad
Adjust	$200 × 22 × .95 − $15 = $4165	$200 × 22 × .95 − $15 = $4165
Don't adjust	$200 × 23 × .95 = $4370	$200 × 23 × .70 = $3220

Then

$$E(\text{income} \mid \text{adjust}) = \$4165(.9243) + \$4165(.0757) = \$4165,$$
$$E(\text{income} \mid \text{don't adjust}) = \$4370(.9243) + \$3220(.0757) = \$4282.95. \Leftarrow$$

Thus, if the first part produced is good, do not adjust.

What if the first part is bad? The posterior distribution is

Good	Bad
.6	.4

The same income table applies, so

$$E(\text{income} \mid \text{adjust}) = \$4165, \Leftarrow$$
$$E(\text{income} \mid \text{don't adjust}) = \$4370(.6) + \$3220(.4) = \$3910.$$

Thus, if the first part produced is bad, adjust.

I now want to compare the expected income under the options of possibly adjusting at the beginning of the day and possibly adjusting after one part has been produced and tested. We can make a tree of possibilities (Fig. 8.4.3). Here

D = don't adjust, A = adjust,
g = good part, b = bad part,
S = sample = see what first part is.

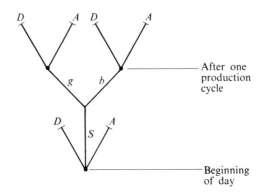

Fig. 8.4.3

We put in the numbers we already have (Fig. 8.4.4). I have broken the paths we would not follow, and brought down to the upper nodes the values of the preferred paths. That is, if we get a good first part, the future expected income under proper choice is $4282.95, etc. Now we must evaluate the sampling option. First we need the probabilities of going along the two branch paths (compound probability again):

$$P(g) = P(g \mid G)P_0(G) + P(g \mid B)P_0(B) = .95 \times .9 + .70 \times .1 = .925,$$
$$P(b) = 1 - .925 = .075.$$

Also, remember that if the first part is good, we get $200 credit for it. We now can calculate the value of the S-branch:

E(income | look at first part and make proper decision)
$$= .925(\$4282.95 + \$200) + .075(\$4165 + \$0) = \$4459.09.$$

This is greater than either $4440 or $4355, so it is the preferred policy. Even if

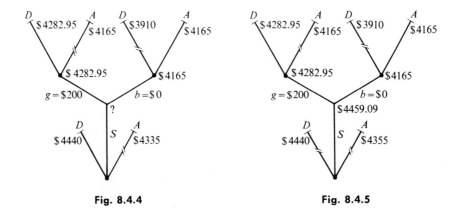

Fig. 8.4.4   Fig. 8.4.5

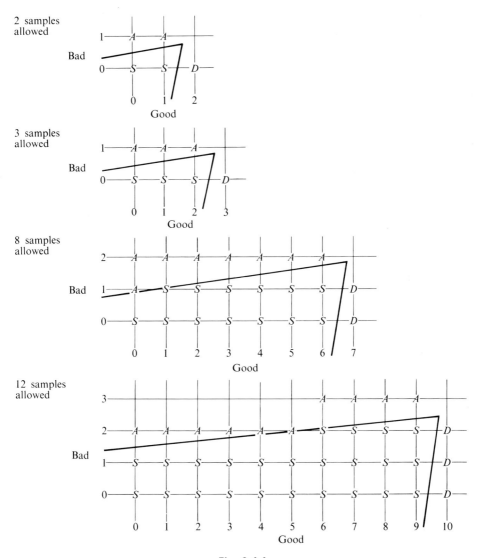

Fig. 8.4.6

we had to pay the mechanic another $15 to hang around, it would still be the best method by a slight amount. The complete decision structure is shown in Fig. 8.4.5.

A presentation like this becomes much too unwieldy if additional sampling and delay in decision is contemplated. The computations get even more complicated; generally a computer is required. Yet again we can plot our two kinds of observations on a simple diagram. The course of the sampling is shown by a path which goes *right* or *up* until it enters a region that says to stop sampling and do something. I worked out on a computer the decision schemes for your being willing to take 2, 3, 8, or even 12 samples before definitely deciding whether to adjust the machine or send the mechanic away; I assumed that the mechanic received his pay only if he worked on the machine. The resulting sampling plans are shown in Fig. 8.4.6.

These wedge-shaped regions seem to be typical of problems like this, in which the cost of an observation is taken into account. There has to be an upper bound to the number of observations allowed because the possible gain from sampling is bounded. Contrast this with the parallel boundaries of Example 8.4.1. There you could give no specific upper bound for the number of observations taken before a decision is made.

PROBLEM SET 8.4

1. If you plot the sequence

$$f, s, f, s, s, s, f, s, s, f, s, s$$

   on Fig. 8.4.1, what decision do you reach?
2. Use Fig. 8.4.1. Take 10 random sequences with $p = .7$, and sample until a decision is made. Do the same with $p = .4$. Record the number of decisions each way and the time it took you to get there. Collect for the class.
3. Plan a sequential probability ratio test for

$$p_I = .3, \quad \alpha = .05,$$
$$p_{II} = .1, \quad \beta = .03.$$

4. In a sequential probability ratio test, what are the approximate probabilities of the two kinds of error if you continue sampling until you get a factor one way or the other of $\frac{10}{1}$?

ANSWERS TO PROBLEM SET 8.4

1. Choose I.
3. $A \doteq \dfrac{.95}{.03} = 31.67; \quad B \doteq \dfrac{.05}{.97} = \dfrac{1}{19.4}$

   The boundary lines are

   $$-1.288 = .477s - .109f \quad \text{and} \quad .477s - .109f = 1.501.$$

CHAPTER **9**

Handling Several Variables

Up till now we have been taking observations on a single variable and making inferences about the parameters of its distribution. More often, however, an experimenter is faced with two or ten or a hundred variables to handle at once, and what to do is not so obvious. Multivariate analysis is a complicated field, and not many very satisfactory results have emerged. Though point estimates can readily be obtained, the accompanying uncertainty is quite perplexing. Most people cannot visualize or imagine the simultaneous variation of many variables. In desperation they resort to all sorts of ad hoc simplifications whose consequences are imperfectly understood. In this chapter I will content myself with short descriptions of how variables are compared, how preplanning an experiment may yield increased precision for the same effort, and how straight-line problems are analyzed. The procedure followed in a particular problem depends strongly on the little details, but our general principles still apply.

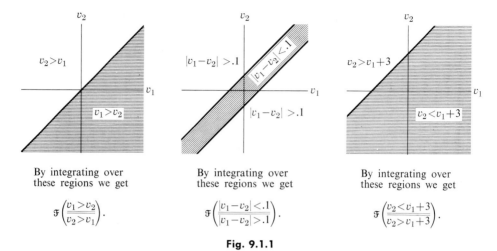

Fig. 9.1.1

9.1 COMPARING TWO VALUES

I have mentioned that experimental practice is greatly concerned with comparing values. Often, as in biology, absolute values have no particular meaning, even when they are quite accurately known. We are interested in how these values change when other characteristics change. The well-established practice of using *controls* testifies to this. Moreover, an experimenter usually has some idea of the magnitude of the differences to be expected, and most scientific experiments are arranged so that the differences sought will be perceptible.

Suppose we have independently acquired information about two variables, and want to compare their values. What procedure do we use and what questions can we ask? Basically we want to manipulate the two posterior distributions to acquire information about the difference in the variate values. We can then ask how sure we are of the *sign of the difference*, what the factor is *that the difference is of a certain minimum size*, etc. Sometimes there are abbreviated ways of obtaining this information, but fundamentally we form the joint density (the product of the two marginal densities if the variables are truly independent) and then integrate over the regions of concern. Figure 9.1.1 shows how to answer several questions. The results are basically factors, not odds, since they express the information that comes from these experiments alone. But if this is all the information that concerns the difference between the variables, the factors become odds.

When the posterior distributions are familiar types, this process can be carried out symbolically, and the numerical work is greatly reduced. It happens that *the sum or difference of two Gaussian variables is also a Gaussian variable*. If

$$v_1 \sim \text{GAU}(* \mid \mu_1, \sigma_1^2), \qquad v_2 \sim \text{GAU}(* \mid \mu_2, \sigma_2^2),$$

and they are independent, then

$$v_2 - v_1 \sim \text{GAU}(* \mid \mu_2 - \mu_1, \sigma_1^2 + \sigma_2^2)$$

and we can get all the answers we want from the good old GAU $(* \mid 0, 1)$ tables.

Example 9.1.1. $x \sim \text{GAU}(* \mid 0, 1)$, $y \sim \text{GAU}(* \mid 1, 1)$, and they are independent. Then

$$y - x \sim \text{GAU}(* \mid 1, 2)$$

and, for example, $P(y > x) = P(y - x > 0) = 1 - \text{GAU}(-.71 \mid 0, 1) = .76$.

Example 9.1.2. Our only knowledge of the strength of alloy A is that it is $\sim \text{GAU}(* \mid 94, 7)$; our only knowledge of the strength of alloy B is that it is $\sim \text{GAU}(* \mid 96, 5)$. What can we say about their relative strength? We have strength$_B$ − strength$_A \sim \text{GAU}(* \mid 2, 12)$. We can then give

95% HDR for difference in strength is $2 \pm 1.96\sqrt{12} = (-4.8, 8.8)$

or

$$\mathcal{F}\left(\frac{\text{strength}_B - \text{strength}_A > 0}{\text{strength}_B - \text{strength}_A < 0}\right) = \frac{1 - \text{GAU}(0 \mid 2, 12)}{\text{GAU}(0 \mid 2, 12)} = \frac{.71}{.29} = 2.5,$$

etc. If our posterior distributions are not GAU but can be closely approximated by GAU, we can use the same method provided we don't go too far into the tails.

If, as it happens more frequently, the posterior distributions are STU, the situation is a little more complicated. Several tactics can be used:

1. You can work numerically.
2. You can treat the distribution as GAU when the numbers of degrees of freedom are large.
3. You can optimistically assume that the underlying variances of the two variables are the same. Then you can get the distribution of the difference into STU form.
4. You can find a good approximation to the distribution of the difference when you know nothing about the relative sizes of the variances.

The third option goes like this: In carrying out our Bayesian analysis on one Gaussian variable, we found that

$S_{\bar{x}\bar{x}}$ had $n - 1$ degrees of freedom; $E(S_{\bar{x}\bar{x}}) = (n - 1)\sigma^2$,

$\dfrac{S_{\bar{x}\bar{x}}}{n - 1}$ estimated the variance of one observation,

$\dfrac{S_{\bar{x}\bar{x}}}{n(n - 1)}$ estimated the variance of the mean, since the variance of a mean is $1/n$ the variance of one observation. If we now have two variables and assume that the variance is the same in the underlying populations, these things happen:

$$\frac{S_{\bar{x}\bar{x}}^{(1)} + S_{\bar{x}\bar{x}}^{(2)}}{n_1 + n_2 - 2} \qquad \text{estimates the common variance of one observation,}$$

$$\frac{S_{\bar{x}\bar{x}}^{(1)} + S_{\bar{x}\bar{x}}^{(2)}}{n_1 + n_2 - 2}\left(\frac{1}{n_1} + \frac{1}{n_2}\right) \qquad \text{estimates the variance of the difference of the two means.}$$

This is quite reasonable intuitively:

$S_{\bar{x}\bar{x}}^{(1)}$ has $n_1 - 1$ degrees of freedom; $\qquad E(S_{\bar{x}\bar{x}}^{(1)}) = (n_1 - 1)\sigma^2$,
$S_{\bar{x}\bar{x}}^{(2)}$ has $n_2 - 1$ degrees of freedom; $\qquad E(S_{\bar{x}\bar{x}}^{(2)}) = (n_2 - 1)\sigma^2$,

$(S_{\bar{x}\bar{x}}^{(1)} + S_{\bar{x}\bar{x}}^{(2)})$ has $n_1 + n_2 - 2$ degrees of freedom; $\quad E(S_{\bar{x}\bar{x}}^{(1)} + S_{\bar{x}\bar{x}}^{(2)}) = (n_1 + n_2 - 2)\sigma^2$.

We just carry through the ordinary STU analysis but use this new estimate of the variance.

Example 9.1.3. We have this data on the first variable:

$$x_1: \quad n_1 = 7, \quad S_x^{(1)} = 5.6, \quad \bar{x}_1 = .8, \quad S_{\bar{x}\bar{x}}^{(1)} = 10.50.$$

Then

$$\frac{10.50}{6} = 1.75 \text{ is the estimated variance of one observation,}$$

$$\frac{10.50}{6 \times 7} = .25 \text{ is the estimated variance of the mean,}$$

so analysis of x_1 by itself gives

$$\mu_1 \sim \text{STU}\,(* \mid .8, .25, 6).$$

We have this data on the second variable:

$$x_2: \quad n_2 = 4, \quad S_x^{(2)} = 6.4, \quad \bar{x}_2 = 1.6, \quad S_{\bar{x}\bar{x}}^{(2)} = 9.72.$$

Then

$$\frac{9.72}{3} = 3.24 \text{ is the estimated variance of one observation,}$$

$$\frac{9.72}{3 \times 4} = .81 \text{ is the estimated variance of the mean,}$$

so analysis of x_2 by itself gives

$$\mu_2 \sim \text{STU}\,(*\mid 1.6, .81, 3).$$

If we now assume that $\sigma_1^2 = \sigma_2^2$, then

$$S_{\bar{x}\bar{x}}^{(1)} + S_{\bar{x}\bar{x}}^{(2)} = 20.22, \quad n_1 + n_2 - 2 = 9 \text{ degrees of freedom,}$$

$$\frac{20.22}{9} = 2.247 \text{ is the estimated common variance of one observation,}$$

$2.247(\frac{1}{7} + \frac{1}{4}) = .884$ is the estimated variance of the difference of the means,

$\bar{x}_2 - \bar{x}_1 = 1.6 - .8 = .8,$

and

$$\mu_2 - \mu_1 \sim \text{STU}\,(*\mid .8, .884, 9).$$

That's if you assume that the underlying variances are equal. If you don't assume this, but let the observations speak for themselves, you have a problem which can be done numerically, but which also can be very well approximated by a formula. [This is called the Behrens-Fisher problem in the literature.] By equating some moments, Welch (1937) produced this result:

Let

$$u_1 = \text{estimated variance of } \mu_1 = \frac{S_{\bar{x}\bar{x}}^{(1)}}{n_1(n_1 - 1)},$$

$$u_2 = \text{estimated variance of } \mu_2 = \frac{S_{\bar{x}\bar{x}}^{(2)}}{n_2(n_2 - 1)},$$

$$b = \frac{u_1}{u_1 + u_2} \qquad (u_1 + u_2 = \text{estimated variance of } \mu_2 - \mu_1).$$

Then

$$\mu_2 - \mu_1 \stackrel{.}{\sim} \text{STU}\,(*\mid \bar{x}_2 - \bar{x}_1, u_1 + u_2, \nu),$$

where

$$\frac{1}{\nu} = \frac{b^2}{n_1 - 1} + \frac{(1 - b)^2}{n_2 - 1}.$$

That is, a fake number of degrees of freedom is constructed to describe the uncertainty. Its size depends on how well the two estimated variances agree. Obviously

$$\text{minimum}\,(n_1 - 1, n_2 - 1) \leq \nu \leq n_1 + n_2 - 2$$

and is usually *not* an integer. You have to interpolate between the rows of the "*t*"-table.

Example 9.1.4

$$n_1 = 10, \quad S_{\tilde{x}\tilde{x}}^{(1)} = 12.08, \quad \frac{S_{\tilde{x}\tilde{x}}^{(1)}}{n_1 - 1} = 1.34,$$

$$n_2 = 3, \quad S_{\tilde{x}\tilde{x}}^{(2)} = 25.14, \quad \frac{S_{\tilde{x}\tilde{x}}^{(2)}}{n_2 - 1} = 13.07,$$

so the estimates of the two variances are quite disparate. The Welch approximation goes:

$$u_1 = \frac{1.34}{10} = .134,$$

$$u_2 = \frac{13.07}{3} = 4.357,$$

$$b = \frac{.134}{.134 + 4.357} = .0299,$$

$$\frac{1}{\nu} = \frac{.0009}{9} + \frac{.9411}{2} = .4706, \quad \nu = 2.12,$$

$$u_1 + u_2 = 4.491,$$

so

$$\mu_2 - \mu_1 \overset{\cdot}{\sim} \text{STU} (* \mid \bar{x}_2 - \bar{x}_1, 4.491, 2.12).$$

If you could have assumed beforehand that the underlying variances were equal, you would have gotten

$$\text{estimated variance of one observation} = \frac{12.08 + 26.14}{11} = 3.47,$$

$$\text{estimated variance of difference} = 3.47(\tfrac{1}{10} + \tfrac{1}{3}) = 1.50,$$

and

$$\mu_2 - \mu_1 \sim \text{STU} (* \mid \bar{x}_2 - \bar{x}_1, 1.50, 11),$$

a much more concentrated distribution.

Figure 9.1.2 shows plots of the posterior densities of the differences of STU variables. The higher curve in each case is the curve which assumes that the variances are equal; with more degrees of freedom to stabilize the estimated variance, the distribution is tighter. The lower curve in each case assumes that you know nothing beforehand about the relative size of the two variances. I calculated both the exact distribution and the Welch approximation in each case, and verified that the approximation is excellent. Figure 9.1.2(c) shows the curves for Example 9.1.4. The estimates of the two individual variances differ by a factor of 10; the posterior densities under the two assumptions are quite different. In this case you'd have to think carefully if you had any evidence that the two variances were the same. In Fig. 9.1.2(a) the two curves are almost superimposed.

9.1 Comparing Two Values

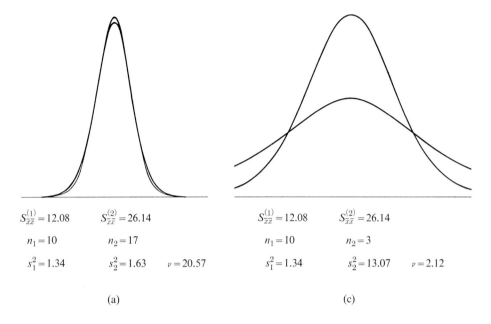

$S_{\bar{x}\bar{x}}^{(1)} = 12.08$ $S_{\bar{x}\bar{x}}^{(2)} = 26.14$ $S_{\bar{x}\bar{x}}^{(1)} = 12.08$ $S_{\bar{x}\bar{x}}^{(2)} = 26.14$
$n_1 = 10$ $n_2 = 17$ $n_1 = 10$ $n_2 = 3$
$s_1^2 = 1.34$ $s_2^2 = 1.63$ $\nu = 20.57$ $s_1^2 = 1.34$ $s_2^2 = 13.07$ $\nu = 2.12$

(a) (c)

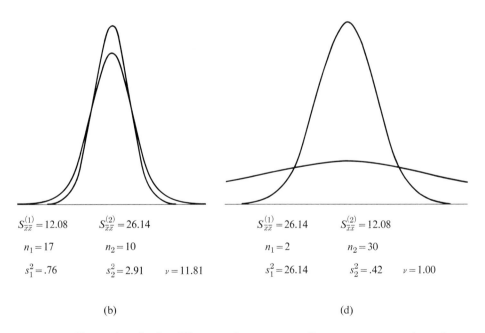

$S_{\bar{x}\bar{x}}^{(1)} = 12.08$ $S_{\bar{x}\bar{x}}^{(2)} = 26.14$ $S_{\bar{x}\bar{x}}^{(1)} = 26.14$ $S_{\bar{x}\bar{x}}^{(2)} = 12.08$
$n_1 = 17$ $n_2 = 10$ $n_1 = 2$ $n_2 = 30$
$s_1^2 = .76$ $s_2^2 = 2.91$ $\nu = 11.81$ $s_1^2 = 26.14$ $s_2^2 = .42$ $\nu = 1.00$

(b) (d)

Fig. 9.1.2 Uncertainty in the difference of means according to two assumptions about the equality of variances.

It seldom makes any difference in your conclusions whether the number of degrees of freedom is 20.57 or 25. Figure 9.1.2(b) is an intermediate case, while (d) is the most extreme.

It is not uncommon to have an estimate of the one variance from more than two sources. The procedure is entirely analogous: A combined estimate of a variance from three separate estimates is

$$\frac{S_{\bar{x}\bar{x}}^{(1)} + S_{\bar{x}\bar{x}}^{(2)} + S_{\bar{x}\bar{x}}^{(3)}}{n_1 - 1 + n_2 - 1 + n_3 - 1} = \frac{\nu_1 s_1^2 + \nu_2 s_2^2 + \nu_3 s_3^2}{\nu_1 + \nu_2 + \nu_3}.$$

This is often called *pooling the variances*. The number of degrees of freedom for the combined estimate of the variance is $\nu_1 + \nu_2 + \nu_3 = n_1 + n_2 + n_3 - 3$.

Example 9.1.5

$$\mu_1 \sim \text{STU} (* \mid 3, .4, 9), \quad \widehat{\sigma^2} = 10 \times .4 = 4.0,$$
$$\mu_2 \sim \text{STU} (* \mid 5, .7, 15), \quad \widehat{\sigma^2} = 16 \times .7 = 11.2,$$
$$\mu_3 \sim \text{STU} (* \mid 4, .5, 3), \quad \widehat{\sigma^2} = 4 \times .5 = 2.0$$

with the same underlying variance. The combined estimate is

$$\widehat{\sigma^2} = \frac{9 \times 4.0 + 15 \times 11.2 + 3 \times 2.0}{9 + 15 + 3} = 7.8$$

with 27 degrees of freedom.

Although this section is supposed to be about comparing *two* variables, I want to make some remarks about comparing more than two. The matter becomes difficult because

1. you don't know exactly what questions to ask,
2. you tend to formulate the questions after seeing the results,
3. an element of competition enters just because there are so many results.

The subject called *multiple comparisons* deals with these questions and provides a variety of methods, none of which apply very universally. I will show you how one type of question can be answered straightforwardly from the posterior distributions. Since you have GAU tables, I'll use GAU posterior distributions, but the same principle applies to any distribution.

Example 9.1.6. We have three sets of measurements on different quantities, all with the same precision:

$$\mu_1 \sim \text{GAU} (* \mid 0, 1), \quad \mu_2 \sim \text{GAU} (* \mid 1, 1), \quad \mu_3 \sim \text{GAU} (* \mid 2, 1).$$

9.1 Comparing Two Values

Table 9.1.1

t	GAU $(t \mid 0, 1)$	GAU $(t \mid 1, 1)$	gau $(t \mid 2, 1)$	Product	Coefficient	Product
−3.5	.000	.000	.0000			
−3.0	.001	.000	.0000			
−2.5	.006	.000	.0000			
−2.0	.023	.001	.0001			
−1.5	.067	.006	.0009			
−1.0	.159	.023	.0044	.000	1	.000
−.5	.309	.067	.0175	.001	4	.004
.0	.500	.159	.0540	.004	2	.008
.5	.691	.309	.1295	.028	4	.112
1.0	.841	.500	.2420	.102	2	.204
1.5	.933	.691	.3521	.227	4	.908
2.0	.977	.841	.3989	.328	2	.656
2.5	.994	.933	.3521	.326	4	1.304
3.0	.999	.977	.2420	.236	2	.472
3.5	1.000	.994	.1295	.129	4	.516
4.0	1.000	.999	.0540	.054	2	.108
4.5	1.000	1.000	.0175	.017	4	.068
5.0	1.000	1.000	.0044	.004	2	.008
5.5	1.000	1.000	.0009	.001	4	.004
6.0	1.000	1.000	.0001	.000	1	.000
						4.372

$$4.372 \times \frac{.5}{3} = .729$$

What is the probability that the mean which appears to be the largest actually is the largest?

Table 9.1.1 shows the calculation. For every possible value, t, we want

$$P(\mu_3 = t) P(\mu_1 < t) P(\mu_2 < t).$$

When we integrate (add) this over all values of t, we get the probability we want. Now

$$P(\mu_3 = t) \propto \text{gau}(t \mid 2, 1),$$
$$P(\mu_1 < t) = \text{GAU}(t \mid 0, 1),$$
$$P(\mu_2 < t) = \text{GAU}(t \mid 1, 1).$$

So I form the product of these values (obtained from the GAU tables in the back of the book) and use Simpson's rule to integrate. I get

$$P(\mu_3 \text{ actually is the largest}) = .73.$$

PROBLEM SET 9.1

1. In a chemistry laboratory the variance of specific gravity (sg) determinations for watery substances with specific gravities near 1 is known to be .0027. Three determinations of the specific gravity of solution A are 1.16, 1.19, 1.12. Three determinations of the specific gravity of solution B are 1.22, 1.19, 1.23. What is the distribution of $sg_B - sg_A$?

2. In Problem 1, what is the 95% HDR for the difference?

3. Out of 7 dirty sweatshirts washed with the detergent Sunburst, 5 were judged satisfactorily clean by an impartial panel. Out of 13 washed with Radiant, 8 were judged clean. If we rate the cleaning power of the detergents by their action on dirty sweatshirts, and if we have no other information about these detergents, what are the odds that Sunburst is better than Radiant?

4. Here is a probability problem using the same mechanism: One automatic machine bores out cylinders in an engine block while another shapes pistons. Neither machine is perfectly accurate:

    the cylinder holes $\sim$ GAU $(*\mid 3.875, .000015)$,
    the pistons $\sim$ GAU $(*\mid 3.855, .000010)$.

    Pistons are randomly selected and assembled into the cylinders. The fit is deemed satisfactory only if the difference in diameters is at least .010 and at most .030. What fraction of all cylinder-piston combinations can be expected to pass inspection?

5. In Problem 4, if an engine block has four cylinders, what fraction of all engine blocks will have all pistons fitting properly?

6. In late 1966 a report was issued about the effect of radiation on inoperable lung cancer patients. In an experiment, about half the patients were given radiation treatments and about half were merely given placebos (sugar pills). Both groups were given transfusions, antibiotics, or whatever palliatives were needed. The numbers surviving after a year were:

Treatment	Number of cases	Number surviving
Radiation	308	56
Placebos	246	34

    What are the approximate posterior odds that the one-year survival rate of irradiated patients is at least .01 greater than the one-year survival rate of nonirradiated patients?

7. A new but expensive diet ingredient for chickens has been proposed. A farmer is toying with the idea of using it, but he wants to be sure that it will be worth the extra cost. He runs a comparative experiment with 50 chickens as controls and 50 as experimentees. He selects the chickens by a random process to avoid any possible bias. [He might subconsciously want the ingredient to be proved beneficial and select the chickens he thought would benefit most.] Here are the summary figures for the

weight increases during the experimental period:

	Control group	Group given new diet
S_x	127.4	151.5
S_{xx}	363.8	514.2
n	49	50
	[One died of natural causes.]	

Compare the analyses with and without the assumption that the underlying variances are the same.

8. $x \sim$ POIS $(*\,|\,5.0)$, $y \sim$ POIS $(*\,|\,7.5)$. What is the approximate distribution of $y - x$?

9. If four equally precise measurements on different populations all have the same value, what is the probability that the first observation actually came from the population with the greatest mean?

10. The posterior distributions for three different means are identical; a fourth is shifted one standard deviation of the mean higher. All have the same variance. What is the probability that the mean which appears largest actually is largest?

ANSWERS TO PROBLEM SET 9.1

1. $\bar{x}_A = 1.157$, $\bar{x}_B = 1.213$

   $sg_A \sim$ GAU $(*\,|\,1.157, .0009)$, $sg_B \sim$ GAU $(*\,|\,1.213, .0009)$

   $sg_B - sg_A \sim$ GAU $(*\,|\,.056, .0018)$

4. diff $\sim$ GAU $(*\,|\,.020, .000025)$

   $P(.010 < \text{diff} < .030) =$ GAU $(.030\,|\,.020, .000025) -$ GAU $(.010\,|\,.020, .000025)$
   $= .9545$

6. Here P_{rad} stands for the patients who received radiation treatments and P_{pla} for the patients who received placebos.

   $p_{\text{rad}} \sim$ BETA $(*\,|\,56, 252)$, $\qquad p_{\text{pla}} \sim$ BETA $(*\,|\,34, 212)$,

   $E(p_{\text{rad}}) = .184$ $\qquad\qquad\qquad\qquad E(p_{\text{pla}}) = .141$,

   $D^2(p_{\text{rad}}) = .000483$ $\qquad\qquad\qquad D^2(p_{\text{pla}}) = .000563$,

   $E(p_{\text{rad}} - p_{\text{pla}}) = .043$, $\qquad\qquad\quad D^2(p_{\text{rad}} - p_{\text{pla}}) = .001046$,

   $P(p_{\text{rad}} - p_{\text{pla}} > .01) \doteq 1 -$ GAU $(.01\,|\,.043, .001046) = .84$,

   $\mathcal{F}\left(\dfrac{p_{\text{rad}} - p_{\text{pla}} > .01}{p_{\text{rad}} - p_{\text{pla}} < .01}\right) \doteq \dfrac{.84}{.16} = 5.3, \qquad \mathcal{O} = ?$

   This calculation assumes that the outcomes of the two treatments each were a series of Bernoulli trials on a random selection of patients from the population of inoperable lung cancer patients available at the hospitals in question. It is very difficult—to say the least—to arrange such a setup in medical trials. The persons running the experiment often have a stake in, or at least a strong personal opinion about, the outcome of the experiment, and unconscious bias may play a part. Were the patients above selected at random? Were the patients themselves allowed to choose their treatment or to decline the treatment offered? Were those given radiation the ones the doctors felt would benefit from the treatment and the others regarded as more or less hopeless? Money spent in such cases is no guarantee of proper methods. The $5,000,000 Salk polio vaccine trials in 1954 were poorly planned in several ways; Brownlee (1955) discusses it.

In medical experiments with drugs, it is the practice to do the experiment *double-blind* if possible; that is, neither the patient nor the attending doctor knows which of the two drugs is being given. [In *blind* experiments the patient does not know.] In *radiation* versus *placebo*, everybody knows what is happening, and psychological responses may have an effect. It is important that the subconscious mechanisms of the body also be treated equally. When an operation is performed on laboratory animals with some part of the brain being destroyed to see what debility results, the control animals are also operated on so that surgical shock does not constitute a difference between control and experimental subjects.

It has been most statisticians' experience that a "haphazard" grab or assignment tends to pick things more alike than the general population under study. Thus the principle of actual mechanical *randomization* seems obvious. Randomization certainly helps avoid unconscious biases and consequently gives added confidence to the person who will use the conclusions. *Underlying factors that you don't know about may still throw you off*, but you have a little hope of averaging them out. So there is a "but": if obvious classes are formed by chance—like all males in the experimental group and all females in the control group—you had better randomize again. We are trying desperately to prevent these other variables from having an effect.

Thus our conclusion is conditional. If there was random choice and Bernoulli trial conditions prevailed, we get a factor of 5+, as stated. What the odds are, I don't know; there is plenty of other information on this subject. But if the experimental selection was not really random, you can conclude practically nothing.

7. The usual summary statistics are

	Control group	Group given new diet
$\bar{x}$	2.60	3.03
$S_{\bar{x}\bar{x}}$	32.6	55.2

(a) If we assume that the underlying variances are the same,

$$\widehat{\sigma}^2 = \frac{32.6 + 55.2}{48 + 49} = .905, \quad .905(\tfrac{1}{49} + \tfrac{1}{50}) = .0366,$$

$$\text{new} - \text{old} \sim \text{STU}\,(*\,|\,.43, .0366, 97).$$

(b) If we know nothing about the underlying variances,

$$u_1 = \frac{32.6}{48 \times 49} = .0139, \quad u_2 = \frac{55.2}{49 \times 50} = .0225,$$

$$u_1 + u_2 = .0364, \quad b = \frac{.0139}{.0364} = .38,$$

$$\frac{1}{\nu} = \frac{.1444}{48} + \frac{.3844}{49} = .01086, \quad \nu = 92,$$

$$\text{new} - \text{old} \; \dot\sim \; \text{STU}\,(*\,|\,.43, .0364, 92).$$

Thus there is essentially no difference in the conclusion. That's not the end of the problem for the farmer, however. He must decide whether the weight gain pays for the increased cost of the feed. Or, rather, it ought to do a little better than just compensate for the increased cost of the feed; there's always a cost—psychological or procedural—in changing over to something new.

10. This is equivalent to having three distributions

$$\text{GAU}\,(*\,|\,0, 1)$$

and one

$$\text{GAU}\,(*\,|\,1, 1).$$

We want to integrate $[\text{GAU}\,(t\,|\,0, 1)]^3 \, \text{gau}\,(t\,|\,1, 1)$. By a calculation similar to that of Table 9.1.1, we get a probability of .551.

9.2 INCREASING EFFICIENCY

In this section I'll talk about a few ways to increase the efficiency of your experiments. I refer to the random sampling efficiency, to reducing the *sampling error*. In any large experiment you know at the outset that part of your money and time is going to have little to do with what is usually thought of as statistical error. You will be spending money getting your instruments in shape, or, in survey sampling, training your investigators. You are going to spend time planning what observations to make or what questions to ask. Perhaps you will plan for spot checks here and there by an independent method just to make sure you're not deceiving yourself. Survey people often assign slightly overlapping areas to two interviewing teams just to get a check on them (*interpenetrating samples*).

The data must be processed, probably punched on cards for entry into a computer. And then you must check that the punching is correct. All this contributes nothing to the reduction-of-sampling-error column. Yet you know that if you don't do it, there will be systematic error, and you'll be hard-pressed to compensate for it. This transcribing-and-punching process is always a pain in the neck. The Census Bureau knows that it misses about $2\frac{1}{2}\%$ of the population anyway, but punching errors cause additional trouble. Coale and Stephan (1962) tell an amusing tale of tracing some glaring discrepancies in the census data to cards getting punched in the wrong column. You're lucky if you have glaring discrepancies. It's the errors you don't notice and don't correct that will get you in trouble.

All this is statistics in the sense that you have to assure yourself that you're measuring what you set out to measure. The safeguards you use depend specifically on the problem, and I want to talk about ways of reducing the *sampling error* when these other items are under fair control. I'm referring to *planning* measures, of course. Before you begin an experiment you have only an expectation of how things will go. [If you are able to experiment sequentially, you may be able to modify your plans somewhat after each step. Always taking into account the newest information should produce better results.]

First, if you are investigating the difference between two populations, you may have the time or money only for a certain number of samples. You would like to know how to divide up the samples between the populations. The answer is to allocate them *in proportion to the standard deviations of the population*. This minimizes the expected variance. If you don't know the variances, this isn't much help, but you may be able to get some information from pilot samples.

Example 9.2.1. You want to minimize the variance of $\bar{x} - \bar{y}$ when $\sigma_x = 20$ and $\sigma_y = 30$. If you can make only 100 observations, make 40 of them on X and 60 of them on Y. You can see intuitively that more observations should be allocated to the variable with the larger variance in order to get the variance of its mean down to a comparable size. Allocation in proportion to the standard deviations does it just right. [Please note that this is *not* the problem we discussed in Chapter 6 concerning the weighting of two *already-obtained estimates* of the *same* variable. There we weighted them inversely proportional to the variances. Here the problem is *planning* the number of observations to divide between two or more variables.]

In stratified sampling you already know that the population can be subdivided into more homogeneous groups or *strata*. When you have just a limited number of observations to make, how many should you make on each stratum? The problem is basically the same as before. If stratum i has N_i members and the standard deviation of an observation is σ_i, the number of observations on that stratum should be proportional to $N_i \sigma_i$. If the σ_i's are not too different (or you don't know them), a good first guess is to make the allocation in *proportion to the size of the stratum*.

Example 9.2.2. The population of a town is stratified as follows:

$$\begin{array}{rr} \text{apartment residents:} & 10{,}000, \\ \text{residents of multiple-family homes:} & 7000, \\ \text{residents of single-family homes:} & 20{,}000. \end{array}$$

If you are making an opinion survey and will question 1000 people altogether, allocate the 1000 in proportion to the sizes, 10,000, 7000, and 20,000; that is, 270 : 190 : 540. If the percentages favoring the question are not too different between the strata, this allocation gives minimum variance of the weighted estimate.

Sometimes you want to obtain estimates of several quantities, and there is absolutely no interaction between the quantities. An example is weighing, where in most cases the sum of the weights is the weight of the sum. Experimental designs which exploit this characteristic are called *weighing designs;* they can achieve substantial reduction in variance.

Example 9.2.3. You have a balance and want to determine the weights of A and B. The variance of a weighing is the same whether one or both objects are placed in either pan. What is the most efficient way to utilize two weighings?

If

1. you weigh A and get a, with $D^2(a) = \sigma^2$,
2. you weigh B and get b, with $D^2(b) = \sigma^2$,

then

$$\hat{A} = a \quad \text{and} \quad D^2(\hat{A}) = \sigma^2,$$
$$\hat{B} = b \quad \text{and} \quad D^2(\hat{B}) = \sigma^2.$$

But if

1. you weigh $A + B$ and get c, with $D^2(c) = \sigma^2$,
2. you weigh $A - B$ and get d, with $D^2(d) = \sigma^2$,

then

$$\hat{A} = \frac{c+d}{2} \quad \text{and} \quad D^2(\hat{A}) = \tfrac{1}{4}D^2(c) + \tfrac{1}{4}D^2(d) = \frac{\sigma^2}{2},$$

$$\hat{B} = \frac{c-d}{2} \quad \text{and} \quad D^2(\hat{B}) = \tfrac{1}{4}D^2(c) + \tfrac{1}{4}D^2(d) = \frac{\sigma^2}{2},$$

and the variance of the estimates is halved.

In biology many animal characteristics fluctuate wildly. The difference between the effects of two different drugs can easily be obscured by the natural variation in the animals. If it is possible, and it sometimes is, to test both drugs (or drug and no-drug) on the same animal, a lot of the variation is eliminated. You then are working with a single variable—the difference in response of a single animal— and end up directly with one STU distribution. This approach avoids a few of the hazards encountered when you randomly select some animals to be controls and some to be experimental subjects.

This scheme of *paired comparisons* cannot be used if the test for one variable affects the result of the test for the other. The animal may change significantly in the time interval between the two tests, or the first test may leave some residual effects. The effects may even vary with the order of administration. Naturally, if the first test is destructive—the animal dies or the part under examination is destroyed—this method is inapplicable.

You may be able to salvage some of the benefits of paired comparisons if you can match two animals or materials, characteristic for characteristic, in everything that is likely to count. A comparison of effects between such pairs is very close to the ideal paired comparison. Working with identical twins is best. [See Student (1931) for a discussion of how using a few identical twins in a famous experiment would have made it unnecessary to use thousands of ordinary subjects.] Sometimes, as in the studies of the relationship between smoking and lung cancer, you have to settle for matching age, weight, residence, etc.

Example 9.2.4. We are investigating how much a certain drug depresses systolic blood pressure in gerbils. We measure blood pressure before and after application of the drug in each animal:

Animal	Before	After
Harpo	91	84
Zeppo	63	55
Groucho	109	99
Liberté	76	70
Egalité	59	49
George	77	72

To find the posterior distribution of the change in blood pressure induced by the drug, we first compute the difference for each animal. The *reductions* in pressure are 7, 8, 10, 6, 10, 5.

$$S_x = 46, \quad S_{xx} = 374, \quad \bar{x} = 7.7, \quad S_{\bar{x}\bar{x}} = 21.3,$$
$$\text{reduction} \sim \text{STU}\, (* \mid 7.7, .71, 5).$$

This is a statement about our knowledge of the mean reduction in blood pressure. If someone asked, "What is the probability that in another animal the reduction would be more than 10?" we would calculate

$$1 - \text{STU}\,(10 \mid 7.7, 4.97, 5) = 1 - \text{STU}\,(1.03 \mid 0, 1, 5) \doteq .18.$$

[Here $4.97 = .71 \times 6 \times \tfrac{7}{6} = s^2/n \times n \times (n+1)/n$.] We also have to remember the following usual assumptions.

1. The underlying variance was constant.
2. We had no prior knowledge of mean or variance.
3. The initial taking of blood pressure had no permanent effect on the little beasties. [Their psychology is beyond my ken.]

When you compare analyses run within a laboratory with analyses made in another laboratory, you are often startled by the differences. The National Bureau of Standards devotes great efforts toward getting measurement techniques standardized, but the situation improves slowly. Some things are difficult to observe; some observations are limited by technology; but other differences are inexplicable. McGuire, Spangler, and Wong (1961) tell the story of how the measurement of the distance from the earth to the sun increased in accuracy as better instrumentation was developed. On the other hand, there is considerable evidence that some scientific results are passed down through the years with uncritical acceptance. For a long time biology texts stated that humans have 48 chromosomes. This is a difficult count to make, but no one complained about it until rather recently; then one man said he had found only 46. Lots of people rushed to count them again, and, sure enough, there are only 46. It seems doubtful that the genetic structure of man changed suddenly! Bridgman (1960) gives an interesting story:

> ...in a new field in which the foundations are not well laid, human psychology may play a surprisingly large role. Rutherford used to enjoy telling how the early measurements of e, the charge on the electron, clustered around the value 3×10^{-10}, until he made the psychological break-through by publishing the value 4.8×10^{-10}, around which all subsequent values have clustered.

It is also well known that there are differences within a laboratory as well as differences between laboratories. Some experimenters are more careful than others, some equipment is better, and human reaction times differ. All this has to be accounted for before you publish results. It is a good practice to try to report separately your sampling errors and any estimates of corrections for possible systematic errors.

PROBLEM SET 9.2

1. We have
$$X \sim \text{GAU}(*\mid \mu_x, 16), \qquad Y \sim \text{GAU}(*\mid \mu_y, 25).$$
We want to spend 9 observations measuring the difference between μ_x and μ_y. How should we allocate them to observations on X and observations on Y?

2. In Problem 1, calculate the variance of $\bar{x} - \bar{y}$ for all allocations from (8, 1) to (1, 8) and graph the result.

3. The popular mayor of Exville hires a public-opinion specialist to see whether his appeal is waning. Six months ago 71% of the voters interviewed approved of his actions:

	Approved	Disapproved
Democrats (60% of voters)	75%	25%
Republicans (40% of voters)	65%	35%

$$.71 = .60 \times .75 + .40 \times .65$$

There are 100,000 voters in Exville. The office of mayor is a political one, so the pollster believes stratification by political party may give improved precision. If n_D Democratic voters are interviewed and n_R Republicans are interviewed, what is an estimate of the variance of the calculated overall approval?

4. In Problem 3, we can minimize $D^2(p)$ by taking

$$n_D \propto \sqrt{.0675} = .260, \qquad n_R \propto \sqrt{.0364} = .191.$$

If we interview n people altogether, what are n_D and n_R under this allocation?

5. *Continuation of Problem 4.* How large must n be for the expected variance of $\hat{p}$ to be about .0001. [This makes the standard deviation about 1%.]

6. *Continuation of Problem 3.* If the allocation was made in proportion to the size of the strata, what would the variance of the estimate be?

7. Calculate the variances of the estimates in these two weighing designs; we are trying to find the weights of A, B, C, and D:

(a) Weigh A, B, C, and D four times each (16 weighings in all).
(b) Weigh

$$A + B + C + D \quad \text{once,}$$
$$A + B - C - D \quad \text{once,}$$
$$A - B + C - D \quad \text{once,}$$
$$A - B - C + D \quad \text{once}$$

(4 weighings in all).

Assume that the variance of every weighing is the same.

8. Fields are notorious for differences in fertility. When trying new strains of grain, one is careful to make paired comparisons in homogeneous areas. [Such areas are called *blocks*.] Suppose we got these yields:

Block	Grain strain 1	Grain strain 2
1	115	118
2	75	74
3	72	76
4	90	92
5	121	126

Analyze this information in two ways:
(a) using paired comparisons,
(b) neglecting your knowledge of the homogeneity of the block structure and treating the two sets of five results separately.

Graph the two posterior distributions on the same axes for comparison.

9. Suppose for the yield Y in a block we have

$$Y = \text{effect of field} + \text{effect of grain}$$

and

$$\text{effect of grain} \sim \text{GAU}(* \mid a, \sigma^2),$$
$$\text{effect of field} \sim \text{GAU}(* \mid b, \tau^2)$$

independently, and neither σ^2 nor τ^2 is known. What are $E(Y)$ and $D^2(Y)$?

10. Suppose we raise n paired crops of grains 1 and 2 under the conditions of Problem 9. The mean yields are a_1 and a_2, and each has variance σ^2. What are E and D^2 of the estimated difference in yield when we do and do not assume the homogeneity of the areas of the field?

ANSWERS TO PROBLEM SET 9.2

1. 4 to X and 5 to Y

3. $\hat{p} = .6\hat{p}_D + .4\hat{p}_R$. If we estimate the variance on the basis of the last returns,

$$D^2(\hat{p}) \doteq .36 \times \frac{.75 \times .25}{n_D} + .16 \times \frac{.65 \times .35}{n_R}$$

$$= \frac{.0675}{n_D} + \frac{.0364}{n_R}.$$

5. $D^2(\hat{p}) = \frac{.0675}{.577n} + \frac{.0364}{.423n} = \frac{.2031}{n} = .0001.$ n should be about 2031.

7. The variance of the mean of each of the first results is $\sigma^2/4$. With the second method,

$$\hat{A} = \tfrac{1}{4} \text{ (sum of four weighings)}$$

and

$$D^2(\hat{A}) = \frac{\sigma^2}{4}, \text{ etc.}$$

We achieve the same variance with only one-fourth as many observations.

8. *Using paired comparisons*, we analyze the five differences

$$3, -1, 4, 2, 5$$

of yield 2 over yield 1. This produces

$$\mu_2 - \mu_1 \sim \text{STU}(* \mid 2.6, 1.06, 4).$$

Using the two sets of measurements separately and the Welch approximation, we get

$$\mu_2 - \mu_1 \mathrel{\dot\sim} \text{STU}\,(*\,|\,2.6,\,225.3,\,8).$$

If we don't pair the comparisons, the great variation in field fertility masks the measurement of the difference in the grains.

10. If we pair,

$$E = a_2 - a_1, \qquad D^2 = \frac{2}{n}\sigma^2.$$

If we don't pair,

$$E = a_2 - a_1, \qquad D^2 = \frac{2}{n}(\sigma^2 + \tau^2).$$

9.3 STRAIGHT-LINE ANALYSIS

A straight-line relationship between two variables lets us summarize the information conveniently and predict values of one variable from the observation of the other. It is common experience that such relationships exist:

> Change in magnitude of a variable star versus logarithm of period.
> Velocity of nerve propagation versus fiber diameter.
> Elongation of a spring versus applied force.

The straight line may be so clear-cut that analysis hardly seems worth while; at other times the line is not so apparent. In any case, the linear relationship usually lasts only over a limited interval; extrapolation to values outside the experimental range can be dangerous.

Example 9.3.1. Kovarik (1910) studied the absorption of β-particles by thin pieces of aluminum. [This was back in the days when radioactive disintegration was relatively new.] In Fig. 9.3.1 the right-hand portion of the diagram shows results which are well explained by a straight line. At the left, however, the line seems to curve up. Kovarik knew that his radioactive source was a mixture of Radium D (Pb^{210}) and Radium E (Bi^{210}), with the latter dominating. Thus he could easily explain the deviation from a straight line of thin pieces of metal as being due to the weaker radiation from the RaD. The straight line was a satisfactory approximation only for the thicker pieces.

It is important to realize that the statistical analysis you carry out again depends rather strongly on the model of the relationship that you assume. Some possibilities are:

a) One variable is relatively errorless compared with the other one. We want to use the first variable to predict or control the second. This is *linear regression*.

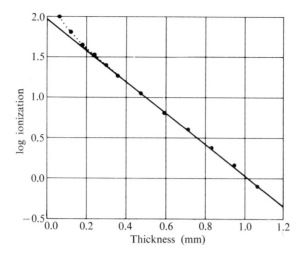

Fig. 9.3.1 Absorption of β-particles. Adapted with permission from A. F. Kovarik (1910), "Absorption and reflexion of the β-particles by matter." *Phil. Mag.*, S6, **20**, 856.

Ionization	log ionization	Thickness (mm)
100.9	2.00	.059
65.9	1.82	.118
46.3	1.67	.177
34.6	1.54	.236
25.7	1.41	.295
19.4	1.29	.354
11.6	1.06	.472
6.70	.83	.590
4.10	.61	.708
2.41	.38	.826
1.52	.18	.944
.80	−.10	1.062

b) Both variables have appreciable error. We believe some strict linear relationship exists between them and want to find it. This is sometimes called the problem of a *linear functional relationship*.

c) The two variables are the simultaneously observed characteristics from a bivariate distribution. We know the type of bivariate distribution and want to estimate the parameters. This is the *bivariate* or *correlation* case.

Here I'll talk only about model (a), linear regression. See Acton (1959) for an introduction to (b), (c), and still other models.

In physics and engineering there are some true linear functional relationships, but regression is a more common situation in other fields. We know, for example, that the vitamin content of a diet helps determine the luster of a mink's coat, but it does not account for all the variation. The academic record of a pupil in high school gives a good indication of what his college record will be, but it cannot predict it exactly. The temperature in a chemical system is one of the variables controlling the speed of a reaction, but it is not the only one. Sometimes the other factors are really unknown, and we cloak our ignorance by invoking random variation; sometimes we are interested in seeing how well a simple model will do. Our aims are:

a) to find out how much of the variation can be accounted for by knowledge of one other variable,

b) to determine the coefficients of the line and their uncertainties,

c) to measure the uncertainty of a future prediction.

I'll treat only the simplest case. Call the predict*or* or controll*ing* variable X and the predict*ed* or controll*ed* variable Y. [The terms *independent variable* for X and *dependent variable* for Y are also used; they are confusing since they have little to do with the notion of statistical independence and dependence.]

a) We have a number of X-values; they are known exactly, or, at least, exactly compared with the Y-values; they may have been chosen systematically or arbitrarily or by some random process; they may also repeat. All that matters is that they be known. Thus we can set our instrument on exact scale divisions, or pick values from a random number table, or let our process proceed until our cup of tea is finished. In Example 9.3.1, the experimenter obviously had some aluminum foil .059 mm thick and put layer after layer together to achieve a variety of thicknesses.

b) For each X-value there is a Y-value. We assume that the distribution of Y depends on the associated value of X in a special way:

$$E(Y) = a + bX,$$

with the variance of Y *constant for all values of* X. See Fig. 9.3.2.

First, we have to get estimates for a and b, the parameters of the line. The usual procedure is to choose them so that

$$\sum [y_i - (a + bx_i)]^2$$

is a *minimum*. This is the method of *least squares*. Each $[y_i - (a + bx_i)]^2$ is the square of the *vertical distance* from the line we are drawing; the "best" line is determined by the condition that the sum of the squares of the vertical distances between observations and the line be a minimum. See Fig. 9.3.3. You may be a

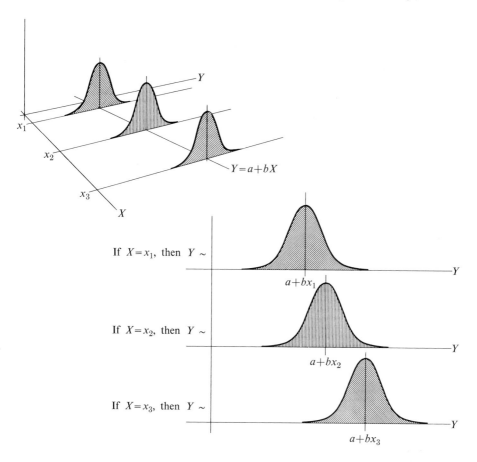

Fig. 9.3.2 The distribution of Y is conditional on the value of X.

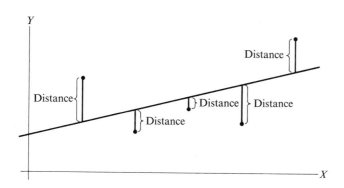

Fig. 9.3.3

little surprised that the distances are measured vertically rather than along perpendiculars to the line. The reason is that our assumptions placed all the variation in the Y's and none in the X's. Therefore we measure only in the direction of the Y's.

Least squares is often put forward as a basic principle. It certainly is the choice which minimizes the expected squared error, just as our choice of $\hat{\mu} = \bar{x}$ did in the univariate case. You can also immediately get estimates of the variance of these estimates. But neither estimates nor variances of estimates have probabilistic meaning until a probability distribution is assumed.

Not necessary, but convenient, is the assumption that

$$Y \sim \text{GAU}\,(* \mid a + bX, \sigma^2),$$

with σ^2 a constant. When we assume no prior knowledge, that is

$$\text{pd}_0\,(a) = \text{FLAT}, \qquad \text{pd}_0\,(b) = \text{FLAT}, \qquad \text{pd}_0\,(\sigma) = \text{LFLAT},$$

the sum of the squares

$$\sum [y_i - (a + bx_i)]^2$$

becomes the main part of the exponent in the joint posterior probability density of the variables:

$$\text{pd}\,(a, b, \sigma \mid \mathbf{x}, \mathbf{y}) \propto \frac{1}{\sigma^{n+1}} \exp \left\{ -\frac{1}{2\sigma^2} \sum [y_i - (a + bx_i)]^2 \right\}.$$

Minimizing the sum of the squares means finding the *maximum* of the posterior density because

$$\exp\,(-\text{minimum}) = \text{maximum}.$$

By now you're quite familiar with the analysis of GAU distributions, so I'll just exhibit the solution. I'll also try to point out analogies with things we've seen before to make it easier to understand. The notation will be an extension of what we used earlier:

$$S_x = \sum x, \qquad S_y = \sum y,$$
$$S_{xx} = \sum x^2, \qquad S_{\bar{x}\bar{x}} = \sum (x - \bar{x})^2,$$
$$S_{yy} = \sum y^2, \qquad S_{\bar{y}\bar{y}} = \sum (y - \bar{y})^2,$$
$$S_{xy} = \sum xy, \qquad S_{\bar{x}\bar{y}} = \sum (x - \bar{x})(y - \bar{y}),$$
$$S_{EE} = \sum \text{residual}^2 = \text{least squares}.$$

Then we have

$$\hat{b} = \frac{S_{\bar{x}\bar{y}}}{S_{\bar{x}\bar{x}}}.$$

When $\hat{b} > 0$, the least-squares line runs SW–NE; when $\hat{b} = 0$, it is horizontal; when $\hat{b} < 0$, it runs NW–SE.

$$\hat{a} = \bar{y} - \hat{b}\bar{x}, \qquad \widehat{\sigma^2} = \frac{S_{EE}}{n-2},$$

where

$$S_{EE} = S_{\tilde{y}\tilde{y}} - \frac{S_{\tilde{x}\tilde{y}}^2}{S_{\tilde{x}\tilde{x}}}.$$

We have $n - 2$ degrees of freedom, since two location constants have been estimated. Formerly just one location constant ($\bar{x}$) was estimated, and we had $n - 1$ degrees of freedom.

The marginal distributions are

$$a \sim \text{STU}\left(* \mid \hat{a}, \frac{S_{EE}}{n(n-2)} \frac{S_{xx}}{S_{\tilde{x}\tilde{x}}}, n-2\right),$$

$$b \sim \text{STU}\left(* \mid \hat{b}, \frac{S_{EE}}{n-2} \frac{1}{S_{\tilde{x}\tilde{x}}}, n-2\right),$$

$$\sigma \sim \text{IGAM}\left(* \mid \sqrt{\frac{S_{EE}}{n-1}}, n-2\right).$$

Many old friends appear! There's nothing intuitive about the forms of the estimates or variances of a and b, but they have STU distributions as you would expect. And σ again has an IGAM distribution. The number of degrees of freedom is $n - 2$, as I remarked above.

Example 9.3.2. My X and Y actually have the relationship

$$E(Y \mid X) = 10 + 1.2X.$$

For my own convenience I make observations at

$$X = 1, 2, 3, 4, \text{ and } 5.$$

If Y had no random variation I would see the values

$$Y = 11.2, 12.4, 13.6, 14.8, \text{ and } 16.$$

However, I postulate that there is Gaussian variation about the mean values with $\sigma^2 = 6$. I actually observe

$$y = 14.1, 11.6, 13.2, 13.2, \text{ and } 17.2,$$

which are

$$+1.18\sigma, \quad -.31\sigma, \quad -.15\sigma, \quad -.60\sigma, \quad \text{and} \quad +.51\sigma$$

from the mean values (see Fig. 9.3.4).

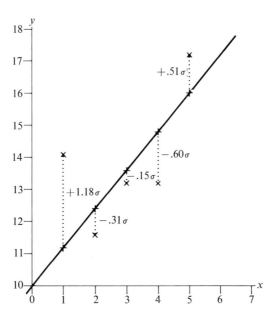

Fig. 9.3.4

This is how the least-squares analysis goes. Do not be surprised that we come out with a line that is "better" than the "true" one. This always happens.

x	y	x^2	xy	y^2
1	14.1	1	14.1	198.81
2	11.6	4	23.2	134.56
3	13.2	9	39.6	174.24
4	13.2	16	52.8	174.24
5	17.2	25	86.0	295.84

$S_x = 15$, $S_y = 69.3$, $S_{xx} = 55$, $S_{xy} = 215.7$, $S_{yy} = 977.69$,
$\bar{x} = 3$, $\bar{y} = 13.86$, $n = 5$,

$$S_{\bar{x}\bar{x}} = S_{xx} - \frac{S_x^2}{n} = 55 - \frac{15^2}{5} = 10,$$

$$S_{\bar{x}\bar{y}} = S_{xy} - \frac{S_x S_y}{n} = 215.7 - \frac{15 \times 69.3}{5} = 7.8,$$

$$S_{\bar{y}\bar{y}} = S_{yy} - \frac{S_y^2}{n} = 977.69 - \frac{69.3^2}{5} = 17.192,$$

$$\hat{b} = \frac{S_{\bar{x}\bar{y}}}{S_{\bar{x}\bar{x}}} = \frac{7.8}{10} = .78,$$

$$\hat{a} = \bar{y} - \hat{b}\bar{x} = 13.86 - .78 \times 3 = 11.52.$$

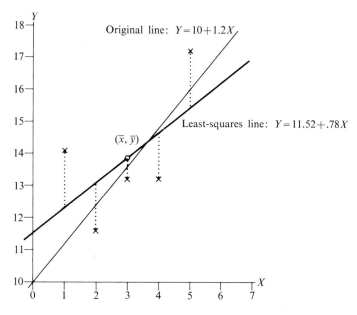

Fig. 9.3.5

Thus the least-squares line is $Y = 11.52 + .78X$. It is also called the *regression line of Y on X*.

Compare the original *y*-values with the *y*-values on this least-squares line:

x	y	$\hat{y} = 11.52 + .78x$	$y - \hat{y}$ = residual	$(y - \hat{y})^2$ = residual²
1	14.1	12.30	+1.80	3.2400
2	11.6	13.08	−1.48	2.1904
3	13.2	13.86	− .66	.4356
4	13.2	14.64	−1.44	2.0736
5	17.2	15.42	+1.78	3.1684
			.00	11.1080

We have the check that

$$S_{\bar{y}\bar{y}} - \frac{S_{\bar{x}\bar{y}}^2}{S_{\bar{x}\bar{x}}} = 17.192 - \frac{7.8^2}{10} = 11.108.$$

Usually we are spared the agony of comparing the "true" line with the least-squares line since we never learn the exact relation! See Fig. 9.3.5.

Note several points:

a) The fitted line goes through $(\bar{x}, \bar{y})$.
b) $\sum \hat{y} = \sum y$; therefore $\sum$ residual $= \sum (y - \hat{y}) = 0$.
c) $\sum \text{residual}^2 = \sum (y - \hat{y})^2 = S_{\bar{y}\bar{y}} - S_{\bar{x}\bar{y}}^2/S_{\bar{x}\bar{x}} = S_{EE}$.

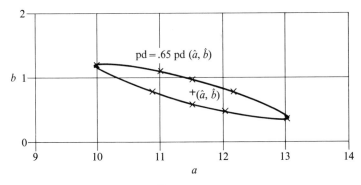

Fig. 9.3.6

These are all useful arithmetic checks. Of course, rounding can produce some inexactness. Since you should always examine residuals, you should compute $\sum \text{residual}^2$ in preference $S_{\hat{y}\hat{y}} - S_{\bar{x}\hat{y}}^2/S_{\bar{x}\bar{x}}$. A computer calculates all this easily and quickly.

The marginal distributions are

$$a \sim \text{STU}(* \mid 11.52, 4.08, 3),$$
$$b \sim \text{STU}(* \mid .78, .371, 3),$$
$$\sigma \sim \text{IGAM}(* \mid 1.67, 3).$$

The joint posterior distribution of a and b is not the product of the marginals; that is, a and b are dependent:

$$\text{pd}(a, b \mid \mathbf{x}, \mathbf{y}) \propto \frac{1}{[S_{EE} + n(a - \hat{a})^2 + 2S_x(a - \hat{a})(b - \hat{b}) + S_{xx}(b - \hat{b})^2]^{n/2}}.$$

If we want to make inferences about the *line*, we have to use this joint distribution.

Example 9.3.3 (*Continuation of Example 9.3.2*). Seven other (a, b)-values which have the same density as the "true line" are:

a (10.00)	b (1.20)
13.04	.36
11.52	.58
11.52	.98
10.87	.78
12.17	.78
12.03	.48
11.01	1.08

These values are plotted in Fig. 9.3.6. You might find it interesting to plot the lines they specify.

The *correlation coefficient* is

$$r = \frac{S_{\bar{x}\bar{y}}}{\sqrt{S_{\bar{x}\bar{x}}S_{\bar{y}\bar{y}}}}.$$

Its square, r^2, tells what portion of the original variance of Y has been accounted for by the least-squares line. The relation is a little deceptive; Table 9.3.1 gives some values. From this we see, for example, that when the correlation coefficient is $\pm.9$, the standard deviation still remaining is 44% of the original.

Table 9.3.1

$|r|$	fitted variance / original variance	fitted s.d. / original s.d.
.0	1.00	1.00
.1	.99	.995
.2	.96	.98
.3	.91	.95
.4	.84	.92
.5	.75	.87
.6	.64	.80
.7	.51	.71
.8	.36	.60
.9	.19	.44
.95	.10	.32
.99	.02	.14
1.00	.00	.00

Example 9.3.4 (*Continuation of Example 9.3.2*)

$$r = \frac{7.8}{\sqrt{10 \times 17.192}} = .595, \quad r^2 = .354.$$

Fitting the least-squares line to the data has removed 35.4% of the original variation. We can check this by

$$\frac{17.192 - 11.108}{17.192} = .354.$$

When the correlation is positive, the observations give the impression of Fig. 9.3.7(a). When the correlation is negative, the observations give the impression of Fig. 9.3.7(b). The closer the correlation is to ± 1, the more the observations resemble a straight line rather than a cloud. [$r = -.999$ in Example 9.3.1.] For algebraic reasons, the correlation coefficient is always between -1 and $+1$.

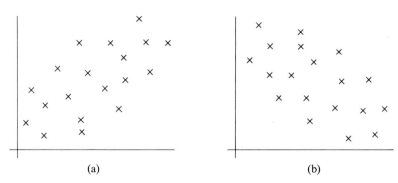

Fig. 9.3.7

An observed correlation different from zero does not imply causality or direct linear relationship; it may just be the result of random fluctuation. At other times, of course, a relationship does exist. It is of great interest to know how strong any particular observed correlation really is. Jeffreys has investigated reasonable alternative distributions for the correlation and has arrived at this factor:

$$\mathcal{F}\left(\frac{\text{uncorrelated}}{\text{correlated}}\bigg|r^2\right) \doteq \sqrt{\frac{2n-1}{\pi}}(1-r^2)^{(n-3)/2}.$$

A factor in favor of correlation does not necessarily mean that the underlying correlation is as large as the observed one. People generally overestimate the strength of observed correlations.

Example 9.3.5 (*Continuation of Example 9.3.2*)

$$n = 5, \qquad r^2 = .354,$$

$$\mathcal{F}\left(\frac{\text{uncorrelated}}{\text{correlated}}\bigg|r^2\right) \doteq \sqrt{\frac{9}{\pi}}(.646)^1 = 1.09.$$

On this evidence it's just about even whether there is any linear relationship between X and Y or not. We know that there actually is—we generated the problem—but the number of observations is quite small.

Example 9.3.6 (*Continuation of Example 9.3.1*)

$$n = 7, \qquad r^2 = .998,$$

$$\mathcal{F}\left(\frac{\text{uncorrelated}}{\text{correlated}}\bigg|r^2\right) \doteq \sqrt{\frac{13}{\pi}}(.002)^2 = \frac{1}{123{,}000}.$$

Obviously strong evidence!

What if we are asked: Here is a new X, call it x_0, one that wasn't used in the regression analysis. What can you predict about the y_0 that is associated with this x_0? If the linear relationship holds over the interval containing the original X-values used in the analysis and the new x_0, the probability density of y_0 is found by the usual compound probability argument. The exclusive and exhaustive possibilities are the various combinations of a, b, and σ; we have to sum (integrate) over all of them:

$$\text{pd}(y_0 \mid \mathbf{x}, \mathbf{y}, x_0) = \iiint \underbrace{\text{gau}(y_0 \mid a + bx_0, \sigma^2)}_{\text{pd }(y_0 \mid \text{line})} \underbrace{\text{pd}(a, b, \sigma \mid \mathbf{x}, \mathbf{y})}_{\text{pd (line} \mid \mathbf{x},\mathbf{y})} da\, db\, d\sigma.$$

It turns out that

$$y_0 \sim \text{STU}\left(* \mid \hat{a} + \hat{b}x_0, \frac{S_{EE}}{n-2}\left[\frac{(x_0 - \bar{x})^2}{S_{\bar{x}\bar{x}}} + \frac{n+1}{n}\right], n - 2\right).$$

The

$$\left[\frac{n+1}{n} \frac{S_{EE}}{n-2}\right]$$

is what we would naturally expect for the distribution of a new observation. The other term shows that things get worse the farther we get from the center of the x's. People warn of the dangers of extrapolation, and quite rightly. Even if the model of a linear relation is correct, the position of the line gets more and more uncertain as $(x_0 - \bar{x})^2$ increases.

Example 9.3.7 (*Continuation of Example 9.3.2*)

$$\sqrt{\frac{11.108}{3}\left[\frac{(x_0 - 3)^2}{10} + \frac{6}{5}\right]} = \sqrt{.37(x_0 - 3)^2 + 4.44}.$$

For convenience, let's let $\sqrt{.37(x_0 - 3)^2 + 4.44} = \Psi$.

$\|x_0 - 3\|$	Ψ	$\Psi \times 2.35$
0	2.11	4.96
1	2.19	5.15
2	2.43	5.71
3	2.78	6.53
4	3.22	7.57
5	3.70	8.69

The last column shows the second column multiplied by 2.35 in order to give the 90% HDR for a STU variable with three degrees of freedom. We have to go this distance on either side of the least-squares line. See Fig. 9.3.8.

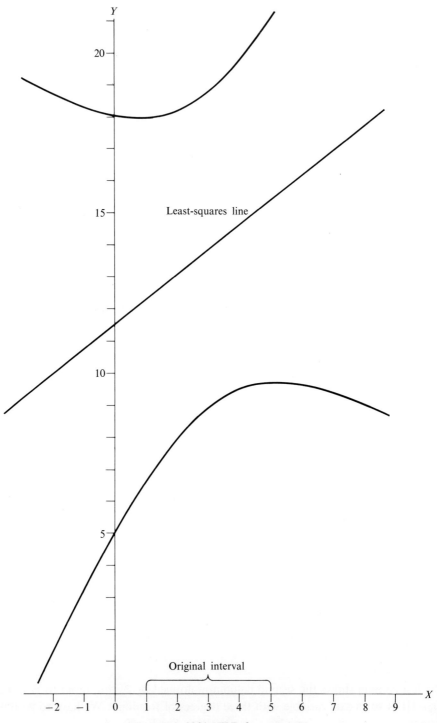

Fig. 9.3.8 90% HDR for a new Y.

Lastly, your regression problem may have arisen like this: You have a very accurate instrument which is slow or expensive to use. You use it (X) to calibrate a faster but less accurate second instrument (Y). Now, by the quicker method, you observe a value y_0. What is the estimate for the true value, x_0—the one that would have been indicated by the accurate instrument? Surprisingly, the Bayesian calculation tells us *not* to solve

$$x_0 = \frac{y_0 - \hat{a}}{\hat{b}},$$

but to do the least-squares calculation *with the variables interchanged*, just as if Y were the errorless variable and X the variable with error! We then predict x_0 using this other least-squares line. The line we get this time is called the *regression line of X on Y*. [This line minimizes the sum of squares of the *horizontal* deviations.]

Example 9.3.8 (*Continuation of Example 9.3.2*). Call the other line

$$X = \alpha + \beta y.$$

We use the statistics we have already calculated in Example 9.3.2 to find

$$\hat{\beta} = \frac{S_{\bar{x}\bar{y}}}{S_{\bar{y}\bar{y}}} = \frac{7.8}{17.192} = .454,$$

$$\hat{\alpha} = \bar{x} - \hat{\beta}\bar{y} = 3 - .454 \times 13.86 = -3.29.$$

This regression line of X on Y,

$$X = -3.29 + .454 Y,$$

is plotted in Fig. 9.3.9, together with the first line we calculated. If we now observe $y_0 = 12$, our estimate of the x_0 corresponding to it is

$$\hat{x}_0 = -3.29 + .454 \times 12 = 2.16 \quad \text{not} \quad \frac{12 - 11.52}{.78} = .62.$$

If we call the new $\sum$ residual^2

$$S_{HH} = S_{\bar{x}\bar{x}} - \frac{S_{\bar{x}\bar{y}}^2}{S_{\bar{y}\bar{y}}} = 6.46,$$

then x_0 has the distribution

$$\text{STU}\left(* \mid \hat{\alpha} + \hat{\beta} y_0, \frac{S_{HH}}{n-2}\left[\frac{(y_0 - \bar{y})^2}{S_{\bar{y}\bar{y}}} + \frac{n+1}{n}\right], n-2\right),$$

which in this case is

$$\text{STU}(* \mid 2.16, 3.01, 3).$$

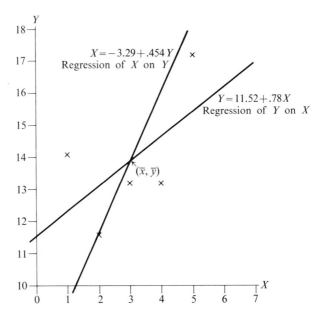

Fig. 9.3.9 The two regression lines.

We get this STU distribution only if we have no prior knowledge of the distribution of the x's. If we do have information—and we might, because the x's we used in the calculation may have been random samples—we have to multiply that prior by this STU density and calculate the expected value of the result.

PROBLEM SET 9.3

1. $n = 10$, $S_x = 56$, $S_y = 3.81$, $S_{\bar{x}\bar{x}} = 82.5$, $S_{\bar{x}\bar{y}} = 5.73$, and $S_{\bar{y}\bar{y}} = .622$. Find and plot the regression lines of Y on X and X on Y. Calculate r.

2. Fit a least-squares line to the right-hand seven points of the Kovarik data (Fig. 9.3.1).

3. What are the marginal distributions of a and b in Problem 2?

4. Produce any consistent set of values for S_x, S_y, $S_{\bar{x}\bar{x}}$, $S_{\bar{x}\bar{y}}$, and $S_{\bar{y}\bar{y}}$, so that, with $n = 10$, you have

$$\hat{a} = 1.0, \quad \hat{b} = -.5, \quad r = -.7.$$

5. What choice of X-values gives the most accurate regression line of Y on X? Is there a drawback to this choice?

6. Suppose that we had the data of Example 9.3.2 but knew that Y was independent of X. What is a 90% HDR for the next observation of Y? Compare with Fig. 9.3.8.

7. The contour lines of pd $(a, b \mid \mathbf{x}, \mathbf{y})$ are ellipses whose axes are not parallel to the X- and Y-axes. This makes them awkward to calculate and contemplate. What can be done to swing them around?

8. If $n = 50$ and $r = -.8$, what is
$$\mathcal{F}\left(\frac{\text{uncorrelated}}{\text{correlated}}\right)?$$

9. If $n = 25$, what is r so that
$$\mathcal{F}\left(\frac{\text{uncorrelated}}{\text{correlated}}\right) = \frac{1}{10}?$$

10. All the analysis so far has assumed that the variation of Y about the line $a + bX$ is independent of X. Sometimes this is not so. How do we handle the case in which, for example, the variation is proportional to X? [That is, $Y \sim \text{GAU}(* \mid a + bX, kX)$.]

11. An experimenter has a record of what fraction of a fly population is killed by what concentration of an insecticide. [The varying doses are administered in separate experiments.] The so-called *probit analysis* may apply. We make the transformations

$$X = \text{logarithm of concentration},$$
$$Y = \text{probit of percentage},$$

and hope that a straight line results. [*Probits* were defined in Section 7.4.] Apply these transformations to the data given below and calculate the least-squares line of Y on X.

Dosage in gm/l	10^{-5}	10^{-4}	10^{-3}	10^{-2}	10^{-1}	1
Percentage killed	6	14	51	81	93	99

Make two plots:
(a) percentage versus concentration,
(b) Y versus X.

12. The time series you see in economics often display highly correlated values. A simple model that explains some of them is that each value depends on the value that precedes it:
$$\text{pd}(y_t \mid y_{t-1}) = \text{gau}(y_t \mid a + by_{t-1}, \sigma^2).$$

The symbol y_t means the value of the economic variable at time t; y_{t-1} means the value at one unit of time before t, say one day before. In Fig. 9.3.10 I've generated and plotted the sequence

$$y_1 = 100, \quad \text{pd}(y_t \mid y_{t-1}) = \text{gau}(y_t \mid 1.1 y_{t-1} - 10, 9).$$

Assume that you have been given the values of Y in the second column of Fig. 9.3.10. How would you estimate a and b?

306 Handling Several Variables 9.3

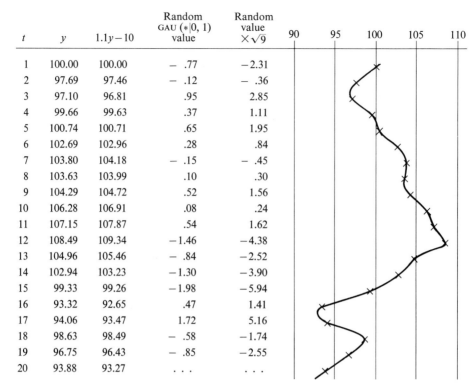

t	y	1.1y − 10	Random GAU (∗|0, 1) value	Random value X√9
1	100.00	100.00	− .77	−2.31
2	97.69	97.46	− .12	− .36
3	97.10	96.81	.95	2.85
4	99.66	99.63	.37	1.11
5	100.74	100.71	.65	1.95
6	102.69	102.96	.28	.84
7	103.80	104.18	− .15	− .45
8	103.63	103.99	.10	.30
9	104.29	104.72	.52	1.56
10	106.28	106.91	.08	.24
11	107.15	107.87	.54	1.62
12	108.49	109.34	−1.46	−4.38
13	104.96	105.46	− .84	−2.52
14	102.94	103.23	−1.30	−3.90
15	99.33	99.26	−1.98	−5.94
16	93.32	92.65	.47	1.41
17	94.06	93.47	1.72	5.16
18	98.63	98.49	− .58	−1.74
19	96.75	96.43	− .85	−2.55
20	93.88	93.27	. . .	. . .

Fig. 9.3.10

ANSWERS TO PROBLEM SET 9.3

2. The thickness of the metal can be measured more accurately than the ionization, so we want the regression line of *log ionization* on *thickness*. Computation gives

$$S_x = 4.956, \quad S_y = 4.25,$$
$$S_{xx} = 3.898720, \quad S_{yy} = 4.0355,$$
$$S_{xy} = 2.25616.$$

From those we derive

$$\bar{x} = .708, \quad \bar{y} = .607, \quad S_{\bar{x}\bar{x}} = .389872, \quad S_{\bar{y}\bar{y}} = 1.4551,$$

and

$$S_{\bar{x}\bar{y}} = -.75284, \quad \hat{b} = -1.931, \quad \hat{a} = 1.974.$$

Let's look at the residuals:

x	y	$\hat{y} = 1.974 - 1.931x$	$y - \hat{y}$	$(y - \hat{y})^2$
.354	1.29	1.290	.000	.000000
.472	1.06	1.063	−.003	.000009
.590	.83	.835	−.005	.000025
.708	.61	.607	+.003	.000009
.826	.38	.381	−.001	.000001
.944	.18	.151	+.029	.000841
1.062	− .10	− .077	−.023	.000529
			.000	.001414

5. We get the best line by putting half the x's at each end of the interval of interest and none in the middle. The drawback is that if we have no observations in the middle, we cannot get any rough check on the linearity of the plot. There is one important advantage, though: We get evidence on the value of σ from *replication* (repeated observations) at fixed values of x; our estimate of σ will not depend on the correctness of the straight-line model.

6. $s^2 = \dfrac{S_{\bar{y}\bar{y}}}{n-1} = \dfrac{17.192}{4} = 4.298$

Then a new observation

$$y_0 \sim \text{STU}\left(* \mid \bar{y}, \frac{n+1}{n} s^2, n-1\right) = \text{STU}(* \mid 13.86, 5.37, 4),$$

90% HDR $= 13.86 \pm 2.13 \times 2.32 = (8.92, 18.80)$.

This coincides with the curves in Fig. 9.3.9 at $X = 3$. Elsewhere it provides a *narrower* interval than that coming from the regression analysis. The more complicated model exacts a real penalty in uncertainty when there are so few observations.

8. $\sqrt{\dfrac{99}{\pi}} (.36)^{47/2} = 2.07 \times 10^{-10} \doteq \dfrac{1}{5{,}000{,}000{,}000}$

10. We want to minimize the sum of *weighted* squares:

$$\sum \left[\frac{y_i - (a + bx_i)}{\sqrt{kx_i}}\right]^2.$$

This is the quantity appearing in the exponent of the likelihood and posterior distribution. There are standard ways of finding a and b by forming and solving the set of *normal equations*. Look these up in any text on regression.

11. Plot (a) should look something like an S-curve. Plot (b) should look something like a straight line. The probit transformation goes:

$$.06 = \text{left tail } (-1.56); \quad \text{probit} = -1.56 + 5 = +3.44,$$
$$.14 = \text{left tail } (-1.08); \quad \text{probit} = -1.08 + 5 = +3.92,$$
$$.51 = \text{left tail } (+ .02); \quad \text{probit} = .02 + 5 = +5.02,$$
$$.81 = \text{left tail } (+ .84); \quad \text{probit} = .84 + 5 = +5.84,$$
$$.93 = \text{left tail } (+1.48); \quad \text{probit} = 1.48 + 5 = +6.48,$$
$$.99 = \text{left tail } (+2.33); \quad \text{probit} = 2.33 + 5 = +7.33.$$

The least-squares analysis goes like this if I take log dosages to base 10:

x	y	x^2	xy	y^2	$\hat{y}$	$y - \hat{y}$
-5	3.44	25	-17.20	11.8336	3.34	$+.10$
-4	3.92	16	-15.68	15.3664	4.14	$-.22$
-3	5.02	9	-15.06	25.2004	4.94	$+.08$
-2	5.84	4	-11.68	34.1056	5.74	$+.10$
-1	6.48	1	-6.48	41.9904	6.54	$-.06$
0	7.33	0	0	53.7289	7.34	$-.01$
-15	32.03	55	-66.10	182.2253		$(-.01)$

$$\bar{x} = -2.5, \quad \bar{y} = 5.34, \quad S_{\bar{x}\bar{x}} = 55 - 37.5 = 17.5,$$
$$S_{\bar{x}\bar{y}} = -66.10 - (-80.08) = 13.98,$$
$$S_{\bar{y}\bar{y}} = 182.2253 - 170.9868 = 11.2385,$$
$$\hat{b} = \frac{13.98}{17.5} = .80, \quad \hat{a} = 5.34 - (-2.5)(.80) = 7.34,$$

Least-squares line: $Y = 7.34 + .80X$.

12. When we try to predict some y_t, we already know y_{t-1}; thus we are in the ordinary regression situation (although this case is called *autoregression*). Remember? At the beginning of this section our x's could be obtained systematically or randomly; all that mattered was that they be *known*. Here we have 19 pairs of observations:

$$y_2 = y_2, \quad x_2 = y_1,$$
$$y_3 = y_3, \quad x_3 = y_2,$$
$$\vdots \qquad \vdots$$
$$y_{20} = y_{20}, \quad x_{20} = y_{19}.$$

We are predicting the next-day's value from the *known* present-day's value.

CHAPTER **10**

A Potpourri of Applications

You've now finished a nine-course dinner, and the last five were mostly large servings of the normal distribution. Whether it be a univariate or a multivariate application, the normal distribution and its variants still play a large role. But there are other distributions around too. For dessert I present you with a sampler of unrelated topics to show the power of statistics and the wide variety of fields to which it is applied. Naturally, whatever the problem, I stick to the familiar Bayesian methods which have already brought us so far. Here and there some neat formulas will emerge, but the numerical techniques will always get you by. Some of my friends do much of their work at an altitude of 30,000 feet, and complain that none of the nine stereo channels on a jet can multiply. I sympathize. When you have no computer, a slide rule and rough calculations may have to do. That's when you appreciate the neat formulas!

310   A Potpourri of Applications   *10.1*

10.1   RARE EVENTS

In a yarn mill, if you choose a foot of yarn at random, you will be unlikely to see a defect; yet in a skein of hundreds of yards you might find several defects scattered around. In a chunk of radioactive material the probability of some particular atom disintegrating in the next second is infinitesimal; yet there are so many atoms that a Geiger counter clicks steadily. The chance of any one person being killed on any one day in an auto accident is quite small; yet the daily death toll in a large state is by no means inconsequential.

These three examples have something in common: There are many opportunities for something to happen, but the chance of something happening on any one of them is quite small. Yet we are interested in periods of time or intervals of space large enough to allow a moderate number of occurrences to take place. If we *assume* that the opportunities are independent and that the same probability applies to each one, we are led to the *Poisson distribution*, POIS.

Suppose we trudge back to our yarn mill and take a number of samples of the output of one machine. Perhaps in every foot of yarn the probability of a knot is the same small number; we'll assume this is close to the truth. Now we rapidly scan the samples, all of which are the same length; we record how many have no knots, how many have one, etc. What is the probability distribution of the number of knots in a skein? I think you would agree that the Bernoulli trials model is relevant:

1) There are two outcomes—knot or no-knot—in any foot.
2) The probability of a knot is the same for each foot.
3) The occurrence of a knot in one foot is independent of what happens in other feet.

If we assume that this is a good approximation, we can go on. Now a knot is a very small thing; it can easily fit into a half-foot interval. If we consider half-feet instead of feet, then n, the number of opportunities for a knot to appear, is doubled, and p, the probability of a knot at any one opportunity, is halved. We could go on to inches or half-inches or smaller. A real knot has some finite size, say, one-eighth of an inch, but conceptually we could keep dividing the interval so

$$n \to \infty \quad \text{and} \quad p \to 0,$$

but

$$np = \lambda = \text{mean number of knots/skein}$$

stays a moderate number. [λ is a Greek "lambda."] If we use this limiting process and say that the probability of having more than one knot in any small interval

10.1   Rare Events

is negligible, we get

$$P(s \text{ knots in a skein}) = \text{pois}(s \mid \lambda)$$
$$= \frac{\lambda^s e^{-\lambda}}{s!}$$

for $s = 0, 1, 2, \ldots$ (no upper limit). The probability is just a function of the average number of knots/skein. In the Gallery we find

$$E[\text{pois}(* \mid \lambda)] = \lambda \quad \text{and} \quad D^2[\text{pois}(* \mid \lambda)] = \lambda;$$

that is, the mean and variance of POIS are the same. This distribution idealizes when we assume that $n \to \infty$ and $p \to 0$, since all phenomena in nature seem to occupy finite time or space. However, it provides a remarkably good description of many interesting situations and is worth a little study.

The classic example of POIS concerns the distribution of the number of deaths in the Prussian Army caused by horses' kicks. The large unit, comparable to a skein of yarn, was a regiment-year; there were, on the average, a few deaths per regiment-year due to this bizarre cause. The reasoning went that there were many opportunities to be kicked to death, but only a small chance of being killed on any one opportunity.

Example 10.1.1. A more appetizing investigation (for me, at least) was reported by Rutherford and Geiger (1910). By watching the scintillations on a screen, they counted the number of α-particles coming from a mass of polonium. They chose a period of 7.5 sec in order to have a reasonable number of counts to handle. [The *Geiger* counter was developed later.] Their results are shown in Table 10.1.1. If we assume that $s \sim \text{POIS}(* \mid \lambda)$, we want to assess our uncertainty about λ, the mean number of counts per 7.5 sec. Suppose that LFLAT is the prior distribution of ignorance for λ. The likelihood is

$$\left(\frac{\lambda^0 e^{-\lambda}}{0!}\right)^{57} \left(\frac{\lambda^1 e^{-\lambda}}{1!}\right)^{203} \left(\frac{\lambda^2 e^{-\lambda}}{2!}\right)^{383} \cdots = \prod_s \left(\frac{\lambda^s e^{-\lambda}}{s!}\right)^{f_s} \propto \lambda^{\Sigma s f_s} e^{-\lambda \Sigma f_s}.$$

If we let
$$\sum s f_s = N = \text{total number of counts} = 10097,$$
$$\sum f_s = T = \text{number of intervals} = 2608,$$

then
$$\text{pd}(\lambda \mid \mathbf{f_s}) \propto \lambda^{N-1} e^{-T\lambda}.$$

N and T are thus sufficient statistics for a POIS variable. When you consult the Table of Kernels you will find that $x^a \exp(-bx)$ characterizes a *Gamma distribution;*

Table 10.1.1

s = number of counts in 7.5 sec	f_s = number of times s was observed
0	57
1	203
2	383
3	525
4	532
5	408
6	273
7	139
8	45
9	27
10	10
11	4
12	—
13	1
14	1
≥ 15	—
	2608 intervals
	10097 counts

$$\bar{s} = \frac{10097}{2608} = 3.872, \qquad \frac{S_{\bar{s}\bar{s}}}{2607} = 3.694$$

in this case it is gamma $(\lambda \mid N, T)$. Hence

$$E[\text{gamma}(* \mid N, T)] = \frac{N}{T} = 3.872 \qquad \text{(our intuitive choice)},$$

$$D^2[\text{gamma}(* \mid N, T)] = \frac{N}{T^2} = .00148, \qquad D = .0385,$$

$$\text{mode}[\text{gamma}(* \mid N, T)] = \frac{N}{T-1} \doteq 3.872.$$

GAMMA is another skew distribution, but it is not as bad as IGAM. I would hazard the guess, then, that a 90% HDR for λ would not be far from

$$3.872 \pm 1.645 \times .0385 = [3.809, 3.935],$$

since the mean is many, many standard deviations from zero. [I used a Gaussian approximation.] When we *fit* a Poisson distribution to the data, we find that N/T is the fitted value for λ.

10.1   Rare Events

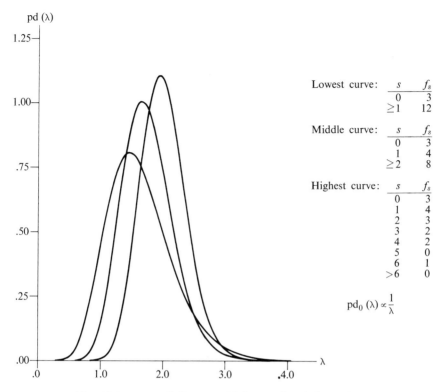

Fig. 10.1.1 POIS: Inference with incomplete information.

In Example 10.1.1 we had the complete distribution of the observed s's. Occasionally this might not be so. Suppose we had been told in a rather minimal report that on 57 occasions the experimenters had detected no α-particles and on 2551 occasions had seen at least one (always referring to a 7.5-sec interval). In this situation no neat formula emerges; we just have to multiply things out. The likelihood would be

$$\left(\frac{\lambda^0 e^{-\lambda}}{0!}\right)^{57} \left(1 - \frac{\lambda^0 e^{-\lambda}}{0!}\right)^{2551},$$

since $P(\geq 1) = 1 - P(0)$. After we multiply this by the prior, which is $\propto 1/\lambda$, we evaluate the product for a scattering of λ-values.

Example 10.1.2. In Fig. 10.1.1 I have worked a small problem to show the effect of different amounts of information. I derived the lowest curve (least information) from knowledge of only f_0 and $f_{\geq 1}$. The middle curve shows my knowledge about λ when I know $f_0, f_1,$ and $f_{\geq 2}$. The highest curve is the one I get when I know the complete distribution of the s's: here it is gamma $(\lambda \mid 30, 15)$,

since
$$3 + 4 + 3 + 2 + 2 + 0 + 1 = 15,$$
$$0 \times 3 + 1 \times 4 + 2 \times 3 + 3 \times 2 + 4 \times 2 + 5 \times 0 + 6 \times 1 = 30.$$

The Poisson distribution is not always applicable to problems like this. Radioactive disintegration seems to be such an ideal case: the number of atoms in a small piece of material is something like 10^{20}, almost infinity! Yet Berkson (1966) has looked at many counts like the one of Rutherford and Geiger and is convinced that the observed variances are almost always slightly less than the means. In a Poisson distribution the two are equal. If this discordant relation should stand up to further investigation, the physicists would have to dig deep for an explanation. I haven't the slightest notion what—if anything—might be wrong here, but in some other situations variances get reduced because dependence exists between the events.

There can also be cases in which the variance is larger than the mean. For example, POIS is definitely not applicable when the probability of the event varies from occasion to occasion. Factory accidents are such a case: some people are more prone to accidents than others. The number of people killed in auto accidents is not ~POIS because deaths tend to occur together: several people from one car may be killed in one accident. Certain assumptions about how accident rates are distributed lead to a form of the *negative binomial distribution*, NEGB, for the number of accidents.

POIS often pops up when counts are made. Consider the problem of dating a rock by the ratio of the abundances of two radioactive elements. Sometimes the abundances are determined by counts: radioactive, mass-spectrometric, or photometric. What do we learn about the ratio from one count on the emission of each element? If the counts are f_1 and f_2, the posterior distributions of the two means are GAMMA $(* \mid f_1, 1)$ and GAMMA $(* \mid f_2, 1)$. We'll let

$$r = \frac{\text{(abundance of element 1)}}{\text{(abundance of element 2)}}.$$

Its distribution is quite skew, but often we can get some information from using

$$\text{mode}(r) = \frac{f_1 - 1}{f_2 + 1},$$

$$E(r) = \frac{f_1}{f_2 - 1},$$

$$D^2(r) = \frac{f_1(f_1 + f_2 - 1)}{(f_2 - 1)^2(f_2 - 2)}.$$

Example 10.1.3. There are 183 emissions from element 1, 874 from element 2. What can we say about the ratio of their abundances? The preceding formulas yield the values

$$\text{mode }(r) = \tfrac{182}{875} = .208,$$
$$E(r) = \tfrac{183}{873} = .210,$$
$$D^2(r) = (183 \times 1056)/(873^2 \times 872) = .00029,$$
$$D(r) = .017.$$

PROBLEM SET 10.1

1. Calculate bin $(s \mid 10, .1)$, bin $(s \mid 100, .01)$, and pois $(s \mid 1)$ for $s = 0, 1$, and 2.
2. On one set of axes plot gamma $(x \mid 10, 4)$ and gamma $(x \mid 1000, 400)$.
3. Calculate pois $(s \mid 3.87)$ for $s = 0, 1, 2, \ldots, 10$. Find $P(\geq 11)$ by subtraction. Multiply each of the probabilities by 2608 to find the expected number of times we would observe s scintillations in the data found by Rutherford and Geiger if $\lambda = 3.87$.
4. During a rabies inoculation drive in a large city, a count was obtained of the number of dogs per dwelling unit:

Dogs/dwelling unit	Dwelling units
0	15779
1	3696
2	347
3	30
4	31
5	11
6	5
7	1
8	1

    Fit a Poisson distribution to this data and evaluate the fit.

5. Suppose you have a population in which

    half the time you draw from POIS $(* \mid 1)$,
    half the time you draw from POIS $(* \mid 1.4)$.

    What are $P(0)$, $P(1)$, $P(2)$? What are the population mean and variance? Is this a Poisson distribution?

6. A geneticist wants to measure what portion of the *E. coli* bacteria in his test tube are penicillin-resistant. He dilutes the culture broth, and puts equal amounts on an ordinary nutrient medium and on a nutrient medium containing penicillin. After incubating the cultures, he counts the number of colonies on each medium, each

colony having developed from one bacterium. There are 512 colonies on the ordinary medium and 7 on the medium containing penicillin. What can he say about the fraction of penicillin-resistant bacteria that were in the test tube when he began his experiment?

7. Linear analysis is not restricted to Gaussian variation. Suppose that the y's are counts, the x's are greater than 0, and the line is known to go through the origin:

$$P(Y \mid X) = \text{pois}(Y \mid bX).$$

How do we estimate b? Do the problem both by analyzing the likelihood and by least squares.

8. A 20 ft by 50 ft plot was divided into 1000 squares. After application of a pesticide, the top inch of soil in each square was examined, and the number of moribund earthworms determined:

x = number dead	Number of squares with x dead
0	514
1	327
2	118
3	31
4	7
5	2
6	1

Calculate the sample mean and variance. Fit a Poisson distribution to the data and show the expected f_x. Discuss the apparent validity of the assumptions required to make a Poisson distribution plausible.

9. A physicist uses a Geiger counter to record the number of particles going through his experimental apparatus. In 5 minutes he counts 2000 particles. If

 (a) the source does not change in intensity,
 (b) the counter counts *all* particles passing through the apparatus,
 (c) the Poisson distribution describes the situation reasonably well,

what is an approximate 95% HDR for the average number of particles emitted by the source in one minute?

ANSWERS TO PROBLEM SET 10.1

3. $\lambda = 3.87$; $e^{-3.87} = .0209 = P(0)$. All the other values can easily be calculated by recursion:

$$\frac{\text{pois}(s \mid \lambda)}{\text{pois}(s-1 \mid \lambda)} = \frac{\lambda^s e^{-\lambda}(s-1)!}{s! \lambda^{s-1} e^{-\lambda}} = \frac{\lambda}{s}.$$

10.1 Rare Events

That is,

$$P(1) = P(0) \times \frac{3.87}{1}, \quad P(2) = P(1) \times \frac{3.87}{2}, \quad \text{etc.}$$

The result is:

s	pois $(s \mid 3.87)$	Expected	Observed
0	.0209	54	57
1	.0807	211	203
2	.1562	408	383
3	.2014	526	525
4	.1949	509	532
5	.1510	394	408
6	.0974	254	273
7	.0539	141	139
8	.0261	68	45
9	.0112	29	27
10	.0043	11	10
≥ 11	.0020	5	6
	1.0000	2608	2608

4. $\lambda = .235$

Dogs/unit	Expected	Observed
0	15715	15779
1	3711	3696
2	438	347
3	34	30
≥ 4	3	49
	19901	19901

The right tail is wagging the distribution excessively. Evidently pairs of dogs lead to 3, 4, or more.

6. We are interested in the ratio of the number that are penicillin-resistant to the total number. Call it r. Therefore

$$E(r) = \tfrac{7}{511} = .0137,$$

$$D^2(r) = \frac{7 \times 518}{511^2 \times 510} = .0000272,$$

$$D(r) = .0052.$$

7. I take $pd_0(b) = \text{LFLAT}$, since b is known to be positive. The likelihood is

$$\prod \left(\frac{(bx_i)^{y_i} e^{-bx_i}}{y_i!} \right) \propto b^{\Sigma y} e^{-b\Sigma x}, \quad \text{so} \quad pd(b) \propto b^{\Sigma y - 1} e^{-b\Sigma x},$$

a GAMMA distribution. In particular,
$$\text{pd}(b) = \text{gamma}(b \mid \sum y, \sum x),$$
so
$$\hat{b} = E(b) = \frac{\sum y}{\sum x}.$$

If we use least squares, we have to take into account the changing variance and minimize

$$\sum \left(\frac{y_i - bx_i}{\sqrt{x_i}}\right)^2 = \sum \left(\frac{y^2}{x}\right) - 2b\sum y + b^2 \sum x$$

$$= (\sum x)\left[b^2 - 2b\left(\frac{\sum y}{\sum x}\right) + \left(\frac{\sum y}{\sum x}\right)^2\right] + \sum \left(\frac{y^2}{x}\right) - \frac{(\sum y)^2}{\sum x}$$

$$= (\sum x)\left(b - \frac{\sum y}{\sum x}\right)^2 + \sum \left(\frac{y^2}{x}\right) - \frac{(\sum y)^2}{\sum x}.$$

Thus again
$$\hat{b} = \frac{\sum y}{\sum x},$$

though we can't analyze the uncertainty until we attribute a distribution to the variables.

9. $400 \pm 2\sqrt{80} = (382, 418)$

10.2 ANGLES

Most of our probability densities have applied to variables having a range from $-\infty$ to $+\infty$. In this section the variable is an angle running from 0° to 360° (0 to 2π radians). But there's no end or beginning; it comes around and sneaks up on itself. Several different densities of this type exist, all described in Batschelet (1965), but I'll only talk about CIRG, the *Circular Normal distribution*. There's a picture of it back in the Gallery.

Where can a distribution like this be used? Wherever angles or cyclic phenomena enter. You may want to describe the orientation of crystal axes, the homing ability of birds, circadian rhythms, color perception, or the isotropy of galactic radio sources. A closely related distribution is used to describe orientation in space.

I will not go deeply into any of these subjects; I will simply show you how to estimate the parameters of CIRG. The only difficulty comes from its bite-its-own-tail feature. Offhand you don't know how to compute a mean because you don't know how to establish an origin. When your observations are as shown in Fig. 10.2.1, those points just above 0° should obviously be called 362° and 370°, not 2° and 10°, if you want to use ordinary arithmetical methods for finding a mean. [Note that here I'm measuring angles *counterclockwise*.]

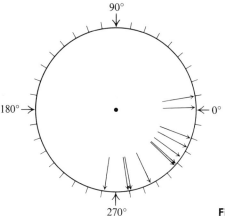

Fig. 10.2.1

The likelihood of a single observation is

$$\text{cirg}\,(\alpha \mid \mu, \kappa) = \frac{e^{\kappa \cos(\alpha - \mu)}}{2\pi I_0(\kappa)}.$$

This looks fearsome, but you can find a tabulation of the density function in Batschelet. The $I_0(\kappa)$ is a *Bessel function;* you don't need to know any of its properties. To some extent the parameter κ plays the role of $1/\sigma^2$. At least, when κ is large,

$$\text{cirg}\,(\alpha \mid \mu, \kappa) \doteq \text{gau}\,(\alpha \mid \mu, 1/\kappa).$$

But this doesn't work for small κ's.

The estimation problem is reasonably straightforward by the usual Bayesian methods. I take

$$\text{pd}_0\,(\mu) \propto 1 \quad \text{and} \quad \text{pd}_0\,(\kappa) = \text{FLAT}.$$

[The choice for $\text{pd}_0\,(\kappa)$ is a compromise; LFLAT can't be used because κ, unlike the σ of GAU, can actually become zero.] The posterior distribution is then

$$\text{pd}\,(\mu, \kappa \mid \boldsymbol{\alpha}) \propto \frac{e^{\kappa \sum \cos(\alpha_i - \mu)}}{I_0^n(\kappa)}.$$

By a trigonometric identity,

$$\sum \cos(\alpha_i - \mu) = R \cos(\mu - \hat{\mu}),$$

where

$$R^2 = \left(\sum \cos \alpha_i\right)^2 + \left(\sum \sin \alpha_i\right)^2$$

and

$$\hat{\mu} = \tan^{-1}\left(\frac{\sum \sin \alpha_i}{\sum \cos \alpha_i}\right).$$

[tan^{-1} x means *the angle whose tangent is x*.] So

$$\text{pd}(\mu, \kappa \mid \boldsymbol{\alpha}) \propto \frac{e^{-\kappa R \cos(\mu - \hat{\mu})}}{I_0^n(\kappa)}.$$

Since the cosine is symmetric about 0, our estimate of μ must be $\hat{\mu}$.

Example 10.2.1. I have nine observations from a CIRG distribution. They are plotted in Fig. 10.2.2. We start the calculation of $\hat{\mu}$ by adding the sines and cosines of the angles:

α°	$\cos \alpha$	$\sin \alpha$
10	.9848	.1736
81	.1564	.9877
17	.9563	.2924
91	−.0175	.9998
29	.8746	.4848
58	.5299	.8480
12	.9781	.2079
36	.8090	.5878
87	.0523	.9986
	5.3239	5.5806

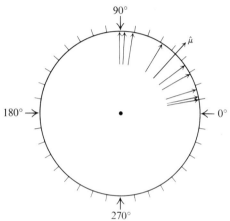

Fig. 10.2.2

$$R^2 = (5.3239)^2 + (5.5806)^2 = 59.4870, \qquad R = 7.713,$$

$$\tan \hat{\mu} = \frac{5.5806}{5.3239} = 1.048, \qquad \hat{\mu} = 46.4^\circ.$$

This is so concentrated a distribution of observations that we might try analyzing it via GAU:

$$\bar{\alpha} = 46.8^\circ, \qquad S_{\bar{\alpha}\bar{\alpha}} = 8752, \qquad \widehat{\sigma^2} = 1094, \qquad \widehat{\sigma_{\bar{\alpha}}^2} = 122 \doteq 11^2,$$

so

$$\mu \overset{\cdot}{\sim} \text{STU}(* \mid 46.8, 122, 8).$$

Example 10.2.2. The nine observations are plotted in Fig. 10.2.3. They are not nearly so concentrated as the ones in Fig. 10.2.2 or Fig. 10.2.1. One might even question whether there was any concentration at all. We carry out the same estimation process for μ:

$\alpha°$	$\cos\alpha$	$\sin\alpha$
183	$-.9986$	$-.0523$
40	.7660	.6428
328	.8480	$-.5299$
244	$-.4384$	$-.8988$
112	$-.3746$	.9272
352	.9903	$-.1392$
245	$-.4226$	$-.9063$
269	$-.0175$	$-.9998$
333	.8910	$-.4540$
	1.2436	-2.4103

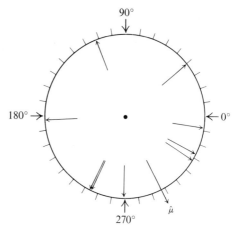

Fig. 10.2.3

$$R^2 = (1.2436)^2 + (2.4103)^2 = 7.3561, \quad R = 2.713,$$

$$\tan\hat{\mu} = \frac{-2.4103}{1.2436} = -1.938, \quad \hat{\mu} = 297°.$$

Estimation of the parameter κ does not go as gracefully. When $\kappa = 0$, cirg is flat around the circle with no directional preference at all. When κ is large, there is a strong directional indication. $E(\kappa)$ must be found numerically, but mode (κ) can be approximated fairly easily. Put $R/n = z$. Then

$$\text{when } z < \tfrac{2}{3}, \quad \hat{\kappa} = \text{mode}(\kappa) \doteq z\left(\frac{2 - z^2}{1 - z^2}\right);$$

$$\text{when } z > \tfrac{2}{3}, \quad \hat{\kappa} = \text{mode}(\kappa) \doteq \frac{z + 1}{4z(1 - z)}.$$

Because of the skewness of the distribution, $E(\kappa) > \text{mode}(\kappa)$.

Example 10.2.3 (*Continuation of Example 10.2.1*)

$$R = 7.713,$$

$$\frac{R}{n} = .857,$$

$$\text{mode}(\kappa) \doteq \frac{1.857}{4(.857)(.143)} = 3.8.$$

Example 10.2.4 (*Continuation of Example 10.2.2*)

$$R = 2.713,$$
$$\frac{R}{n} = .301,$$
$$\text{mode}(\kappa) \doteq .301 \left(\frac{1.91}{.91}\right) = .63.$$

Example 10.2.5. In Example 10.2.1 I was able to use the GAU distribution to get some idea of the precision of $\hat{\mu}$. Can I do anything to find the precision of $\hat{\mu}$ in Example 10.2.2? Of course; I calculate the joint probability density very roughly, scaling the value at $\kappa = 0$ to be 1.0:

		\multicolumn{5}{c}{κ}
		0.00	0.63	1.26	1.89	2.52
	27°	1.0	.4	.0	.0	.0
	342°	1.0	1.4	.4	.0	.0
μ	297°	1.0	2.3	1.2	.0	.0
	252°	1.0	1.4	.4	.0	.0
	207°	1.0	.4	.0	.0	.0

All the other values of the density for $\kappa \neq 0$ are close to 0. Now I integrate out κ by Simpson's rule (not bothering with the interval/3) (Fig. 10.2.4). Next I find the total area so I can normalize (again I don't bother with the interval/3; I just

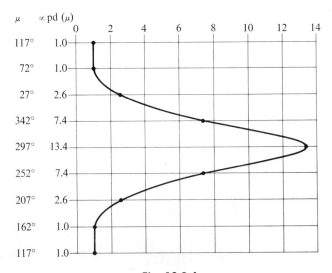

μ	$\propto$ pd (μ)
117°	1.0
72°	1.0
27°	2.6
342°	7.4
297°	13.4
252°	7.4
207°	2.6
162°	1.0
117°	1.0

Fig. 10.2.4

keep everything proportional):

$$\begin{array}{rrcl}
252\text{--}342: & 68.4 & \text{or} & 61\%, \\
162\text{--}252: & 18.8 & \text{or} & 17\%, \\
342\text{--}72: & 18.8 & \text{or} & 17\%, \\
72\text{--}162: & \underline{6.0} & \text{or} & 5\%, \\
& 112.0. &
\end{array}$$

So $297° \pm 45°$ is a 61% HDR for μ. We can also find that $297° \pm 90°$ is a 81% HDR for μ. The conclusion is that μ is not very well determined, that the directional indication is weak.

PROBLEM SET 10.2

1. Here are nine CIRG observations:

$$46°, 54°, 67°, 33°, 47°, 42°, 50°, 40°, 35°.$$

Use the CIRG formulation to estimate μ and κ. In addition, analyze these observations as if they came from GAU.

2. Here are nine CIRG observations:

$$136°, 3°, 105°, 214°, 33°, 312°, 236°, 343°, 212°.$$

Use the CIRG formulation to estimate μ and κ.

3. When the alignment of crystal axes is investigated, there is an ambiguity of 180° in the description; that is, an orientation of 10° is equivalent to 190° because the ends of the axes are not distinguishable. How can this situation be handled by CIRG?

4. Describe how to estimate μ when the observations come from the density

$$\text{pd}\,(\theta\,|\,\mu) = \frac{1}{2\pi}[1 + \cos(\theta - \mu)].$$

ANSWERS TO PROBLEM SET 10.2

1. $\hat{\mu} = 46°$, $\hat{\kappa} = 34.5$, $\mu \stackrel{.}{\sim} \text{STU}\,(*\,|\,46, 12, 8)$
3. If x is the observed angle, work with the variable $y = 2x$, which runs from 0° to 360°.

10.3 RELIABILITY AND FAILURES

In our complex, industrial society the problem of making mechanisms reliable assumes great importance. Statistical techniques help both to model the life span of machinery and to maintain quality on production lines. The variety of

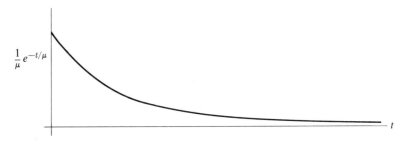

Fig. 10.3.1

factors involved is enormous, and there are few clear-cut opportunities to apply simple standard distributions. Standard distributions are usually based upon theory; at the moment, at least, many procedures in this field are quite empirical. One trouble (or blessing) is that most of the possibly relevant distributions have much the same shape: they begin at 0, have a hump, and then tail off slowly. Often the precise choice does not make too much difference as far as plain description goes.

You remember that POIS appeared when the number of opportunities for something to happen got very large while the probability of it happening at each opportunity grew very small. I now want to discuss a limiting form that describes the *length of time* you have to *wait* for the first occurrence of the event of interest. Under the same conditions—make the number of intervals large but their size small—we get the *Exponential distribution*

$$\text{expd}\,(t \mid \mu) = \frac{1}{\mu}\,e^{-t/\mu}.$$

It looks like Fig. 10.3.1. The probabilities of the short waiting times are the largest; the probability of having to wait a very long time is small. The average waiting time is μ.

The exponential distribution, EXPD, has been applied to problems such as estimating the life expectancy of light bulbs from a test sample. If we assume that this model applies—and all our inferences are based on such an assumption—we use the customary procedure:

posterior $\propto$ prior $\times$ likelihood.

We are, of course, aiming at a description of our uncertainty about μ.

All right, suppose we turn on a number of light bulbs and see how long they burn. The problem is not as clean as you might expect. An unreasonable number of them may immediately flash out; we really have to alter our model to say that some percentage will burn out immediately and that we want to estimate the life expectancy of those that last at least a few seconds. This means that we toss out the obviously bad ones right away; after all, a bulb's lifetime should be in the tens or hundreds of hours. [There is a similar trouble with electronic parts, but it is called "burn-in"! Animal and human populations also show burn-in.]

Okay, the bulbs are burning merrily. Now and then one burns out and we record the time. Sometimes a new one is started in the old socket. Occasionally the night-cleaning force breaks one with a mop handle; the time of this happening is recorded. When we get tired and decide to call it quits, we have three kinds of data: burned-out bulbs, broken bulbs, and bulbs still burning. So long as we record the times carefully, we can treat the last two categories as one: both types of bulbs had not burned out when last seen. Therefore we have

$$\text{times of burnouts:} \quad t_1, t_2, \ldots, t_f,$$
$$\text{periods without burnouts:} \quad u_1, u_2, \ldots, u_s.$$

If we knew nothing about μ, we would assign as a prior of ignorance $\text{pd}_0(\mu) = \text{LFLAT}$, since $\mu > 0$. We would most likely have some idea of the approximate size of μ, but I will neglect it here; you know how to put it in numerically. For burnouts

$$\text{pd}(t_i) = \frac{1}{\mu} \exp\left(-\frac{t_i}{\mu}\right).$$

For nonburnouts the appropriate probability is the right tail of EXPD, the probability that the time to burn out is *at least* some value. In EXPD the right tail has the simple form

$$P(> u_i \mid \mu) = \exp\left(-\frac{u_i}{\mu}\right) = 1 - \text{EXPD}(u_i \mid \mu).$$

Our likelihood is thus a mixture of pd's and P's, but it all works out. We get

$$\text{pd}(\mu \mid \mathbf{t}, \mathbf{u}) \propto \underbrace{\frac{1}{\mu}}_{\text{prior}} \times \underbrace{\frac{1}{\mu^f} \exp\left(-\frac{\sum t}{\mu}\right)}_{\text{burnouts}} \times \underbrace{\exp\left(-\frac{\sum u}{\mu}\right)}_{\text{nonburnouts}}$$

$$= \frac{1}{\mu^{f+1}} \exp\left(-\frac{\sum t + \sum u}{\mu}\right) = \frac{1}{\mu^{f+1}} e^{-T/\mu},$$

where f is the number of burnouts and T is the total burning time of all bulbs, burned-out or not. This is a distribution similar to IGAM, but it has no formal name.

$$E(\mu) = \frac{T}{f - 1},$$

$$D^2(\mu) = \frac{T^2}{(f-1)^2(f-2)},$$

$$\text{mode}(\mu) = \frac{T}{f+1}.$$

Example 10.3.1. Here is a miniature life test which shows all the principles. Note that the mean and the mode of the posterior distribution turn out to be quite far apart; this indicates a skew distribution.

Bulb	Fate	t	u
1	Burned out at 751 hr	751	
2	Burned out at 594 hr	594	
3	Burned out at 1213 hr	1213	
4	Still burning at 1320 hr		1320
5	Burned out at 1126 hr	1126	
6	Burned out at 819 hr	819	
7	Broken at 103 hr		103
8	Still burning at 1208 hr		1208
		4503	2631

[After bulb 7 was accidentally broken, it was replaced by bulb 8, which survived the entire test. The test was to last eight weeks; but the planners forgot the last day was Thanksgiving, so it lasted only 55 days.]

$$f = 5, \quad T = 4503 + 2631 = 7134,$$

$$\text{pd}\,(\mu \mid \mathbf{t}, \mathbf{u}) \propto \frac{1}{\mu^6} e^{-7134/\mu},$$

$$E(\mu) = \frac{7134}{4} = 1783.5,$$

$$D^2(\mu) = \frac{7134^2}{48} = 1{,}060{,}000,$$

$$D(\mu) = 1030,$$

$$\text{mode}\,(\mu) = \frac{7134}{6} = 1189.$$

You can see it's a very skew distribution with a long right tail. A normal approximation seems inapplicable, so to answer questions about the uncertainty in μ you have to integrate areas. As I wrote this *I* happened to be at an altitude of 29,000 feet over Arizona, so I decided to show that I could calculate $P(\mu < 1000)$ without a digital computer. The probability density of μ with all the normalizing coefficients in is

$$\frac{7.7 \times 10^{17} e^{-7134/\mu}}{\mu^6}.$$

With my trusty slide rule, which I carry for calculation and protection, I computed:

μ	pd (μ)	Simpson's coefficient	Product
1000	.00063	1	.00063
900	.00052	4	.00208
800	.00038	2	.00076
700	.00025	4	.00100
600	.00012	2	.00024
500	.000032	4	.00013
400	.0000035	2	.00001
300	.000000050	4	.00000
200		2	
100		4	
000		1	
			.00485

The coefficients are those of Simpson's rule for numerical integration. I started at 1000 and worked down, since I knew the mode was 1189. I stopped when the last decimal place seemed reasonably secure. To get the area, I then multiplied the sum of the products by interval/3: $\frac{100}{3} \times .00485 = .16$. Allowing for a little rounding off because of the rarefied atmosphere, I arrived at

$$P(\mu < 1000) \doteq \tfrac{1}{6} \quad \text{or} \quad \mathfrak{F}\left(\frac{\mu > 1000}{\mu < 1000}\right) \doteq 5.$$

To make this conclusion I assumed that

1) I had no prior information about μ,
2) the EXPD model applied,
3) the bulbs came from the *same* EXPD population,
4) the observations and bulbs were independent.

The conclusion refers only to continuous burning time, since that was the way the experiment was run. What the life expectancy is under frequent on-off switching cannot be determined from this experiment.

The other standard distributions used to describe the length of time before failure are GAMMA, WEIB (the *Weibull distribution*), and sometimes even GAU. It is difficult to produce many good theoretical reasons for using any of them. GAMMA may appear when we look for *several* occurrences of an event, EXPD when we look for just one. Thus GAMMA might be applicable to cases in which several

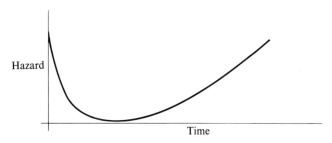

Fig. 10.3.2

subsystems have to be knocked out before the overall mechanism goes; that is, there is some redundancy in the device for safety.

Sometimes it is convenient to think not of the time before failure, but of the *instantaneous danger of failure* to which our device or organism is subjected. This is called the *hazard function*, and is defined as

$$h(t) = \frac{\text{pd}(t)}{\text{right tail}(t)}.$$

If you look back at the definition, you can see that for EXPD

$$h(t) = \frac{\frac{1}{\mu} \exp\left(-\frac{t}{\mu}\right)}{\exp\left(-\frac{t}{\mu}\right)} = \frac{1}{\mu},$$

a constant. Another kind of hazard function is shown in Fig. 10.3.2. Here there is a fairly high attrition rate at the beginning, a slow burn-in phenomenon, then relatively high-quality performance, and finally gradually accelerated aging. This is characteristic of WEIB.

If you think better in terms of hazard functions, you can go to the survival distribution function by

$$\text{pd}(t) = h(t) \exp\left[-\int_0^t h(x)\, dx\right].$$

PROBLEM SET 10.3

1. Plot the posterior distribution of μ in Example 10.3.1.
2. Suppose a system has two identical components, *both* of which must work in order for the system to work. Assume that the time before failure of each one independently is $\sim$ EXPD $(* \mid 1000)$. What is the probability that the system will not fail for at least 100 hr?
3. Solve Problem 2 for general t and μ.

4. The density of the distribution in Problem 3 is

$$\frac{2}{\mu} e^{-2t/\mu}.$$

What is the hazard function? Sketch it.

5. Suppose that the system in Problem 2 will work so long as *either one or both* of the components work. What is the probability that the system will not fail for at least 100 hr?

6. Solve Problem 5 for general t and μ.

7. The density of the distribution in Problem 6 is

$$\frac{2}{\mu} e^{-t/\mu}(1 - e^{-t/\mu}).$$

What is the hazard function? Sketch it.

8. Under proper operating conditions transistors show practically no aging; that is, the hazard is constant. In one computer 1000 transistors of a certain type are used. After one year of operation it was found that 40 new transistors had been installed. What can you infer about the average lifetime of this transistor in this operating environment?

ANSWERS TO PROBLEM SET 10.3

2. $P(\text{one lasts} > 100 \mid \mu = 1000) = e^{-100/1000} = e^{-.1} = .905$
   $P(\text{both last} > 100) = (.905)^2 = .819$

4. $h(t) = 2/\mu$, a horizontal line. This is twice the hazard of a single component.

6. $1 - (1 - e^{-t/\mu})^2 = e^{-t/\mu}(2 - e^{-t/\mu})$

8. Constant hazard means that EXPD is the distribution of time before failure.

$$f = \text{number of failures} = 40,$$
$$T = \text{total operating time} = 1000 \times 365 \times 24 = 8{,}760{,}000 \text{ hr},$$

[I neglect maintenance and downtime.]

$$\text{pd}(\mu \mid f, T) \propto \frac{e^{-8760000/\mu}}{\mu^{41}},$$

[In EXPD, f and T are sufficient for **t** and **u**.]

$$E(\mu) = \frac{8760000}{39} = 224{,}600 \text{ hr},$$

$$D^2(\mu) = \frac{8760000^2}{39^2 \times 38} = 1{,}326{,}000{,}000 \text{ hr}^2,$$

$$D(\mu) = 36{,}400 \text{ hr},$$

$$\text{mode}(\mu) = \frac{8760000}{41} = 213{,}700 \text{ hr}.$$

330   A Potpourri of Applications   10.4

If I use a normal approximation, I get

$$P(\mu > 178{,}000 \text{ hr}) \doteq .90.$$

[If the lifetime is so large, it seems unlikely that the right-hand part of the hazard function has been adequately investigated.]

10.4 RADAR

This last application concerns the detection of a returning radar signal. A little pulse of energy leaves a radar transmitter, bounces off a target, and returns to the radar receiver. From the time interval between transmission and reception,

Table 10.4.1

1	2	3	4	5	6
Position	Noise	a Signal	u Signal + noise	$\sum ua$	$\exp(\sum ua)$
1	−1.16	.0	−1.16	2.186	9
2	−.18	.0	−.18	3.229	25
3	.37	.0	.37	1.480	4
4	2.94	.0	2.94	−.907	
5	−.21	.0	−.21	−1.879	
6	−.15	.0	−.15	−2.928	
7	−.33	.0	−.33	−4.444	
8	.75	.0	.75	−4.276	
9	−.97	.0	−.97	−1.963	
10	.16	.0	.16	1.161	3
11	.78	.0	.78	4.196	67
12	.47	.0	.47	6.296	540
13	.25	.2	.45	6.665	790
14	1.15	.7	1.85	5.339	208
15	1.70	1.0	2.70	2.197	9
16	−.44	.7	.26	−1.019	
17	1.20	.2	1.40	−1.351	
18	−1.90	−.2	−2.10	.218	
19	.30	−.7	−.40	.236	
20	.61	−1.0	−.39	−1.019	
21	−.46	−.7	−1.16	−1.478	
22	−.88	−.2	−1.08	−1.823	
23	.05	.0	.05	−2.393	
24	1.55	.0	1.55	−1.952	
25	−2.77	.0	−2.77	−.455	
26	−.87	.0	−.87	.896	2
27	−.26	.0	−.26	1.299	4
28	.93	.0	.93	.872	2
29	.83	.0	.83	.433	1
30	−.42	.0	−.42	.627	2

10.4 Radar

the range to the target can be deduced. That signal sent out is not large; the return signal is extremely small. Not only has it diminished in size, but *noise* has been added to it—electrical disturbances both from the atmosphere and from the radar receiver itself. The signal is now so submerged in the noise that its position is not obvious. The problem is to find it. Here I will show you a mathematical solution; in practice electronic equipment does the calculation for us. (See Woodward, 1953).

I'll take an extremely simplified version of the process, build up the problem step by step, and then try to solve it. [Since I know there's a solution, I'll try even harder!] First we generate some Gaussian noise to represent the noise in the atmosphere and the receiver. Time is divided up into equal intervals, and a random value from GAU $(* \mid 0, 1)$ is picked for each interval. I choose *independent*

Table 10.4.1 (Cont.)

1	2	3	4	5	6
Position	Noise	a Signal	u Signal + noise	$\sum ua$	$\exp(\sum ua)$
31	− .28	.0	− .28	.689	2
32	− .47	.0	− .47	.059	1
33	.97	.0	.97	− .717	
34	− .40	.0	− .40	−1.329	
35	− .65	.0	− .65	−1.553	
36	.25	.0	.25	−1.246	
37	−1.27	.0	−1.27	−1.100	
38	.99	.0	.99	−1.614	
39	−1.23	.0	−1.23	−1.768	
40	.98	.0	.98	−1.478	
41	− .87	.0	− .87	−1.813	
42	.97	.0	.97	−1.902	
43	.52	.0	.52	− .264	
44	− .73	.0	− .73	2.165	9
45	1.09	.0	1.09	3.658	39
46	1.15	.0	1.15	3.833	46
47	.14	.0	.14	2.946	19
48	1.28	.0	1.28	1.820	6
49	.97	.0	.97	1.072	3
50	1.29	.0	1.29	.638	2
51	− .37	.0	− .37	—	
52	− .93	.0	− .93	—	
53	− .60	.0	− .60	—	
54	− .27	.0	− .27	—	
55	1.13	.0	1.13	—	
56	− .69	.0	− .69	—	
57	−1.89	.0	−1.89	—	
58	.80	.0	.80	—	
59	−2.28	.0	−2.28	—	

values to create *white noise*. When the values are not independent it is called *colored noise* (e.g., green, blue). In Table 10.4.1 the 59 intervals of time are numbered in column 1; the noise is in column 2. [We can analyze a signal by considering equally spaced values so long as the radio frequency has some upper limit, a *band-limited* signal.]

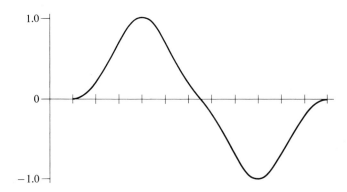

Fig. 10.4.1

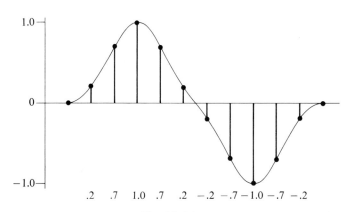

Fig. 10.4.2

Now I add in the returning radar signal, which I will later try to locate. Assume that I know the size and shape of the signal (Fig. 10.4.1). I digitize it by finding the amplitude at equally spaced intervals of time (Fig. 10.4.2):

$$.2, .7, 1.0, .7, .2, -.2, -.7, -1.0, -.7, -.2.$$

I add these ten values to the noise at positions 13 through 22 and add 0 to all the rest of the noise. Column 3 shows the *signal*, and column 4 shows

10.4 Radar

signal + noise. Putting the signal at position 13 was an arbitrary choice; it had to go somewhere.

The values in column 4 are what the radar receiver sees. Where is the signal? To shorten the problem we will consider only 50 hypotheses: the hypotheses that the signal starts in one of the first 50 positions. [When I speak of the position of the signal, I mean the time interval in which the first nonzero value (.2) falls.] I will assign equal prior probabilities to these hypotheses, so the discrimination will depend on the likelihood.

Table 10.4.2

Position	Signal + noise	Signal
1	−1.16	.2
2	− .18	.7
3	.37	1.0
4	2.94	.7
5	− .21	.2
6	− .15	− .2
7	− .33	− .7
8	.75	−1.0
9	− .97	− .7
10	.16	− .2
11	.78	.0
12	.47	.0
⋮	⋮	⋮

What about the likelihood? Each hypothesis assumes that it knows the signal values, that Gaussian noise has been added to them. Therefore at each position we have

$$\text{signal + noise} \sim \text{GAU}(* \mid \text{signal}, 1),$$

because the noise had 0 mean. Let me write out Hypothesis 1 in detail. Table 10.4.2 shows the assumption.

Then the probability density is

$$\text{gau}(-1.16 \mid .2, 1) \quad \text{at position 1,}$$
$$\text{gau}(- .18 \mid .7, 1) \quad \text{at position 2,}$$
$$\text{gau}(.37 \mid 1.0, 1) \quad \text{at position 3, etc.}$$

You can evaluate this numerically, but we can get a meaningful and useful formula.

Let the *signal* value at position i be a_i and the *signal + noise* value be u_i. Then the probability of the assumption at position i is $\text{gau}(u_i \mid a_i, 1)$. The probability of all 59 positions works out to

$$\left(\frac{1}{\sqrt{2\pi}}\right)^{59} \exp\left[-\frac{1}{2}\sum(u_i - a_i)^2\right]$$

$$= \left(\frac{1}{\sqrt{2\pi}}\right)^{59} \exp\left[-\frac{1}{2}\left(\sum u_i^2 - 2\sum u_i a_i + \sum a_i^2\right)\right].$$

This simplifies drastically. All the π's can be discarded since we'll have to normalize anyway. The

$$\exp\left(-\tfrac{1}{2}\sum u_i^2\right)$$

is the same for all 50 hypotheses and certainly can't help us discriminate between them. Similarly, the

$$\exp\left(-\tfrac{1}{2}\sum a_i^2\right)$$

is the same for all 50 hypotheses; the ten nonzero values of the signal are all in there somewhere, and all the other positions are 0. This leads to

$$P(\text{signal starts at this position}) \propto \exp\left(\sum u_i a_i\right).$$

So we have to calculate the appropriate $\sum u_i a_i$ for each starting position of the signal. This means just ten multiplications and additions, since most of the a_i's are 0.

Table 10.4.3

Hypothesis 1		Hypothesis 2		Hypothesis 3	
u	a	u	a	u	a
−1.16	.2	−1.16		−1.16	
− .18	.7	− .18	.2	− .18	
.37	1.0	.37	.7	.37	.2
2.94	.7	2.94	1.0	2.94	.7
− .21	.2	− .21	.7	− .21	1.0
− .15	− .2	− .15	.2	− .15	.7
− .33	− .7	− .33	− .2	− .33	.2
.75	−1.0	.75	− .7	.75	− .2
− .97	− .7	− .97	−1.0	− .97	− .7
.16	− .2	.16	− .7	.16	−1.0
.78		.78	− .2	.78	− .7
.47		.47		.47	− .2
.25		.25		.25	
1.15		1.15		1.15	

If you do this by hand, just write the ten nonzero signal values on a strip of paper and slide it along the column of *signal + noise* values. Table 10.4.3 shows you how: Then multiply the juxtaposed values and add. The sums of cross-products for all 50 possible starting positions are given in column 5 of Table 10.4.1. In column 6 are the values of $\exp(\sum ua)$ which form the unnormalized posterior distribution.

I have not bothered to calculate the normalized posterior distribution since the unnormalized values in Fig. 10.4.3 are sufficient to give the picture. Obviously positions 12, 13, and 14 outweigh all the others; position 13 happens to be the right answer.

You see that the *statistic* which measures the agreement between the hypothesized *signal* and the *signal + noise* is the *sum of cross-products*. We met this in regression. It is a very fruitful statistic in many applications in which you want to match two strings of data, even if the variation is not Gaussian. The thing to remember is that we can go from column 5 to column 6 and interpret those values as (unnormalized) *probabilities* only because we know that the noise is Gaussian. If we don't know the underlying distribution, we can get a *ranking* of the possi-

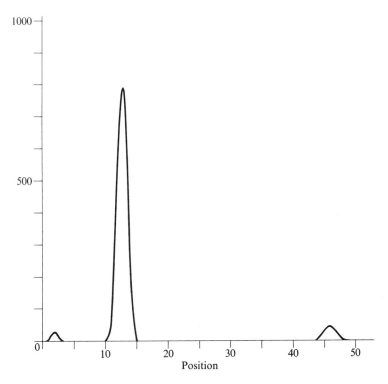

Fig. 10.4.3

336   A Potpourri of Applications   10.4

bilities by computing the *sum of cross-products*, but we don't know what the differences between the values signify. They might mean a lot or not much at all. Nevertheless, a ranking of possibilities is often a great help, especially in exploratory work.

PROBLEM SET 10.4

1. I took a sequence of seven independent digits from a Poisson distribution and added the sequence (1, 2, 1) to three consecutive places. The result was

$$1, 9, 7, 3, 6, 4, 2.$$

   What is the posterior distribution for the position of the (1, 2, 1)? Compare the answer with what you obtain using the sum of cross-products.

2. In the Southwest archeologists study tree rings in order to establish a time scale. In a certain region they have set up a master sequence of *wet* and *dry* years:

   W, W, D, W, D, W, W, D, D, D, D, W, D, W, W, W, D, W, D, W, W, W, W, D, W, W, D.

   When a worker tries to match a particular sample against the master sequence, he finds that there is not perfect agreement because of local climatic variations. The probability that any particular year agrees is .8 (independently of other years). Given a sample sequence

   $$W, W, D, W, D, D, W, W$$

   that could start at any of the first 20 positions of the master sequence, what are the probabilities of each of those 20 positions?

3. Describe how to analyze the radar problem when you know the shape of the return signal, but you have only the probability density of its amplitude, λ.

ANSWERS TO PROBLEM SET 10.4

1. If u means the sequence of observed values, and a means the sequence (1, 2, 1) with 0's surrounding it, then the likelihood for any match is

$$\prod \left(\frac{e^{-\lambda} \lambda^{u-a}}{(u-a)!} \right) = e^{-7\lambda} \lambda^{\Sigma u - \Sigma a} \prod \left(\frac{1}{(u-a)!} \right).$$

   Thus λ makes no difference. [We immediately throw out possibilities where an $a_i > u_i$.] The relative probabilities of the five starting positions are

$$504, 1134, 252, 360, 144.$$

10.4

The corresponding sums of cross-products are

$$26, 26, 19, 19, 16.$$

2. The likelihoods will be

$$.8^{\text{no. of agreements}} .2^{\text{no. of disagreements}}.$$

The relative probabilities for various matches are:

Agree	Disagree	Relative probability
1	7	1
2	6	4
3	5	16
4	4	64
5	3	256
6	2	1024
7	1	4096

Position 15 does the best, but it still has only 53% of all the probability.

EPILOGUE

Alexander wept when he heard from Anaxarchus that there were an infinite number of worlds. His friends asked him if some accident had befallen him, and he replied: "Do you not think it a sorrowful matter that when there are such a vast multitude of worlds, I have not yet conquered one?"

Plutarch

Suggestions for Further Reading

Here is a list of some references which you might find of benefit at this early stage. There are, of course, many excellent standard statistical texts; the ones given here are of a more specialized nature. The names and dates refer to the bibliography which follows this section. I have also listed the major statistical journals; most of them, with the exceptions noted, are pretty heavy going for beginners, but you should know what the resources in this field are.

REFERENCES

1. Bayesian Statistics

Bayes (1763). The first statement of the theorem. For the connoisseur.

Edwards et al. (1963). On an intermediate level; not just for psychologists.

Fry (1965). Contains good discussion of Bayesian point of view; also elementary probability.

Jaynes (1958). Good and readable. The very last part demands a little physics.

Jeffreys (1961). Absolutely the best, but on a more advanced level. You must read this when you have more mathematics.

Laplace (1820). A very readable classic.

Lindley (1965). Advanced and comprehensive.

Schlaifer (1959). Excellent for beginners; oriented towards business decision theory.

Tribus (1969). Upperclass level with engineering orientation.

2. Decision Theory

Raiffa (1968). Excellent.

Schlaifer (1959). Cannot be too highly praised.

Wald (1947 and 1950). The classics. On a more advanced level, but some parts are surprisingly easy.

3. Information Theory

Iaglom and Iaglom (1960). Easy and interesting.

Shannon and Weaver (1949). Slightly above beginner level.

Woodward (1953). Uses mathematical and engineering concepts freely, but very good.

4. Probability Theory

Feller (1968). Delightful as well as instructive. You can read this as a beginner or as an advanced student and get something out of it.

5. Planning Experiments

a) Oriented towards sample surveys

Backstrom and Hursh (1963). Easy but good.

Stephan and McCarthy (1958). Also easy but good.

b) Oriented towards scientific experiments

Cox (1958). Thorough without being too mathematical.

Fisher (1947). A classic but most readable.

6. Descriptive Statistics (Graphical display)

Berry and Marble (1968). Illustrations of new techniques scattered throughout book.

Dickinson (1963). Very good.

Huff and Geis (1954). Entertaining and instructive.

7. General Reference

Kendall and Stuart (1958–1966). An encyclopedia of statistics, mostly on an advanced level. But even these three big volumes can't completely cover fast-growing statistical theory.

8. Tables

Fisher and Yates (1963).

Owen (1962).

STATISTICAL JOURNALS

American Statistician. Mostly a newsletter; some minor discussions.

Annals of Mathematical Statistics. Heavy theory.

Annals of Statistical Mathematics (Japanese journal in English). Theory.

Applied Statistics. Section C of the Journal of the Royal Statistical Society. Up-to-date discussions of modern ideas. Good reading, even for near-beginners.

Biometrics. Mostly applications to biology; some theory.

Biometrika. Originally applications; now mostly theory. Some historical articles.

Journal of the American Statistical Association. Both theory and applications; good reading for near-beginners.

Journal of the Royal Statistical Society. Section A deals with applications. Section B deals with theory, much of it Bayesian, with many comments about applications. The best reading available today, though not much of it is elementary. Good discussion of up-to-date topics. It has been the practice that "points of agreement are not over-emphasized."

Review of the International Statistical Institute. Mixed; much official statistics.

Sankhya (Indian journal in English). Leans toward theory.

Statistical Theory and Method Abstracts. Abstracts only.

Technometrics. Industrial applications with relevant theory. Since this is the fastest growing field in statistics today, the newest things appear here early.

Bibliography

ACTON, F. S. (1959). *Analysis of Straight-line Data.* New York: Wiley. (Now available in Dover edition.)

ARBUTHNOT, J. (1710). An argument for divine providence taken from the constant regularity of the births of both sexes. *Phil. Trans. Roy. Soc. London* **23,** 186–190.

BACKSTROM, C. H., and G. D. HURSH (1963). *Survey Research.* Evanston, Ill.: Northwestern Univ. Press.

BATSCHELET, E. (1965). *Statistical Methods for the Analysis of Problems in Animal Orientation and Certain Biological Rhythms.* American Institute of Biological Sciences Monograph.

BAYES, T. (1763). An essay towards solving a problem in the doctrine of chances. *Phil. Trans. Roy. Soc. London* **53,** 370–418. Reproduced with biography of Bayes in G. A. Barnard (1958), Studies in the history of probability and statistics: IX. *Biometrika* **45,** 293–315.

BERKSON, J. (1966). Examination of randomness of α-particle emissions. In F. N. David (ed.), *Research Papers in Statistics.* New York: Wiley.

BERRY, B. J. L., and D. F. MARBLE, eds. (1968). *Spatial Analysis: A Reader in Statistical Geography.* Englewood Cliffs, N.J.: Prentice-Hall.

BOYD, W. C. (1952). *Genetics and the Races of Man.* Boston: Little, Brown and Co.

BOX, G. E. P., and G. C. TIAO (1962). A further look at robustness via Bayes's Theorem. *Biometrika* **49,** 419–432.

BRACKEN, J. (1966). Percentage points of the beta distribution for use in Bayesian analysis of Bernoulli processes. *Technometrics* **8,** 687–694.

BRIDGMAN, P. W. (1960). Critique of critical tables. *Proc. Nat. Acad. Sci. U.S.* **46,** 1394–1400.

BROWNLEE, K. A. (1955). Statistics of the 1954 polio vaccine trials. *J. Am. Statist. Assoc.* **50,** 1005–1013.

CARROLL, J. B. (1966). An experiment in evaluating the quality of translations. *Mech. Transl.* **9**, 55–66.

COALE, A. J., and F. F. STEPHAN (1962). The case of the Indians and the teen-age widows. *J. Am. Statist. Assoc.* **57**, 338–347.

COOK, E., and W. R. GARNER (1966). *Percentage Baseball* (2nd ed.). Cambridge, Mass.: M.I.T. Press.

CORNFIELD, J. (1966). A Bayesian test of some classical hypotheses—with applications to sequential clinical trials. *J. Am. Statist. Assoc.* **61**, 577–594.

COX, D. R. (1958). *Planning of Experiments.* New York: Wiley.

DICKINSON, G. C. (1963). *Statistical Mapping and the Presentation of Statistics.* London: Edward Arnold.

EDWARDS, W., H. LINDMAN, and L. J. SAVAGE (1963). Bayesian statistical inference for psychological research. *Psychol. Rev.* **70**, 193–242.

EISENHART, C. (1968). Expressions of the uncertainties of final results. *Science* **160**, 1201–1204.

FELLER, W. (1968). *An Introduction to Probability Theory and its Applications,* Vol. I (3rd ed.). New York: Wiley.

FISHER, R. A. (1947). *The Design of Experiments* (4th ed.). New York: Hafner.

FISHER, R. A., and F. YATES (1963). *Statistical Tables for Biological, Agricultural and Medical Research* (6th ed.). Edinburgh: Oliver and Boyd.

FRY, T. C. (1965). *Probability and its Engineering Uses* (2nd ed.). Princeton, N.J.: D. Van Nostrand.

GABBE, J. D., M. B. WILK, and W. L. BROWN (1967). Statistical analysis and modeling of the high-energy proton data from the Telstar 1 satellite. *Bell System Tech. J.* **46**, 1301–1450.

GOOD, I. J. (1950). *Probability and the Weighing of Evidence.* London: Charles Griffin.

——— (1965). *The Estimation of Probabilities.* Cambridge, Mass.: M.I.T. Press.

GRAYSON, J. (1961). Decisions under uncertainty: Drilling decisions by oil and gas operators. Division of Research, Graduate School of Business Administration, Harvard University.

HUFF, D., and I. GEIS (1954). *How to Lie with Statistics.* New York: Norton.

IAGLOM, A. M., and I. M. IAGLOM (1960). *Probability and Information.* Moscow: State Publishing House for Physics and Mathematical Literature. (This is in Russian. There is a German translation, and Dover will put out an English translation.)

JAYNES, E. T. (1958). Probability theory in science and engineering. Colloquium Lectures in Pure and Applied Science, No. 4. Socony-Mobil Oil Co.

JEFFREYS, H. (1961). *Theory of Probability* (3rd ed.). Oxford: Clarendon Press.

KENDALL, M., and A. STUART (1958–1966). *The Advanced Theory of Statistics.* 3 vols. London: Charles Griffin.

KERRIDGE, D. (1963). Bounds for the frequency of misleading Bayes inferences. *Ann. Math. Statist.* **34,** 1109–1110.

KOVARIK, A. F. (1910). Absorption and reflexion of the β-particles by matter. *Phil. Mag. S6,* **20,** 849–866.

LAPLACE, P. S. (1820). *Essai Philosophique sur les Probabilités.* Translated by F. W. Truscott and F. L. Emory. *A Philosophical Essay on Probabilities.* New York: Dover, 1951.

LINDLEY, D. V. (1965). *Introduction to Probability and Statistics from a Bayesian Viewpoint.* 2 vols. New York: Cambridge Univ. Press.

LISCO, T. E. (1967). The value of commuters' travel time: A study in urban transportation. Ph.D. dissertation, University of Chicago.

MCGUIRE, J. B., E. R. SPANGLER, and L. WONG (1961). The size of the solar system. *Sci. Am.* **204** (April), 64–72.

MILLER, R. L., and J. S. KAHN (1962). *Statistical Analysis in the Geological Sciences.* New York: Wiley.

NEWCOMB, S. (1886). A generalized theory of the combination of observations so as to obtain the best results. *Am. J. Math.* **8,** 343–366.

OWEN, D. B. (1962). *Handbook of Statistical Tables.* Reading, Mass.: Addison-Wesley.

PORTER, R. C. (1967). Extra-point strategy in football. *Am. Statistician* **21** (December), 14–15.

RAIFFA, H. (1968). *Decision Analysis.* Reading, Mass.: Addison-Wesley.

RESCIGNO, A., and G. A. MACCACARO (1961). The information content of biological classifications. In *Information Theory: 4th London Symposium,* 437–446. London: Butterworth.

RUTHERFORD, E., and H. GEIGER (1910). The probability variations in the distribution of α-particles (with note by H. Bateman). *Phil. Mag. S6,* **20,** 698–707.

SCHLAIFER, R. (1959). *Probability and Statistics for Business Decisions.* New York: McGraw-Hill.

SHANNON, C. E., and W. WEAVER (1949). *The Mathematical Theory of Communication.* Urbana, Ill.: Univ. of Illinois Press.

STEPHAN, F. F., and P. J. MCCARTHY (1958). *Sampling Opinions.* New York: Wiley. (Now in paperback edition.)

"STUDENT" (1931). The Lanarkshire milk experiment. *Biometrika* **23,** 398–406.

THORNDIKE, E. L., and I. LORGE (1944). *The Teacher's Word Book of 30,000 Words.* New York: Teachers College Press, Columbia University.

TRIBUS, M. (1962). The use of the maximum entropy estimate in the estimation of reliability. In R. E. Machol and P. Gray (eds.), *Recent Developments in Information and Decision Processes.* New York: Macmillan.

_____ (1969). *Rational Descriptions, Decisions, and Designs.* New York: Pergamon.

TUKEY, J. W. (1962). The future of data analysis. *Ann. Math. Statist.* **33,** 1–67.

WALD, A. (1947). *Sequential Analysis.* New York: Wiley.

―――― (1950). *Statistical Decision Functions.* New York: Wiley.

WELCH, B. L. (1937). The significance of the difference between two means when the population variances are unequal. *Biometrika* **29,** 350–362.

WOODWARD, P. M. (1953). *Probability and Information Theory with Applications to Radar.* New York: McGraw-Hill.

Appendixes   A. *Mathematical Details*
B. *Logarithms*
C. *Random Numbers*

Appendix A. Mathematical Details

A1. SUMS

$\sum$ The Greek capital letter "sigma" indicates a sum. In statistics we repeatedly want to add up collections of numbers, and this notation represents the operation concisely. A sum basically says: "Add *that* and *that* and *that* and ... and *that*." The benefit gained lies in our being able to point out all the "thats" without enumerating them.

1. The simple notation

Given a set of numbers called the "x"s, we indicate their sum simply by $\sum x$. This is read aloud as "the sum of the x's," or something similar.

Example. We have four x's: 1, 5, 8, and 6. Their sum in

$$\sum x = 1 + 5 + 8 + 6 = 20.$$

Another familiar operation is summing the squares of a set of numbers. We simply write $\sum x^2$ and read it as "the sum of x-squared."

Example. For the same x's,

$$\sum x^2 = 1 + 25 + 64 + 36 = 126.$$

If an expression in parentheses appears immediately after the $\sum$, we calculate the various values that the expression takes, and then add the results.

Example. For the same x's,

$$\sum (x - 3) = (-2) + 2 + 5 + 3 = 8.$$

Note that this is different from

$$\sum x - 3 = 20 - 3 = 17.$$

If there are n x's altogether, we define the *average*, *mean*, or *arithmetic mean* to be

$$\bar{x} = \frac{\sum x}{n}.$$

This means, of course, that $\sum x = n\bar{x}$.

Example. For the x's above,

$$\bar{x} = \tfrac{20}{4} = 5.$$

Note that

$$\sum (x - \bar{x}) = (-4) + 0 + 3 + 1 = 0.$$

The result that $\sum (x - \bar{x}) = 0$ applies to any set of x's, not just this one. To prove this, note that the result we get when we apply the sum operator to an expression in parentheses is the same as what we get when we apply it to the individual terms in the expression and combine those results:

$$\begin{aligned}\sum (x - \bar{x}) &= \sum x - \sum \bar{x} \\ &= \sum x - n\bar{x} \\ &= \sum x - \sum x \\ &= 0.\end{aligned}$$

[The sum above has n terms. When we break it into the two separate sums, each of these has n terms also. In $\sum \bar{x}$ each of the terms is the same; therefore that sum is $n\bar{x}$.]

In the same way we can derive one of the very important identities in statistics. Suppose we have calculated $\bar{x}$; it is now just a number. What is $\sum (x - \bar{x})^2$?

$$\begin{aligned}\sum (x - \bar{x})^2 &= \sum (x^2 - 2\bar{x}x + \bar{x}^2) \\ &= \sum x^2 - 2\bar{x}\sum x + \sum \bar{x}^2 \\ &= \sum x^2 - 2\bar{x}(n\bar{x}) + n\bar{x}^2 \\ &= \sum x^2 - n\bar{x}^2.\end{aligned}$$

Example. For the x's above, $\bar{x} = 5$. Then

$$\sum (x - \bar{x})^2 = (-4)^2 + 0^2 + 3^2 + 1^2 = 26,$$
$$\sum x^2 - n\bar{x}^2 = 126 - 4 \times 5^2 = 126 - 100 = 26.$$

It also can be shown that for any number a,
$$\sum (x - a)^2 = \sum (x - \bar{x})^2 + n(\bar{x} - a)^2.$$

Example. For the x's above and $a = 1$,
$$\sum (x - 1)^2 = 0^2 + 4^2 + 7^2 + 5^2 = 90,$$
$$\sum (x - \bar{x})^2 + n(\bar{x} - 1)^2 = 26 + 4 \times 4^2 = 90.$$

If we have two sets of numbers, associated in pairs, we can extend our notion of summation.

Example

The x's are 1, 5, 8, and 6;
the y's are 7, 2, 4, and 3.

Then
$$\sum xy = 1 \times 7 + 5 \times 2 + 8 \times 4 + 6 \times 3 = 67.$$

Since $\bar{x} = 5$ and $\bar{y} = 4$, we also have
$$\sum (x - \bar{x})(y - \bar{y}) = (-4) \times 3 + 0 \times (-2) + 3 \times 0 + 1 \times (-1) = -13.$$

By the method used previously we can show that
$$\sum (x - \bar{x})(y - \bar{y}) = \sum xy - n\bar{x}\bar{y}.$$

Example. For the x's and y's above,
$$\sum (x - \bar{x})(y - \bar{y}) = -13,$$
$$\sum xy - n\bar{x}\bar{y} = 67 - 4 \times 5 \times 4 = -13.$$

Note 1. In these examples the means were integers. If the means are not integers, we get less round-off error if we write the formulas as
$$\sum (x - \bar{x})^2 = \sum x^2 - \frac{(\sum x)^2}{n},$$
$$\sum (x - \bar{x})(y - \bar{y}) = \sum xy - \frac{(\sum x)(\sum y)}{n}.$$

Note 2. If we use a digital computer, computing
$$\sum (x - \bar{x})^2 \quad \text{and} \quad \sum (x - \bar{x})(y - \bar{y})$$
directly is probably best; your computer center can give you advice. The idea is that you often want to inspect the *residuals* $(x - \bar{x})$ anyway. We also get even better round-off error control.

2. The S-notation

A few of the sums we discussed occur so many times in statistics that a more compact notation is sometimes used. Usage varies a little between authors, but I will define them as follows:

$$S_x = \sum x, \qquad S_{xx} = \sum x^2, \qquad S_{\bar{x}\bar{x}} = \sum (x - \bar{x})^2,$$
$$S_{xy} = \sum xy, \qquad S_{\bar{x}\bar{y}} = \sum (x - \bar{x})(y - \bar{y}).$$

3. Subscript notation

Sometimes various complications arise and we want to indicate more explicitly what and how many things are to be added. To do this we start by labeling each of our variables with a subscript.

Example. The x's are 1, 5, 8, and 6. We write

$$x_1 = 1, \qquad x_2 = 5, \qquad x_3 = 8, \qquad x_4 = 6.$$

[This is read "the first x is 1" or "x-sub-1 is 1," etc.]. We then pick some letter which does not occur in any part of our expressions; it is used as a *dummy index*, a means of counting off the numbers in our sum. In this case we can choose for a dummy any letter except x. Since the traditional choices are i, j, k, l, or m, I'll pick i. To express

$$1 + 5 + 8 + 6$$

we write

$$\sum_{i=1}^{4} x_i,$$

which is read "the sum of x-sub-i for i equals 1 to 4." Remember that the i is an arbitrary letter which has no significance outside the domain of this particular summation sign. We could just as correctly have written

$$\sum_{j=1}^{4} x_j \quad \text{or} \quad \sum_{k=1}^{4} x_k \quad \text{or} \quad \sum_{l=1}^{4} x_l.$$

We now can do all sorts of complicated things:

$$\sum_{i=1}^{3} x_i = 1 + 5 + 8 = 14,$$
$$\sum_{i=2}^{4} x_i = 5 + 8 + 6 = 19,$$
$$\sum_{i=1}^{2} x_{2i} = 5 + 6 = 11.$$

The dummy variable itself can be used to help evaluate its expression:

$$\sum_{m=1}^{4} (-1)^m x_m = -1 + 5 - 8 + 6.$$

In fact, sometimes there are no subscripted values around, and the successive values of the dummy variable completely determine the expression:

$$\sum_{k=1}^{3} k^2 = 1^2 + 2^2 + 3^2 = 14,$$

$$\sum_{j=2}^{5} j(j-1) = 2 \times 1 + 3 \times 2 + 4 \times 3 + 5 \times 4 = 40.$$

Occasionally we cannot reasonably express our selection of summands by enumerating a sequence of subscripts or by calculating an expression. We then are forced to use things like this:

$$\sum_{x_i \text{ even}} x_i,$$

which means: "Look over the x's and pick out all of them which are even; add these." For the given x's this sum is $8 + 6 = 14$. Other variations which you run across now and then are just local abbreviations which have to be explained on the spot.

Problems

For Problems 1 through 18 use these values:

x's are 5, 2, 4, 8, 6,
y's are 1, 1, 3, 2, 2,
z's are 10, 4, 1.

1. $\sum x$ 2. $\sum y$ 3. $\bar{z}$ 4. $\bar{y}$ 5. S_{xx}
6. S_{yy} 7. $S_{\bar{x}\bar{x}}$ 8. $S_{\bar{y}\bar{y}}$ 9. S_{xy} 10. $S_{\bar{x}\bar{y}}$
11. $\sum (z-3)^2$ 12. $\sum_{i=1}^{5} x_i$ 13. $\sum_{j=2}^{3} y_j z_j$ 14. $\sum_{k=1}^{3} k(k+1)$
15. $\left(\sum_{l=1}^{2} l^2\right)\left(\sum_{j=1}^{2} \sqrt{j}\right)$ 16. $\sum_{x_i \text{ a square}} x_i$ 17. $\sum_{y_i < 3} y_i$ 18. $\sum_{x_i = y_i + 1} x_i$

19. Prove that $\sum (x-a)^2 = \sum (x-\bar{x})^2 + n(\bar{x}-a)^2$.
    [Hint: $(x-a) = (x-\bar{x}) + (\bar{x}-a)$.]

Answers

1. 25   2. 9   3. 5
4. 1.8   5. 145   6. 19

7. 20   8. 2.8   9. 47
10. 2   11. 54   12. 25.
13. 7   14. 20   15. 12.07
16. 4   17. 6   18. 6

A2. PRODUCTS

$\prod$ The Greek capital letter "pi" indicates a *product*. Do not confuse it with the lower-case letter which is the well-known constant $\pi = 3.14159...$. All the conventions used for the summation sign Σ apply to the product sign $\prod$, except that we must multiply instead of add.

Example. The x's are 1, 5, 8, 6.

$$\prod x = 1 \times 5 \times 8 \times 6 = 240.$$

The only time that products have a really convenient representation is when the factors are expressed as powers of a common base. Then we can add the exponents and perhaps use some of the helpful summation formulas.

Example. The x's are 10^5, 10^7, 10^{-1}; the y's are e^2, $e^{5.1}$, e^{-8}, $e^{.7}$.

$$\prod x = 10^{(5+7-1)} = 10^{11},$$
$$\prod y = e^{(2+5.1-8+.7)} = e^{-.2}.$$

Problems

1. $\prod_{i=1}^{4} i$ 2. $\prod_{j=1}^{5} 10^j$ 3. $\prod_{k=1}^{3} e^{(x_k-1)^2}$

Answers

1. 24   2. 10^{15} 3. $e^{\Sigma(x-\bar{x})^2 + 3(\bar{x}-1)^2}$, where $\bar{x} = \frac{1}{3}\Sigma x$

A3. FACTORIALS

! The exclamation point is the sign of the *factorial* function. Unlike almost all other functional names, it is written *after* its argument rather than before. This peculiar feature sometimes requires us to pay special attention to parentheses.

For a positive integer n we define n-factorial to be

$$n! = 1 \times 2 \times 3 \times 4 \times \cdots \times (n-1) \times n,$$

that is, the product of the numbers from 1 up to n. In product notation,

$$n! = \prod_{k=1}^{n} k.$$

Note the relation

$$n! = n \times (n-1)!.$$

To make things come out right for some other purposes, we define $0! = 1$. Note how fast the function increases as n increases.

$$0! = 1, \quad 2! = 2, \quad 4! = 24, \quad 10! = 3{,}628{,}800,$$
$$1! = 1, \quad 3! = 6, \quad 5! = 120, \quad 100! \doteq 10^{158}.$$

The factorial function is defined even if x is not an integer. There is no very easy way to describe how the calculation is done, but the values of $x!$ go smoothly between the values we know for the integers. The relation

$$x! = x \times (x-1)!$$

holds for all x, whether integer or not.

Some of the values of $x!$ for x not an integer turn out to be simple. For example,

$$(\tfrac{1}{2})! = \sqrt{\pi}/2.$$

Values of $x!$ for other than the integers and half-integers are undistinguished infinite decimals. Sometimes the term *gamma function* is used in place of *factorial* for non-integral x. We will stick with the factorial terminology in this book.

Problems

1. $6!$ 2. $7!$ 3. $(\tfrac{3}{2})!$

Answers

1. 720 2. 5040 3. $\tfrac{3}{4}\sqrt{\pi}$

A4. BINOMIAL COEFFICIENTS

$\binom{n}{r}$ This arrangement of two numbers *without a fraction bar* is a "binomial coefficient." Its value is the number of combinations of n things taken r at a time. For example,

$$\binom{5}{2} = 10,$$

since there are 10 ways of picking pairs from a set of 5:

	1	2	3	4	5
	×	×			
	×		×		
	×			×	
	×				×
		×	×		
		×		×	
		×			×
			×	×	
			×		×
				×	×

These numbers have scads of fascinating properties which are well covered in every book on probability. I list only two:

$$\binom{n}{r} = \frac{n!}{r!(n-r)!} = \frac{n!}{(n-r)!r!} = \binom{n}{n-r},$$

$$(a+b)^n = \sum_{r=0}^{n} \binom{n}{r} a^r b^{n-r} \quad \text{for } n \text{ a positive integer.}$$

Examples

$$\binom{5}{2} = \frac{5!}{2!3!} = \frac{120}{2 \times 6} = 10 = \frac{5!}{3!2!} = \binom{5}{3},$$

$$(2+x)^3 = \binom{3}{0} 2^0 x^3 + \binom{3}{1} 2^1 x^2 + \binom{3}{2} 2^2 x^1 + \binom{3}{3} 2^3 x^0$$
$$= x^3 + 6x^2 + 12x + 8.$$

The appearance of the symbols $\binom{n}{r}$ as *coefficients* in the expansion of a *binomial* (an expression with *two* terms) explains their name. They are 0 for r a negative integer or for $r > n$.

Problems

1. $\binom{4}{2}$ 2. $\binom{6}{3}$

3. Show that

$$\binom{6}{2} + \binom{6}{3} = \binom{7}{3}.$$

4. Prove that in general
$$\binom{n}{r} + \binom{n}{r+1} = \binom{n+1}{r+1}.$$

Answers

1. 6   2. 20

A5. THE CONSTANT e

e   Our letter "ee" is the universal mathematical constant $e = 2.71828\ldots$. It's an infinite, never-repeating decimal. Though it happens that the next four decimal places are 1828, the tenth place is 5.

It can be expressed as a series,
$$e = 1 + \frac{1}{1!} + \frac{1}{2!} + \frac{1}{3!} + \frac{1}{4!} + \cdots,$$
as a limit,
$$e = \lim_{n \to \infty} \left(1 + \frac{1}{n}\right)^n,$$
and in a mystical relation,
$$e^{2\pi \sqrt{-1}} = 1.$$

The number *e* appears in many statistical formulas. For most purposes you can forget its mystical properties and treat it as just another numerical constant.

Problem

Sum the series as far as 1/4! and see how close you already are to the final value.

A6. NUMERICAL INTEGRATION

∫   The *integral* sign is actually an elongated *S*; it indicates that we are dealing with something related to a sum. We will treat integrals quite simply here. The emphasis will be on evaluating them numerically rather than on using the methods of calculus to derive closed formulas for them. The trouble is that most of the integrals we run across cannot be expressed in simple formulas. However, they can often be evaluated numerically; with modern computers it is simple, fast, and cheap. I will present here a few methods for computing integrals by hand. They are more for emergencies than for every-day use.

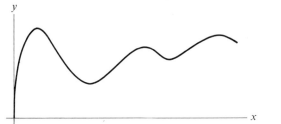

Fig. A6.1

If y is a function of x with a graph like Fig. A6.1, then

$$\int_a^b y(x)\, dx,$$

read "the integral of y with respect to x from $x = a$ to $x = b$" or, more briefly, "the integral of y 'dee' x from a to b," means the area shown in Fig. A6.2.

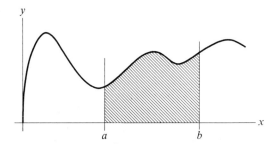

Fig. A6.2

In principle, all we have to do to evaluate such an area is to slice it up into narrow strips and approximate the area of each strip by a rectangle; then we add up all the rectangular areas (Fig. A6.3). As we make the strips narrower and narrower, we can find the area to as high an accuracy as we wish. Much of the work in numerical analysis deals with finding ways of taking into account the little corners and curved lines you neglected when you treated the strips as rectangles. If you can do a good job here, then you don't have to make so many subdivisions.

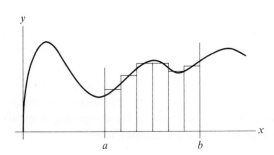

Fig. A6.3

A6 Numerical Integration

Example. The curve $y = 6x(1 - x)$ looks like Fig. A6.4 for $0 \leq x \leq 1$. What is the area enclosed between the curve and the x-axis? According to the above definition, it is $\int_0^1 6x(1 - x)\,dx$. If you know calculus, you can find that this area is 1. How do we find it numerically? First, let's do it according to the definition. We divide the interval $x = 0$ to $x = 1$ into ten strips of equal width (.1), and take as the height of each rectangle the value of $6x(1 - x)$ at the middle of each interval:

Interval	Middle (x)	Height (y)	Area
0.0–0.1	.05	.285	.0285
0.1–0.2	.15	.765	.0765
0.2–0.3	.25	1.125	.1125
0.3–0.4	.35	1.365	.1365
0.4–0.5	.45	1.485	.1485
0.5–0.6	.55	1.485	.1485
0.6–0.7	.65	1.365	.1365
0.7–0.8	.75	1.125	.1125
0.8–0.9	.85	.765	.0765
0.9–1.0	.95	.285	.0285
			1.0050

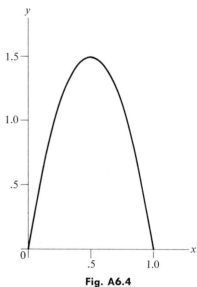

Fig. A6.4

The first line says:

a) the first interval is from $x = 0.0$ to $x = 0.1$;

b) the midpoint of this interval is $x = .05$;

c) the height is $6(.05)(1 - .05) = 6(.05)(.95) = .285$;

d) the area of this little rectangle is height $\times$ width $= .285 \times .1 = .0285$.

We're only off by one-half of one percent; if we had used fewer points it would have been worse.

We can usually do better if we use a formula which approximates the curve at the top of each strip by a parabola. It is called *Simpson's rule:*

a) Divide the interval into an *even* number of subintervals. Call the width of the subinterval h.

b) Calculate the values of the function at the endpoints and all the dividing points.

c) Multiply these values by the coefficients

$$1, 4, 2, 4, 2, \ldots, 2, 4, 1,$$

that is, 1 for the endpoints, and alternating 4's and 2's inbetween. For three points, the coefficients are 1, 4, and 1.

d) Sum these products and multiply the sum by $h/3$.

Let's see how it goes for the curve we have been working with:

Point (x)	Height (y)	Coefficient	Product
.0	.00	1	.00
.1	.54	4	2.16
.2	.96	2	1.92
.3	1.26	4	5.04
.4	1.44	2	2.88
.5	1.50	4	6.00
.6	1.44	2	2.88
.7	1.26	4	5.04
.8	.96	2	1.92
.9	.54	4	2.16
1.0	.00	1	.00
			30.00

and since $h = .1$,

$$30.00 \times \frac{.1}{3} = 1.00. \quad \text{Right on the head!}$$

You shouldn't be impressed by this; Simpson's rule always gives exact answers for parabolas. In fact, I can do it with only two subintervals (three points):

Point (x)	Height (y)	Coefficient	Product
.0	.00	1	.00
.5	1.50	4	6.00
1.0	.00	1	.00
			6.00

and since $h = .5$,

$$6.00 \times \frac{.5}{3} = 1.00.$$

Let's try it with the quadrant of a circle. The equation is $y = \sqrt{1 - x^2}$, and you know the area is $\pi/4 = .7854$ (Fig. A6.5).

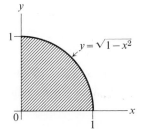

Fig. A6.5

Point (x)	Height (y)	Coefficient	Product
.00	1.000	1	1.000
.25	.968	4	3.872
.50	.866	2	1.732
.75	.661	4	2.644
1.00	.000	1	.000
			9.248

($h = .25$)

$$9.248 \times \frac{.25}{3} = .771 \quad (1.8\% \text{ error}).$$

If I use only two subintervals, I get:

Point (x)	Height (y)	Coefficient	Product
.0	1.000	1	1.000
.5	.866	4	3.464
1.0	.000	1	.000
			4.464

($h = .5$)

$$4.464 \times \frac{.5}{3} = .744 \quad (5.2\% \text{ error}).$$

Problems

1. Find $\int_4^5 x^3 \, dx$ by using two subintervals.
2. Find $\int_0^1 x \log_{10}(1 + x) \, dx$.

Answers

1. Use points 4, 4.5, 5; cubes are 64, 91.125, 125; coefficients are 1, 4, 1; products are 64, 364.5, 125. Hence

$$\int_4^5 x^3 \, dx = 553.5 \times \frac{.5}{3} = 92.25.$$

2. .10857...

Appendix B. Logarithms

B1. USING LOGARITHMS

I have a number N. What is $\log_{10} N$?

N cannot be 0. N cannot be negative. Handle your signs separately and work only with positive numbers.

Steps to follow	Example 1	Example 2
	$N = 22.61$	$N = .0139$
Rewrite N as $A \times 10^B$, where $1.0 \leq A < 10.0$.	$N = 2.261 \times 10^1$	$N = 1.39 \times 10^{-2}$
Find the table entry M corresponding to A. You may have to interpolate. [A decimal point should go before each table entry.]	$\begin{array}{ll} A & M \\ 2.26 \to .3541 \\ 2.27 \to .3560 \\ \left[\begin{array}{l}\text{diff} = .0019 \\ .1 \times \text{diff} = .0002\end{array}\right] \\ 2.261 \to .3543 \end{array}$	$\begin{array}{ll} A & M \\ 1.39 \to .1430 \end{array}$
$\log_{10} N = M + B$	$\log_{10} 22.61 = 1.3543$	$\log_{10} .0139 = .1430 - 2$ $= -1.8570$

B1   Using Logarithms

I have $\log_{10} N$. What is N?

Steps to follow	Example 1	Example 2
	$\log_{10} N = 4.9162$	$\log_{10} N = -2.1416$
Rewrite $\log_{10} N$ as $M + B$, where $0 \leq M < 1.0$	$.9162 + 4$	$.8584 - 3$
Find M in the inside of the table. Find the coordinates and call them A. You may have to interpolate. Here again $1.0 \leq A < 10.0$.	$\begin{array}{cc} M & A \\ \hline .9159 \to 8.24 \\ .9165 \to 8.25 \\ \left[\begin{array}{l}.0006 = \text{diff} \\ .9162 - .9159 = .0003 \\ .0003/.0006 = .5\end{array}\right] \\ .9162 \to 8.245 \end{array}$	$\begin{array}{cc} M & A \\ \hline .8579 \to 7.21 \\ .8585 \to 7.22 \\ \left[\begin{array}{l}.0006 = \text{diff} \\ .8584 - .8579 = .0005 \\ .0005/.0006 = .8\end{array}\right] \\ .8584 \to 7.218 \end{array}$
$N = A \times 10^B$	$N = 8.245 \times 10^4$ $= 82{,}450$	$N = 7.218 \times 10^{-3}$ $= .007218$

Rules

1. [Here x and y must be positive, but a may be positive, zero, or negative.]

$$\log_{10}(x \times y) = \log_{10} x + \log_{10} y,$$
$$\log_{10}(x/y) = \log_{10} x - \log_{10} y,$$
$$\log_{10} x^a = a \times \log_{10} x,$$
$$\log_{10} \sqrt[y]{x} = (1/y) \times \log_{10} x.$$

You cannot use logarithms to add or subtract numbers.

2. Logarithms to base e are written $\ln N$:

$$\ln N = 2.3026 \times \log_{10} N \quad \left(2.3026 = \frac{1}{\log_{10} e}\right).$$

3. Logarithms to base 2 are written $\lg N$:

$$\lg N = 3.3219 \times \log_{10} N \quad \left(3.3219 = \frac{1}{\log_{10} 2}\right).$$

The expression $\log N$ is a little ambiguous. Sometimes it means base 10 and sometimes it means base e.

B2. LOGARITHMS TO BASE 10

	0	1	2	3	4	5	6	7	8	9
1.0	0000	0043	0086	0128	0170	0212	0253	0294	0334	0374
1.1	0414	0453	0492	0531	0569	0607	0645	0682	0719	0755
1.2	0792	0828	0864	0899	0934	0969	1004	1038	1072	1106
1.3	1139	1173	1206	1239	1271	1303	1335	1367	1399	1430
1.4	1461	1492	1523	1553	1584	1614	1644	1673	1703	1732
1.5	1761	1790	1818	1847	1875	1903	1931	1959	1987	2014
1.6	2041	2068	2095	2122	2148	2175	2201	2227	2253	2279
1.7	2304	2330	2355	2380	2405	2430	2455	2480	2504	2529
1.8	2553	2577	2601	2625	2648	2672	2695	2718	2742	2765
1.9	2788	2810	2833	2856	2878	2900	2923	2945	2967	2989
2.0	3010	3032	3054	3075	3096	3118	3139	3160	3181	3201
2.1	3222	3243	3263	3284	3304	3324	3345	3365	3385	3404
2.2	3424	3444	3464	3483	3502	3522	3541	3560	3579	3598
2.3	3617	3636	3655	3674	3692	3711	3729	3747	3766	3784
2.4	3802	3820	3838	3856	3874	3892	3909	3927	3945	3962
2.5	3979	3997	4014	4031	4048	4065	4082	4099	4116	4133
2.6	4150	4166	4183	4200	4216	4232	4249	4265	4281	4298
2.7	4314	4330	4346	4362	4378	4393	4409	4425	4440	4456
2.8	4472	4487	4502	4518	4533	4548	4564	4579	4594	4609
2.9	4624	4639	4654	4669	4683	4698	4713	4728	4742	4757
3.0	4771	4786	4800	4814	4829	4843	4857	4871	4886	4900
3.1	4914	4928	4942	4955	4969	4983	4997	5011	5024	5038
3.2	5051	5065	5079	5092	5105	5119	5132	5145	5159	5172
3.3	5185	5198	5211	5224	5237	5250	5263	5276	5289	5302
3.4	5315	5328	5340	5353	5366	5378	5391	5403	5416	5428
3.5	5441	5453	5465	5478	5490	5502	5514	5527	5539	5551
3.6	5563	5575	5587	5599	5611	5623	5635	5647	5658	5670
3.7	5682	5694	5705	5717	5729	5740	5752	5763	5775	5786
3.8	5798	5809	5821	5832	5843	5855	5866	5877	5888	5899
3.9	5911	5922	5933	5944	5955	5966	5977	5988	5999	6010
4.0	6021	6031	6042	6053	6064	6075	6085	6096	6107	6117
4.1	6128	6138	6149	6160	6170	6180	6191	6201	6212	6222
4.2	6232	6243	6253	6263	6274	6284	6294	6304	6314	6325
4.3	6335	6345	6355	6365	6375	6385	6395	6405	6415	6425
4.4	6435	6444	6454	6464	6474	6484	6493	6503	6513	6522
4.5	6532	6542	6551	6561	6571	6580	6590	6599	6609	6618
4.6	6628	6637	6646	6656	6665	6675	6684	6693	6702	6712
4.7	6721	6730	6739	6749	6758	6767	6776	6785	6794	6803
4.8	6812	6821	6830	6839	6848	6857	6866	6875	6884	6893
4.9	6902	6911	6920	6928	6937	6946	6955	6964	6972	6981
5.0	6990	6998	7007	7016	7024	7033	7042	7050	7059	7067
5.1	7076	7084	7093	7101	7110	7118	7126	7135	7143	7152
5.2	7160	7168	7177	7185	7193	7202	7210	7218	7226	7235
5.3	7243	7251	7259	7267	7275	7284	7292	7300	7308	7316
5.4	7324	7332	7340	7348	7356	7364	7372	7380	7388	7396

$$\log_{10} e = .4343, \quad \log_{10} \pi = .4971$$

Logarithms to base 10 (Cont'd)

	0	1	2	3	4	5	6	7	8	9
5.5	7404	7412	7419	7427	7435	7443	7451	7459	7466	7474
5.6	7482	7490	7497	7505	7513	7520	7528	7536	7543	7551
5.7	7559	7566	7574	7582	7589	7597	7604	7612	7619	7627
5.8	7634	7642	7649	7657	7664	7672	7679	7686	7694	7701
5.9	7709	7716	7723	7731	7738	7745	7752	7760	7767	7774
6.0	7782	7789	7796	7803	7810	7818	7825	7832	7839	7846
6.1	7853	7860	7868	7875	7882	7889	7896	7903	7910	7917
6.2	7924	7931	7938	7945	7952	7959	7966	7973	7980	7987
6.3	7993	8000	8007	8014	8021	8028	8035	8041	8048	8055
6.4	8062	8069	8075	8082	8089	8096	8102	8109	8116	8122
6.5	8129	8136	8142	8149	8156	8162	8169	8176	8182	8189
6.6	8195	8202	8209	8215	8222	8228	8235	8241	8248	8254
6.7	8261	8267	8274	8280	8287	8293	8299	8306	8312	8319
6.8	8325	8331	8338	8344	8351	8357	8363	8370	8376	8382
6.9	8388	8395	8401	8407	8414	8420	8426	8432	8439	8445
7.0	8451	8457	8463	8470	8476	8482	8488	8494	8500	8506
7.1	8513	8519	8525	8531	8537	8543	8549	8555	8561	8567
7.2	8573	8579	8585	8591	8597	8603	8609	8615	8621	8627
7.3	8633	8639	8645	8651	8657	8663	8669	8675	8681	8686
7.4	8692	8698	8704	8710	8716	8722	8727	8733	8739	8745
7.5	8751	8756	8762	8768	8774	8779	8785	8791	8797	8802
7.6	8808	8814	8820	8825	8831	8837	8842	8848	8854	8859
7.7	8865	8871	8876	8882	8887	8893	8899	8904	8910	8915
7.8	8921	8927	8932	8938	8943	8949	8954	8960	8965	8971
7.9	8976	8982	8987	8993	8998	9004	9009	9015	9020	9025
8.0	9031	9036	9042	9047	9053	9058	9063	9069	9074	9079
8.1	9085	9090	9096	9101	9106	9112	9117	9122	9128	9133
8.2	9138	9143	9149	9154	9159	9165	9170	9175	9180	9186
8.3	9191	9196	9201	9206	9212	9217	9222	9227	9232	9238
8.4	9243	9248	9253	9258	9263	9269	9274	9279	9284	9289
8.5	9294	9299	9304	9309	9315	9320	9325	9330	9335	9340
8.6	9345	9350	9355	9360	9365	9370	9375	9380	9385	9390
8.7	9395	9400	9405	9410	9415	9420	9425	9430	9435	9440
8.8	9445	9450	9455	9460	9465	9469	9474	9479	9484	9489
8.9	9494	9499	9504	9509	9513	9518	9523	9528	9533	9538
9.0	9542	9547	9552	9557	9562	9566	9571	9576	9581	9586
9.1	9590	9595	9600	9605	9609	9614	9619	9624	9628	9633
9.2	9638	9643	9647	9652	9657	9661	9666	9671	9675	9680
9.3	9685	9689	9694	9699	9703	9708	9713	9717	9722	9727
9.4	9731	9736	9741	9745	9750	9754	9759	9763	9768	9773
9.5	9777	9782	9786	9791	9795	9800	9805	9809	9814	9818
9.6	9823	9827	9832	9836	9841	9845	9850	9854	9859	9863
9.7	9868	9872	9877	9881	9886	9890	9894	9899	9903	9908
9.8	9912	9917	9921	9926	9930	9934	9939	9943	9948	9952
9.9	9956	9961	9965	9969	9974	9978	9983	9987	9991	9996

$$\log_{10} \frac{1}{\sqrt{2\pi}} = -.3991, \quad \log_{10} \sqrt{\frac{\pi}{2}} = .0981$$

B3. $\log_{10} N!$

N	$\log_{10} N!$	N	$\log_{10} N!$	N	$\log_{10} N!$	N	$\log_{10} N!$
0	.0000	40	47.9116	80	118.8547	120	198.8254
1	.0000	41	49.5244	81	120.7632	121	200.9082
2	.3010	42	51.1477	82	122.6770	122	202.9945
3	.7782	43	52.7811	83	124.5961	123	205.0844
4	1.3802	44	54.4246	84	126.5204	124	207.1779
5	2.0792	45	56.0778	85	128.4498	125	209.2748
6	2.8573	46	57.7406	86	130.3843	126	211.3751
7	3.7024	47	59.4127	87	132.3238	127	213.4790
8	4.6055	48	61.0939	88	134.2683	128	215.5862
9	5.5598	49	62.7841	89	136.2177	129	217.6967
10	6.5598	50	64.4831	90	138.1719	130	219.8107
11	7.6012	51	66.1906	91	140.1310	131	221.9280
12	8.6803	52	67.9066	92	142.0948	132	224.0485
13	9.7943	53	69.6309	93	144.0632	133	226.1724
14	10.9404	54	71.3633	94	146.0364	134	228.2995
15	12.1165	55	73.1037	95	148.0141	135	230.4298
16	13.3206	56	74.8519	96	149.9964	136	232.5634
17	14.5511	57	76.6077	97	151.9831	137	234.7001
18	15.8063	58	78.3712	98	153.9744	138	236.8400
19	17.0851	59	80.1420	99	155.9700	139	238.9830
20	18.3861	60	81.9202	100	157.9700	140	241.1291
21	19.7083	61	83.7055	101	159.9743	141	243.2783
22	21.0508	62	85.4979	102	161.9829	142	245.4306
23	22.4125	63	87.2972	103	163.9958	143	247.5860
24	23.7927	64	89.1034	104	166.0128	144	249.7443
25	25.1906	65	90.9163	105	168.0340	145	251.9057
26	26.6056	66	92.7359	106	170.0593	146	254.0700
27	28.0370	67	94.5619	107	172.0887	147	256.2374
28	29.4841	68	96.3945	108	174.1221	148	258.4076
29	30.9465	69	98.2333	109	176.1595	149	260.5808
30	32.4237	70	100.0784	110	178.2009	150	262.7569
31	33.9150	71	101.9297	111	180.2462	151	264.9359
32	35.4202	72	103.7870	112	182.2955	152	267.1177
33	36.9387	73	105.6503	113	184.3485	153	269.3024
34	38.4702	74	107.5196	114	186.4054	154	271.4899
35	40.0142	75	109.3946	115	188.4661	155	273.6803
36	41.5705	76	111.2754	116	190.5306	156	275.8734
37	43.1387	77	113.1619	117	192.5988	157	278.0693
38	44.7185	78	115.0540	118	194.6707	158	280.2679
39	46.3096	79	116.9516	119	196.7462	159	282.4693

For larger N,

$$\log_{10} N! \doteq .3991 + (N + \tfrac{1}{2}) \log_{10} N - .4343N.$$

B4. $-p \log_{10} p$

	0	1	2	3	4	5	6	7	8	9
.000	0000	0004	0007	0011	0014	0017	0019	0022	0025	0027
.00		0030	0054	0076	0096	0115	0133	0151	0168	0184
.0		0200	0340	0457	0559	0651	0733	0808	0878	0941
.1	1000	1054	1105	1152	1195	1236	1273	1308	1341	1370
.2	1398	1423	1447	1468	1487	1505	1521	1535	1548	1559
.3	1569	1577	1584	1589	1593	1596	1597	1598	1597	1595
.4	1592	1588	1582	1576	1569	1561	1551	1541	1530	1518
.5	1505	1491	1477	1461	1445	1428	1410	1392	1372	1352
.6	1331	1309	1287	1264	1240	1216	1191	1165	1139	1112
.7	1084	1056	1027	0998	0968	0937	0906	0874	0842	0809
.8	0775	0741	0707	0672	0636	0600	0563	0526	0489	0450
.9	0412	0373	0333	0293	0253	0212	0170	0129	0086	0043
.99		0039	0035	0030	0026	0022	0017	0013	0009	0004
1.0	0000									

Each entry above should be preceded by a decimal point.
If you want logarithms to base e, multiply the entries by 2.3026.
If you want logarithms to base 2, multiply the entries by 3.3219.

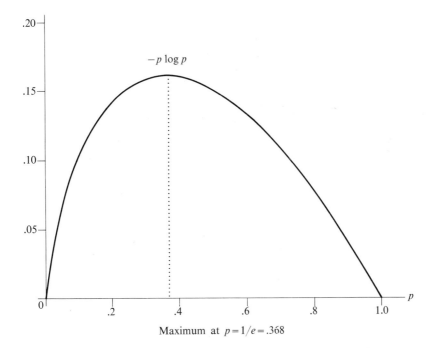

Maximum at $p = 1/e = .368$

Appendix C. Random Numbers

C1. FLAT RANDOM DIGITS

	00 01 02 03 04	05 06 07 08 09	10 11 12 13 14	15 16 17 18 19
00	76 60 40 33 45	33 80 40 85 91	43 65 50 92 08	86 79 61 39 89
01	69 14 45 24 33	34 59 79 67 03	98 85 34 93 60	15 37 49 16 70
02	78 47 90 12 16	88 17 31 26 34	69 35 81 15 09	31 29 99 92 80
03	69 81 12 95 30	09 24 85 84 33	71 01 47 76 45	38 42 03 64 64
04	99 92 94 90 22	01 39 79 20 15	80 62 14 97 89	47 35 09 58 50
05	45 81 07 99 49	98 49 17 75 78	16 99 67 47 68	80 32 81 24 15
06	89 13 92 24 58	49 66 79 68 73	40 53 32 90 83	04 45 79 51 47
07	73 64 39 27 14	19 34 21 81 03	97 38 60 44 92	49 53 16 66 56
08	05 51 14 85 67	39 88 58 42 09	75 15 59 50 76	40 87 30 06 12
09	95 46 18 11 19	68 23 65 55 63	94 30 29 78 23	37 95 44 03 56
10	09 02 89 92 44	66 85 86 70 21	99 91 88 60 71	74 39 92 68 30
11	87 29 76 77 69	06 15 00 17 44	96 41 47 74 77	75 56 72 43 45
12	71 62 53 20 00	38 55 91 62 59	56 11 03 35 22	93 89 02 05 09
13	96 93 47 04 92	78 98 33 24 34	16 22 07 79 74	65 23 86 22 06
14	03 55 90 58 71	21 33 27 42 69	70 08 93 72 37	02 16 04 88 73
15	25 82 45 50 03	88 57 60 49 97	32 17 48 76 71	09 45 10 07 31
16	25 96 54 63 39	18 09 22 57 69	30 70 03 73 36	52 80 11 68 64
17	45 78 84 34 95	10 71 44 60 12	74 27 77 86 94	17 25 90 90 79
18	09 74 23 67 52	33 41 62 04 25	39 97 74 52 06	81 36 85 38 45
19	82 90 55 22 16	14 21 34 04 99	44 69 55 72 01	04 83 85 48 09
20	37 08 94 37 00	22 07 99 12 78	15 53 32 15 02	08 50 16 28 99
21	56 38 95 58 49	59 63 31 31 78	89 93 83 89 59	04 45 31 15 26
22	02 29 60 58 52	70 17 82 82 58	40 93 09 14 24	04 58 51 27 63
23	40 73 28 10 69	92 52 11 04 51	29 06 06 42 78	89 56 73 51 73
24	00 83 46 22 02	20 66 62 21 16	33 94 00 64 84	95 70 46 42 47
25	12 20 83 46 42	48 48 30 03 96	69 68 74 82 33	48 13 93 49 66
26	00 07 48 38 89	57 18 20 39 90	57 88 79 19 47	64 93 64 56 36
27	51 36 76 18 20	36 14 48 54 51	05 40 92 79 57	17 56 56 22 88
28	53 53 81 82 03	08 27 77 53 46	67 85 95 06 34	56 02 46 93 03
29	88 37 84 18 98	87 96 79 43 45	12 03 65 23 35	36 49 48 50 49
30	38 13 15 73 74	01 83 57 67 00	50 74 29 01 17	98 02 05 45 87
31	93 27 95 09 06	81 91 10 14 94	03 11 52 08 84	72 43 24 31 62
32	50 30 51 87 08	72 12 33 02 82	97 53 85 77 79	11 06 55 54 14
33	69 24 81 33 60	14 44 82 75 00	59 00 19 56 45	76 62 11 09 64
34	50 25 60 94 70	15 42 19 71 25	40 21 61 57 20	12 35 36 88 92
35	68 80 30 67 36	30 89 17 34 94	17 50 53 31 68	11 83 32 61 56
36	70 93 34 28 32	15 49 37 93 08	59 94 46 66 92	45 24 32 86 17
37	63 88 12 09 59	02 93 76 94 98	49 42 19 09 78	58 02 65 41 16
38	41 67 79 86 36	52 59 14 80 89	26 29 35 86 95	97 87 41 84 40
39	89 30 68 31 64	18 93 34 13 42	25 26 57 45 19	20 85 87 46 12
40	86 66 29 55 70	98 28 98 11 27	99 75 21 22 06	89 00 25 26 73
41	10 09 74 99 11	74 23 62 11 27	69 88 03 38 25	40 79 29 24 79
42	74 67 58 19 59	36 41 83 33 08	41 16 77 97 19	51 67 30 32 49
43	76 71 98 34 91	41 30 22 57 68	56 03 57 96 52	09 89 73 92 34
44	91 24 19 46 36	93 34 00 83 20	21 78 53 13 79	27 70 27 34 95
45	19 81 94 17 03	92 63 26 08 22	00 62 07 93 37	59 14 07 81 21
46	72 74 09 28 34	59 83 46 25 74	88 17 64 39 36	80 39 05 49 13
47	27 19 08 62 88	75 69 19 57 46	17 24 03 93 00	49 61 98 74 63
48	94 51 95 75 76	10 98 27 36 94	92 39 39 43 77	91 66 39 97 66
49	32 80 07 16 06	35 27 71 64 44	38 19 15 74 15	69 15 08 31 79

C2. NORMAL DEVIATES
Random values from GAU (* | 0, 1)

	00	01	02	03	04	05	06	07	08	09
00	.31	-.51	-1.45	-.35	.18	.09	.00	.11	-1.91	-1.07
01	.90	-.36	.33	-.28	.30	-2.62	-1.43	-1.79	-.99	-.35
02	.22	.58	.87	-.02	.04	.12	-.17	.78	-1.31	.95
03	-1.00	.53	-1.90	-.77	.67	.56	-.94	.16	2.22	-.08
04	-.12	-.43	.69	.75	-.32	-.71	-1.13	-.79	-.26	-.86
05	.01	.37	-.36	.68	.44	.43	1.18	-.68	-.13	-.41
06	.16	-.83	-1.88	.89	-.39	.93	-.76	-.12	.66	2.06
07	1.31	-.82	-.36	.36	.24	-.95	.41	-.77	.78	-.27
08	-.38	-.26	-1.73	.06	-.14	1.59	.96	-1.39	.51	-.05
09	.38	.42	-1.39	-.22	-.28	-.03	2.48	1.11	-1.10	.40
10	1.07	2.26	-1.68	-.04	.19	1.38	-1.53	-1.41	.09	-1.91
11	-1.65	-1.29	-1.03	.06	2.18	-.55	-.34	-1.07	.80	1.77
12	1.02	-.67	-1.11	.08	-1.92	-.97	-.70	-.04	-.72	-.47
13	.06	1.43	-.46	-.62	-.11	.36	.64	-.27	.72	.68
14	.47	-1.84	.69	-1.07	.83	-.25	-.91	-1.94	.96	.75
15	.10	1.00	-.54	.61	-1.04	-.33	.94	.56	.62	.07
16	-.71	.04	.63	-.26	-1.35	-1.20	1.52	.63	-1.29	1.16
17	-.94	-.94	.56	-.09	.63	-.36	.20	-.60	-.29	.94
18	.29	.62	-1.09	1.84	-.11	.19	-.45	.23	-.63	-.06
19	.57	.54	-.21	.09	-.57	-.10	-1.25	-.26	.88	-.26
20	.24	.19	-.67	3.04	1.26	-1.21	.52	-.05	.76	-.09
21	-1.47	1.20	.70	-1.80	-1.07	.29	1.18	.34	-.74	1.75
22	-.01	.49	1.16	.17	-.48	.81	1.40	-.17	.57	.64
23	-.63	-.26	.55	-.21	-.07	-.37	.47	-1.69	.05	-.96
24	.85	-.65	-.94	.12	-1.67	.28	-.42	.14	-1.15	-.41
25	1.07	-.36	1.10	.83	.37	-.20	-.75	-.50	.18	1.31
26	1.18	-2.09	-.61	.44	.40	.42	-.61	-2.55	-.09	-1.33
27	.47	.88	.71	.31	.41	-1.96	.34	-.17	1.73	-.33
28	.26	.90	.11	.28	.76	-.12	-1.01	1.29	-.71	2.15
29	.39	-.88	-.15	-.38	.55	-.41	-.02	-.74	-.48	.46
30	-1.01	-.89	-1.23	.07	-.07	-.08	-.08	-1.95	-.34	-.29
31	1.36	.18	.85	.55	.00	-.43	.27	-.39	.25	.69
32	1.02	-2.49	1.79	.04	-.03	.85	-.29	-.77	.28	-.33
33	-.53	-1.13	.75	-.39	.43	.10	-2.17	.37	-1.85	.96
34	.76	1.21	-.68	.26	.93	.99	1.12	-1.72	-.04	-.73
35	.07	-.23	-.88	-.23	.68	.24	1.38	-2.10	-.79	-.27
36	.27	.61	.43	-.38	.68	-.72	.90	-.14	-1.61	-.88
37	.93	.72	-.45	2.80	-.12	.74	-1.47	.39	-.61	-2.77
38	1.03	-.43	.95	-1.49	-.63	.22	.79	-2.80	-.41	.61
39	-.32	1.41	-.23	-.36	.60	-.59	.36	.63	.73	.81
40	1.41	.64	.06	.25	-1.75	.39	1.84	1.23	-1.27	-.75
41	.25	-.70	.33	.12	.04	1.03	-.64	.08	1.63	.34
42	-1.15	.57	.34	-.32	2.31	.74	.85	-1.25	-.17	.14
43	.72	.01	.50	-1.42	.26	-.74	-.55	1.86	-.17	-.10
44	-.92	.15	-.66	.83	.50	.24	-.40	1.90	.35	.69
45	-.42	.62	.24	.55	-.06	.14	-1.09	-1.53	.30	-1.56
46	-.54	1.21	-.53	.29	1.04	-.32	-1.20	.01	.05	.20
47	-.13	-.70	.07	.69	.88	1.18	.61	-.46	-1.54	.50
48	-.29	.36	1.44	-.44	.53	-.14	.66	.00	.33	-.36
49	1.90	-1.21	-1.87	-.27	-1.86	-.49	.25	.25	.14	1.73

Random values from GAU (∗ | 0, 1) (Cont'd)

	10	11	12	13	14	15	16	17	18	19
00	-.73	.25	-2.08	.17	-1.04	-.23	.74	.23	.70	-.79
01	-.87	-.74	1.44	-.79	-.76	-.42	1.93	.88	.80	-.53
02	1.18	.05	.10	-.15	.05	1.06	.82	.90	-1.38	.51
03	-2.09	1.13	-.50	.37	-.18	-.16	-1.85	-.90	1.32	-.83
04	-.32	1.06	1.14	-.23	.49	1.10	-.27	-.64	.47	-.05
05	.90	-.86	.63	-1.62	-.52	-1.55	.78	-.54	-.29	.19
06	-.16	-.22	-.17	-.81	.49	.96	.53	1.73	.14	1.21
07	.15	-1.12	.80	-.30	-.77	-.91	.00	.94	-1.16	.44
08	-1.87	.72	-1.17	-.36	-1.42	-.46	-.58	.03	2.08	1.11
09	.87	.95	.05	.46	-.01	.85	1.19	-1.61	-.10	-.87
10	.52	.12	-1.04	-.56	-.91	-.13	.17	1.17	-1.24	-.84
11	-1.39	-1.18	1.67	2.88	-2.06	.10	.05	-.55	.74	.33
12	-.94	-.46	-.85	-.29	.54	.71	.90	-.42	-1.30	.50
13	-.51	.04	-.44	-1.87	-1.06	1.18	-.39	.22	-.55	-.54
14	-1.50	-.21	-.89	.43	-1.81	-.07	-.66	-.02	1.77	-1.54
15	-.48	1.54	1.88	.66	-.62	.28	-.34	2.42	-1.65	2.06
16	.89	-.23	.57	.23	1.81	1.02	.33	1.23	1.31	.06
17	.38	1.52	-1.32	2.13	-.14	.28	-.46	.25	.65	1.18
18	-.53	.37	.19	-2.41	.16	.36	-.15	.14	-.15	-.73
19	.15	.62	-1.29	1.84	.80	-.65	1.72	-1.77	.07	.46
20	-.81	-.22	1.16	1.09	-.73	-.15	.87	-.88	.92	-.04
21	-1.61	2.51	-2.17	.49	-1.24	1.16	.97	.15	.37	.18
22	.26	-.48	-.43	-2.08	.75	1.59	1.78	-.55	.85	-1.87
23	-.32	.75	-.35	2.10	-.70	1.29	.94	.20	-1.16	.89
24	-1.00	1.37	.68	.00	1.87	-.14	.77	-.12	.89	-.73
25	.66	.04	-1.73	.25	.26	1.46	-.77	-1.67	.18	-.92
26	-.20	-1.53	.59	-.15	-.15	-.11	.68	-.14	-.42	-1.51
27	1.01	-.44	-.20	-2.05	-.27	-.50	-.27	-.45	.83	.49
28	-1.81	.45	.27	.67	-.74	-.17	-1.11	.13	-1.18	-1.41
29	-.40	1.34	1.50	.57	-1.78	.08	.95	.69	.38	.71
30	-.01	.15	-1.83	1.18	.11	.62	1.86	.42	.03	-.14
31	-.23	-.19	-1.08	.44	-.41	-1.32	.14	.65	-.76	.76
32	-1.27	.13	-.17	-.74	-.44	1.67	-.07	-.99	.51	.76
33	-1.72	1.70	-.61	.18	.48	-.26	-.12	-2.83	2.35	1.25
34	.78	1.55	-.19	.43	-1.53	-.76	.83	-.46	.48	-.43
35	1.86	1.12	-2.09	1.82	-.71	-1.76	-.20	-.38	.82	-1.08
36	-.50	-.93	-.68	-1.62	-.88	.05	-.27	.23	-.58	-.24
37	1.02	-.81	-.62	1.46	-.31	-.37	.08	.59	-.27	.37
38	-1.57	.10	.11	-1.48	1.02	2.35	.27	-1.22	-1.26	2.22
39	2.27	-.61	.61	-.28	-.39	-.45	-.89	1.43	-1.03	-.01
40	-2.17	-.69	1.33	-.26	.15	-.10	-.78	.64	-.70	.14
41	.05	-1.71	.21	.55	-.60	-.74	-.90	2.52	-.07	-1.11
42	-.38	1.75	.93	-1.36	-.60	-1.76	-1.10	.42	1.44	-.58
43	.40	-1.50	.24	-.66	.83	.37	-.35	.16	.96	.79
44	.39	.66	.19	-2.08	.32	-.42	-.53	.92	.69	-.03
45	-.12	1.18	-.08	.30	-.21	.45	-1.84	.26	.90	.85
46	1.20	-.91	-1.08	-.99	1.76	-.80	.51	.25	-.11	-.58
47	-1.04	1.28	2.50	1.56	-.95	-1.02	.45	-1.90	-.02	-.73
48	-.32	.56	-1.03	.11	-.72	.53	-.27	-.17	1.40	1.61
49	1.08	.56	.34	-.28	-.37	.46	.03	-1.13	.34	-1.08

A Gallery   *Table of Kernels / Notation / BETA / BIGAU /*
of Standard   *BIN / CIRG / EXPD / GAMMA / GAU / IGAM /*
Distributions   *NEGB / POIS / RECT / STU / WEIB*

A Gallery of Standard Distributions

Here I have collected the commonest statistical distributions. I have given the formulas for the density, mean, variance, mode, and other useful information. I have put tables and graphs of the distributions with their descriptions. The main purpose of this section is reference; few people ever have a detailed working knowledge of it. Hence the distributions are given in alphabetical order, not in the order in which they are discussed in the book.

TABLE OF KERNELS

This table is designed to help you identify posterior densities. In the expressions, x represents the variable under study. Multiplicative constants not involving x have been discarded since normalization will take care of them. The letters a, b, c, and d represent the constants as demanded by the problem.

1	RECT, FLAT	$\exp\{-bx\}$	EXPD
$1/x$	LFLAT	$x^a \exp\{-bx\}$	GAMMA
$x^a(1-x)^b$	BETA	$x^a \exp\{-bx^c\}$	WEIB
$\exp\{-(ax^2 + bx + c)\}$	GAU	$\exp\{-a(x-b)^2\}$	GAU
$\dfrac{1}{[a + b(x-c)^2]^d}$	STU	$\dfrac{1}{x^a}\exp\left\{-\dfrac{b}{x^2}\right\}$	IGAM

375

NOTATION

[I use the first distribution on the list, BETA, to help exhibit the notation.]

- P — probability of
- $|$ — given
- pd — probability density (in general)
- PD — cumulative distribution (in general)
- $\sim$ — is distributed according to
- $\dot\sim$ — is approximately distributed according to
- $\doteq$ — is approximately equal to
- $\propto$ — is proportional to
- BETA — the beta distribution
- $p \sim$ BETA — p is distributed according to the beta distribution
- beta $(p \mid s, f)$ — the value of the beta density at the point p when the parameters are s and f
- beta $(* \mid s, f)$ — the beta density with parameters s and f
- BETA $(p \mid s, f)$ — the cumulative beta distribution at the point p; that is, $P(\leq p \mid s, f)$
- BETA $(* \mid s, f)$ — the cumulative beta distribution
- $E[\text{beta}(* \mid s, f)]$ — the expected value (mean) of a beta variable with parameters s and f
- $D^2[\text{beta}(* \mid s, f)]$ — the variance of a beta variable with parameters s and f
- mode $[\text{beta}(* \mid s, f)]$ — the value of p for which beta $(p \mid s, f)$ is largest
- $\hat{p}$ — an estimate of p
- $\mathcal{F}$ — factor
- $\mathcal{O}$ — odds

BETA: The beta distribution

$$\text{beta}(p \mid s, f) = \frac{(s + f + 1)!}{s!\,f!} p^s (1 - p)^f \quad \text{for} \quad \begin{cases} 0 \leq p \leq 1, \\ -1 < s, \\ -1 < f, \end{cases}$$

$$E[\text{beta}(* \mid s, f)] = \frac{s + 1}{s + f + 2},$$

$$D^2[\text{beta}(* \mid s, f)] = \frac{(s + 1)(f + 1)}{(s + f + 2)^2 (s + f + 3)},$$

$$\text{mode}[\text{beta}(* \mid s, f)] = \frac{s}{s + f}.$$

This distribution is common both as a prior and as a posterior distribution for the p of the binomial variable bin $(* \mid n, p)$.

Caution: Some texts use the definition

$$\text{Be}(p \mid a, b) = \frac{(a+b-1)!}{(a-1)!(b-1)!} p^{a-1}(1-p)^{b-1}.$$

If you hear "beta function," check the definition.

You can find percentile tables of BETA in Bracken (1966). These tables use the alternative definition for the beta distribution. You can use them with our definition by setting

his $n =$ our $s + f + 2$ and   his $r =$ our $s + 1$.

Example. The median of BETA $(* \mid 2, 10)$ is .200. Look in Bracken's table for

$n = 14,$ $r = 3,$ .50 percentile

to find .2004.

Three tables of BETA appear on the following pages.

1. *Highest density regions:* intervals containing a specified amount of probability and having the property that the density inside is larger than the density outside.

Example. $s = 2, f = 10$. 95% of the probability is contained in (.032, .421), and the density inside this interval is larger than the density outside.

2. *Ratio of tail areas*

$$\frac{\text{right tail }(p)}{\text{left tail }(p)} = \frac{1 - \text{BETA}(p \mid s, f)}{\text{BETA}(p \mid s, f)}.$$

Example. $s = 2, f = 10$. At $p = .085$ the right tail area is 10 times as large as the left tail area. Note that the column headed 1/1 gives the *median*.

3. *Factor that* $p = \frac{1}{2}$

$$\mathcal{F}\left(\frac{p = \frac{1}{2}}{p \neq \frac{1}{2}} \,\middle|\, s, f\right) = \text{beta}(\tfrac{1}{2} \mid s, f).$$

Contour lines of this function are plotted.

Example. $s = 50, f = 25$. The odds are about $\frac{1}{10}$, that is, 10 to 1 *against* equality.

This calculation assumes that the alternatives to $p = \frac{1}{2}$ have a flat prior density.

BETA $(* \mid s, f)$
Highest density regions

s	f	.90	.95	.99		s	f	.90	.95	.99
0	0	.000	.684	(flat distribution)	.000 .776	.000 .900	3	0	.562 1.000	.473 1.000	.316 1.000
0	1	.000 .536	.000 .632	.000 .785		3	1	.395 .957	.330 .974	.220 .991
0	2	.000 .438	.000 .527	.000 .684		3	2	.289 .862	.239 .895	.156 .943
0	3	.000 .369	.000 .451	.000 .602		3	3	.225 .775	.184 .816	.118 .882
0	4	.000 .319	.000 .393	.000 .536		3	4	.184 .701	.149 .746	.093 .823
0	5	.000 .280	.000 .348	.000 .482		3	5	.155 .639	.124 .685	.077 .768
0	6	.000 .250	.000 .312	.000 .438		3	6	.134 .586	.107 .633	.065 .718
0	7	.000 .226	.000 .283	.000 .401		3	7	.117 .542	.093 .588	.056 .674
0	8	.000 .189	.000 .238	.000 .342		3	8	.105 .503	.083 .548	.050 .634
0	10	.000 .189	.000 .238	.000 .342		3	10	.086 .440	.068 .483	.040 .567
0	20	.000 .104	.000 .133	.000 .197		3	20	.045 .270	.035 .302	.020 .366
0	50	.000 .044	.000 .057	.000 .086		3	50	.018 .125	.014 .141	.008 .176
1	0	.316 1.000	.224 1.000	.100 1.000		4	0	.631 1.000	.549 1.000	.398 1.000
1	1	.135 .865	.094 .906	.041 .959		4	1	.475 .970	.409 .982	.292 .995
1	2	.068 .712	.044 .772	.016 .867		4	2	.368 .892	.315 .919	.223 .958
1	3	.043 .605	.026 .670	.009 .780		4	3	.299 .816	.254 .851	.177 .907
1	4	.030 .525	.018 .591	.005 .708		4	4	.251 .749	.212 .788	.146 .854
1	5	.023 .464	.013 .527	.004 .647		4	5	.216 .690	.182 .732	.124 .804
1	6	.019 .416	.011 .476	.003 .593		4	6	.190 .639	.159 .682	.107 .758
1	7	.015 .376	.009 .433	.002 .548		4	7	.169 .595	.141 .638	.094 .716
1	8	.013 .343	.007 .398	.002 .508		4	8	.152 .556	.126 .599	.084 .677
1	10	.010 .293	.006 .341	.001 .443		4	10	.127 .492	.105 .533	.069 .611
1	20	.005 .168	.002 .199	.001 .268		4	20	.069 .311	.056 .342	.036 .405
1	50	.002 .074	.001 .088	.000 .122		4	50	.029 .147	.024 .164	.015 .200
2	0	.464 1.000	.368 1.000	.215 1.000		5	0	.681 1.000	.607 1.000	.464 1.000
2	1	.288 .932	.228 .956	.133 .984		5	1	.536 .977	.473 .987	.353 .996
2	2	.189 .811	.147 .853	.083 .917		5	2	.432 .912	.379 .935	.283 .967
2	3	.138 .711	.105 .761	.057 .844		5	3	.361 .845	.315 .876	.232 .923
2	4	.108 .632	.081 .685	.042 .777		5	4	.310 .784	.268 .818	.196 .876
2	5	.088 .568	.065 .621	.033 .717		5	5	.271 .729	.234 .766	.169 .831
2	6	.074 .515	.054 .568	.027 .665		5	6	.241 .680	.207 .719	.149 .788
2	7	.064 .471	.046 .522	.023 .619		5	7	.217 .637	.186 .677	.133 .748
2	8	.056 .434	.041 .484	.020 .579		5	8	.197 .599	.168 .639	.120 .711
2	10	.045 .375	.032 .421	.015 .511		5	10	.167 .535	.142 .574	.100 .646
2	20	.022 .223	.016 .255	.007 .320		5	20	.094 .346	.079 .377	.054 .439
2	50	.009 .100	.006 .116	.003 .150		5	50	.040 .168	.034 .185	.023 .222

BETA $(* \mid s, f)$
Highest density regions (Cont'd)

s	f	.90	.95	.99	s	f	.90	.95	.99
6	0	.720	.652 1.000	.518 1.000	10	0	.811 1.000	.762 1.000	.658 1.000
6	1	.584 .981	.524 .989	.407 .997	10	1	.707 .990	.659 .994	.557 .999
6	2	.485 .926	.432 .946	.335 .973	10	2	.625 .955	.579 .968	.489 .985
6	3	.414 .866	.367 .893	.282 .935	10	3	.560 .914	.517 .932	.433 .960
6	4	.361 .810	.318 .841	.242 .893	10	4	.508 .873	.467 .895	.389 .931
6	5	.320 .759	.281 .793	.212 .851	10	5	.465 .833	.426 .858	.354 .900
6	6	.287 .713	.251 .749	.189 .811	10	6	.429 .796	.392 .823	.324 .870
6	7	.260 .672	.227 .708	.170 .774	10	7	.398 .762	.363 .790	.299 .840
6	8	.238 .635	.208 .672	.154 .739	10	8	.372 .730	.339 .759	.277 .811
6	10	.204 .571	.177 .608	.130 .676	10	10	.328 .672	.298 .702	.242 .758
6	20	.118 .378	.101 .409	.073 .469	10	20	.207 .478	.186 .507	.149 .562
6	50	.052 .187	.044 .205	.031 .242	10	50	.098 .254	.088 .273	.069 .311
7	0	.750 1.000	.688 1.000	.562 1.000	20	0	.896 1.000	.867 1.000	.803 1.000
7	1	.624 .985	.567 .991	.452 .998	20	1	.832 .995	.801 .998	.732 .999
7	2	.529 .936	.478 .954	.381 .977	20	2	.777 .978	.745 .984	.680 .993
7	3	.458 .883	.412 .907	.326 .944	20	3	.730 .955	.698 .965	.634 .980
7	4	.405 .831	.362 .859	.284 .906	20	4	.689 .931	.658 .944	.595 .964
7	5	.363 .783	.323 .814	.252 .867	20	5	.654 .906	.623 .921	.561 .946
7	6	.328 .740	.292 .773	.226 .830	20	6	.622 .882	.591 .899	.531 .927
7	7	.300 .700	.266 .734	.205 .795	20	7	.593 .859	.563 .877	.504 .908
7	8	.276 .664	.244 .699	.188 .762	20	8	.567 .836	.538 .855	.480 .889
7	10	.238 .602	.210 .637	.160 .701	20	10	.522 .793	.493 .814	.438 .851
7	20	.141 .407	.123 .437	.092 .496	20	20	.374 .626	.351 .649	.308 .692
7	50	.064 .205	.055 .223	.040 .261	20	50	.204 .378	.190 .397	.164 .433
8	0	.774 1.000	.717 1.000	.599 1.000	50	0	.956 1.000	.943 1.000	.914 1.000
8	1	.657 .987	.602 .993	.492 .998	50	1	.926 .998	.912 .999	.878 1.000
8	2	.566 .944	.516 .959	.421 .980	50	2	.900 .991	.884 .994	.850 .997
8	3	.497 .895	.452 .917	.366 .950	50	3	.875 .982	.859 .986	.824 .992
8	4	.444 .848	.401 .874	.323 .916	50	4	.853 .971	.836 .976	.800 .985
8	5	.401 .803	.361 .832	.289 .880	50	5	.832 .960	.815 .966	.778 .977
8	6	.365 .762	.329 .792	.261 .846	50	6	.813 .948	.795 .956	.758 .969
8	7	.336 .724	.301 .756	.238 .812	50	7	.795 .936	.777 .945	.739 .960
8	8	.311 .689	.278 .721	.219 .781	50	8	.778 .925	.759 .934	.722 .950
8	10	.270 .629	.241 .661	.189 .723	50	10	.746 .902	.727 .912	.689 .931
8	20	.164 .433	.145 .462	.111 .520	50	20	.622 .796	.603 .810	.567 .836
8	50	.075 .222	.066 .241	.050 .278	50	50	.419 .581	.404 .596	.375 .625

BETA $(* \mid s, f)$

p for selected values of $\frac{\text{right tail}}{\text{left tail}}$

s	f	100/1	10/1	1/1	1/10	1/100	s	f	100/1	10/1	1/1	1/10	1/100
6	0	.517	.710	.906	.986	.999	10	0	.657	.804	.939	.991	.999
6	1	.409	.584	.799	.935	.980	10	1	.560	.705	.864	.957	.987
6	2	.343	.501	.714	.875	.947	10	2	.493	.633	.800	.915	.964
6	3	.297	.440	.645	.818	.907	10	3	.443	.576	.744	.873	.936
6	4	.262	.393	.588	.766	.866	10	4	.403	.529	.695	.833	.906
6	5	.234	.355	.540	.718	.826	10	5	.370	.490	.653	.795	.875
6	6	.213	.324	.500	.676	.788	10	6	.342	.456	.615	.759	.845
6	7	.194	.298	.465	.638	.752	10	7	.318	.427	.582	.726	.816
6	8	.179	.276	.435	.603	.718	10	8	.298	.402	.552	.696	.788
6	10	.155	.241	.385	.544	.658	10	10	.264	.359	.500	.641	.736
6	20	.093	.147	.244	.363	.459	10	20	.169	.235	.341	.458	.547
6	50	.043	.068	.116	.180	.237	10	50	.082	.116	.174	.244	.304
7	0	.562	.741	.917	.988	.999	20	0	.803	.892	.967	.995	1.000
7	1	.455	.623	.820	.942	.983	20	1	.734	.830	.925	.977	.992
7	2	.388	.542	.741	.889	.952	20	2	.681	.780	.885	.953	.980
7	3	.339	.481	.676	.836	.917	20	3	.638	.737	.849	.928	.964
7	4	.302	.433	.621	.787	.879	20	4	.602	.700	.816	.902	.946
7	5	.272	.395	.575	.742	.841	20	5	.570	.667	.785	.877	.927
7	6	.248	.362	.535	.702	.806	20	6	.541	.637	.756	.853	.907
7	7	.228	.335	.500	.665	.772	20	7	.516	.610	.729	.829	.888
7	8	.211	.312	.469	.631	.740	20	8	.493	.586	.704	.807	.868
7	10	.184	.274	.418	.573	.682	20	10	.453	.542	.659	.765	.831
7	20	.112	.171	.271	.390	.484	20	20	.325	.398	.500	.603	.675
7	50	.052	.080	.132	.198	.255	20	50	.177	.222	.290	.364	.423
8	0	.599	.766	.926	.989	.999	50	0	.913	.954	.986	.998	1.000
8	1	.495	.655	.838	.948	.984	50	1	.879	.925	.968	.990	.997
8	2	.427	.577	.764	.899	.957	50	2	.850	.900	.950	.979	.991
8	3	.377	.517	.702	.851	.924	50	3	.825	.878	.932	.968	.984
8	4	.339	.469	.650	.805	.889	50	4	.803	.857	.915	.956	.976
8	5	.308	.430	.605	.762	.854	50	5	.782	.838	.898	.944	.967
8	6	.282	.397	.565	.724	.821	50	6	.763	.820	.884	.932	.957
8	7	.260	.369	.531	.688	.789	50	7	.745	.802	.868	.920	.948
8	8	.242	.344	.500	.656	.758	50	8	.728	.786	.854	.907	.938
8	10	.212	.304	.448	.598	.702	50	10	.696	.756	.826	.884	.918
8	20	.132	.193	.296	.414	.507	50	20	.577	.636	.710	.778	.823
8	50	.062	.093	.146	.214	.272	50	50	.386	.434	.500	.566	.614

BETA   381

BETA $(* \mid s, f)$

p for selected values of $\dfrac{\text{right tail}}{\text{left tail}}$ (Cont'd)

s	f	100/1	10/1	1/1	1/10	1/100		s	f	100/1	10/1	1/1	1/10	1/100
0	0	.010	.091	.500	.909	.990		3	0	.315	.549	.841	.976	.998
0	1	.005	.047	.293	.699	.901		3	1	.221	.405	.686	.894	.967
0	2	.003	.031	.206	.550	.785		3	2	.173	.324	.579	.807	.915
0	3	.002	.024	.159	.451	.685		3	3	.142	.270	.500	.730	.858
0	4	.002	.019	.129	.381	.603		3	4	.121	.232	.440	.664	.802
0	5	.002	.016	.109	.329	.537		3	5	.105	.204	.393	.608	.750
0	6	.001	.014	.094	.290	.483		3	6	.093	.182	.355	.560	.703
0	7	.001	.012	.083	.259	.438		3	7	.083	.164	.324	.519	.661
0	8	.001	.011	.074	.234	.401		3	8	.076	.149	.298	.483	.623
0	9	.001	.009	.061	.196	.343		3	10	.064	.127	.256	.424	.557
0	20	.000	.005	.033	.108	.197		3	20	.036	.072	.151	.263	.362
0	50	.000	.002	.014	.046	.087		3	50	.016	.032	.068	.122	.175
1	0	.099	.301	.707	.953	.995		4	0	.397	.619	.871	.981	.998
1	1	.059	.186	.500	.814	.941		4	1	.294	.479	.736	.912	.973
1	2	.042	.135	.386	.691	.860		4	2	.236	.394	.636	.837	.929
1	3	.033	.106	.314	.595	.779		4	3	.198	.336	.560	.768	.879
1	4	.027	.088	.264	.521	.706		4	4	.171	.293	.500	.707	.829
1	5	.023	.075	.229	.463	.644		4	5	.150	.260	.452	.653	.782
1	6	.020	.065	.201	.416	.591		4	6	.134	.234	.412	.607	.738
1	7	.017	.058	.180	.377	.545		4	7	.121	.213	.379	.567	.698
1	8	.016	.052	.162	.345	.505		4	8	.111	.195	.350	.531	.661
1	9	.013	.043	.136	.295	.440		4	10	.094	.167	.305	.471	.597
1	20	.008	.023	.075	.170	.266		4	20	.054	.098	.184	.300	.398
1	50	.003	.010	.032	.075	.121		4	50	.024	.044	.085	.143	.197
2	0	.215	.450	.794	.969	.997		5	0	.463	.671	.891	.984	.998
2	1	.140	.309	.614	.865	.958		5	1	.356	.537	.771	.925	.977
2	2	.105	.238	.500	.762	.895		5	2	.293	.453	.679	.859	.939
2	3	.085	.193	.421	.676	.827		5	3	.250	.392	.607	.796	.895
2	4	.071	.163	.364	.606	.764		5	4	.218	.347	.548	.740	.850
2	5	.061	.141	.321	.547	.707		5	5	.194	.311	.500	.689	.806
2	6	.053	.125	.286	.499	.657		5	6	.174	.282	.460	.645	.766
2	7	.048	.111	.259	.458	.612		5	7	.159	.258	.425	.605	.728
2	8	.043	.101	.236	.423	.573		5	8	.146	.238	.395	.570	.692
2	9	.036	.085	.200	.367	.507		5	10	.125	.205	.347	.510	.630
2	20	.020	.047	.115	.220	.319		5	20	.073	.123	.215	.333	.430
2	50	.009	.021	.050	.100	.150		5	50	.033	.056	.101	.162	.218

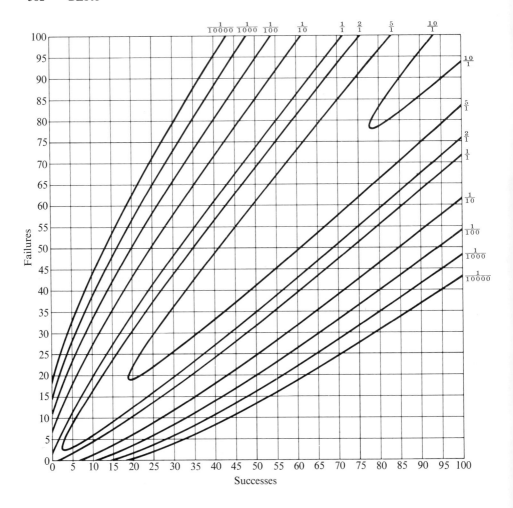

$$\mathfrak{F}\left(\frac{p = \tfrac{1}{2}}{p \neq \tfrac{1}{2}} \bigg| s, f\right)$$

assuming flat distribution of alternatives

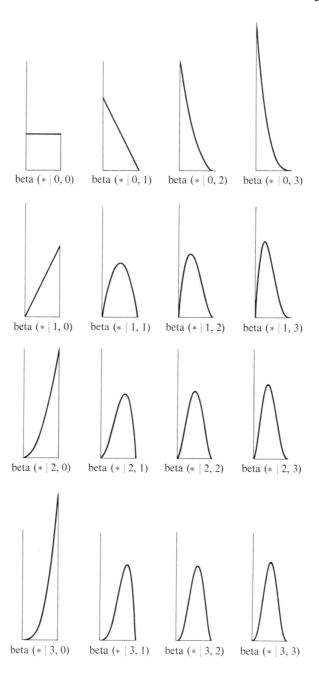

BIGAU: The bivariate Gaussian or bivariate normal distribution

$$\text{bigau}(x, y \mid \mu_x, \mu_y, \sigma_x^2, \sigma_y^2, \rho) = \frac{1}{2\pi\sigma_x\sigma_y\sqrt{1-\rho^2}}$$

$$\times \exp\left\{-\frac{1}{2(1-\rho^2)}\left[\left(\frac{x-\mu_x}{\sigma_x}\right)^2 - 2\rho\left(\frac{x-\mu_x}{\sigma_x}\right)\left(\frac{y-\mu_y}{\sigma_y}\right) + \left(\frac{y-\mu_y}{\sigma_y}\right)^2\right]\right\}$$

for

$$-\infty < x < \infty, \quad -\infty < y < \infty, \quad -\infty < \mu_x < \infty,$$
$$-\infty < \mu_y < \infty, \quad 0 < \sigma_x < \infty, \quad 0 < \sigma_y < \infty, \quad -1 < \rho < +1.$$

ρ is the correlation coefficient. The marginal densities are

$$\text{gau}(x \mid \mu_x, \sigma_x^2) \quad \text{and} \quad \text{gau}(y \mid \mu_y, \sigma_y^2).$$

If $\rho = 0$, x and y are independent, and the joint density equals the product of the marginal densities.

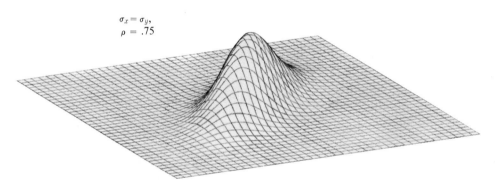

$\sigma_x = \sigma_y$, $\rho = .75$

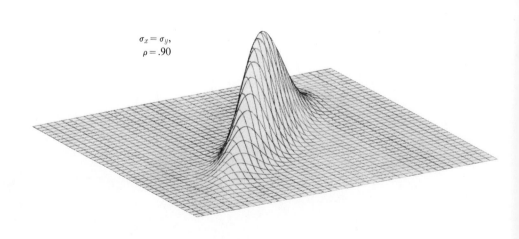

$\sigma_x = \sigma_y$, $\rho = .90$

BIN: The binomial distribution

$$\text{bin}(s \mid n, p) = \binom{n}{s} p^s (1-p)^{n-s} \quad \text{for} \quad \begin{cases} s = 0, 1, 2, \ldots, n, \\ 0 \leq p \leq 1, \end{cases}$$

$$E[\text{bin}(* \mid n, p)] = np,$$
$$D^2[\text{bin}(* \mid n, p)] = np(1-p),$$
$$\text{mode}[\text{bin}(* \mid n, p)] \text{ is near } (n+1)p - \tfrac{1}{2}.$$

bin $(s \mid n, p)$ is the probability of s successes in n trials when

1. the only possibilities are *success* and *failure*,
2. the probability of success on each trial is p,
3. the trials are independent.

The distribution is symmetric only when $p = \tfrac{1}{2}$. Then $n/2$ is the center.

GAU can provide a good approximation near the middle of BIN, especially when p is not close to 0 or 1. However, GAU generally underestimates the extreme tails; there POIS is a better approximator.

When $n \to \infty$ and $p \to 0$ but np stays reasonable,

$$\text{bin}(s \mid n, p) \to \text{pois}(s \mid np).$$

CIRG: The circular normal distribution

$$\text{cirg}(\alpha \mid \mu, \kappa) = \frac{e^{\kappa \cos(\alpha - \mu)}}{2\pi I_0(\kappa)} \quad \text{for} \quad \begin{cases} \mu - \pi < \alpha < \mu + \pi \text{ (in radians)}, \\ 0 \leq \mu \leq 2\pi \text{ (in radians)}, \\ 0 \leq \kappa. \end{cases}$$

This is a distribution on a circle. When $\kappa = 0$, the distribution is flat. When $\kappa > 0$,

$$E[\text{cirg}(* \mid \mu, \kappa)] = \mu,$$
$$D^2[\text{cirg}(* \mid \mu, \kappa)] \text{ has no convenient formula},$$
$$\text{mode}[\text{cirg}(* \mid \mu, \kappa)] = \mu.$$

$I_0(\kappa)$ is a Bessel function; Bessel functions always pop up when problems involve circular symmetry.

When κ is large,

$$\text{cirg}(\alpha \mid \mu, \kappa) \to \text{gau}(\alpha \mid \mu, 1/\kappa).$$

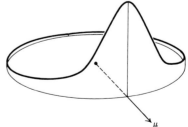

EXPD: The exponential distribution

$$\text{expd}\,(x \mid \mu) = \frac{1}{\mu} e^{-x/\mu} \quad \text{for} \quad \begin{cases} 0 \le x < \infty, \\ 0 < \mu, \end{cases}$$

$$E[\text{expd}\,(* \mid \mu)] = \mu,$$
$$D^2[\text{expd}\,(* \mid \mu)] = \mu^2,$$
$$\text{mode}\,[\text{expd}\,(* \mid \mu)] = 0,$$
$$\text{EXPD}\,(x \mid \mu) = 1 - e^{-x/\mu},$$
$$1 - \text{EXPD}\,(x \mid \mu) = e^{-x/\mu}.$$

If μ is the average length of time until an occurrence of interest, $\text{EXPD}\,(* \mid \mu)$ is the distribution of the time until the first occurrence.

$$\text{expd}\,(x \mid \mu) = \text{gamma}\,(x \mid 1, 1/\mu).$$

Caution: You sometimes see EXPD written with $\tau = 1/\mu$, where τ = average number of occurrences per unit time.

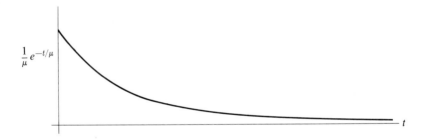

GAMMA: The gamma distribution

$$\text{gamma}(x \mid n, \tau) = \frac{\tau(\tau x)^{n-1} e^{-\tau x}}{(n-1)!} \quad \text{for} \quad \begin{cases} 0 \leq x < \infty, \\ 0 < n, \\ 0 < \tau, \end{cases}$$

$$E[\text{gamma}(* \mid n, \tau)] = n/\tau,$$
$$D^2[\text{gamma}(* \mid n, \tau)] = n/\tau^2,$$
$$\text{mode}[\text{gamma}(* \mid n, \tau)] = \begin{cases} (n-1)/\tau & \text{for } 1 \leq n, \\ 0 & \text{for } 0 < n \leq 1. \end{cases}$$

This is a "waiting-time" distribution. If τ is the average number of occurrences per unit time, GAMMA $(* \mid n, \tau)$ is the distribution of the length of time until n occurrences have taken place. NEGB is its discrete relative.

If $u \sim$ GAMMA $(* \mid n_1, \tau)$ and $v \sim$ GAMMA $(* \mid n_2, \tau)$, then

$$u + v \sim \text{GAMMA}(* \mid n_1 + n_2, \tau).$$

Caution: There are several other definitions of a gamma distribution.

GAU: The Gaussian or normal distribution

$$\text{gau}(x \mid \mu, \sigma^2) = \frac{1}{\sigma\sqrt{2\pi}} \exp\left[-\frac{1}{2}\left(\frac{x-\mu}{\sigma}\right)^2\right] \quad \text{for} \quad \begin{cases} -\infty < x < +\infty, \\ -\infty < \mu < +\infty, \\ 0 < \sigma < \infty, \end{cases}$$

$$E[\text{gau}(* \mid \mu, \sigma^2)] = \mu,$$
$$D^2[\text{gau}(* \mid \mu, \sigma^2)] = \sigma^2,$$
$$\text{mode}[\text{gau}(* \mid \mu, \sigma^2)] = \mu.$$

The distribution is symmetric about μ. GAU $(* \mid \mu, \sigma^2)$ is often referred to as $\mathcal{N}(\mu, \sigma^2)$. The unit normal distribution

$$\text{gau}(x \mid 0, 1) = \frac{1}{\sqrt{2\pi}} e^{-x^2/2}$$

is the one universally tabulated. It is sometimes called $\phi(x)$, while GAU $(x \mid 0, 1)$ is called $\Phi(x)$.

If $u \sim$ GAU $(* \mid \mu_1, \sigma_1^2)$ and $v \sim$ GAU $(* \mid \mu_2, \sigma_2^2)$ independently, then

$$u + v \sim \text{GAU}(* \mid \mu_1 + \mu_2, \sigma_1^2 + \sigma_2^2).$$

GAU (∗ | 0, 1)

Areas

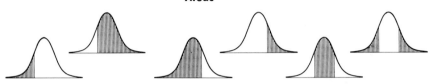

z	LEFT TAIL	RIGHT TAIL	z	LEFT TAIL	RIGHT TAIL	CENTER	BOTH TAILS
-4.0	.0000317	.9999683	0.0	.500	.500	.0000	1.0000
-3.9	.0000481	.9999519	0.1	.540	.460	.0797	.9203
-3.8	.0000723	.9999277	0.2	.579	.421	.159	.841
-3.7	.000108	.999892	0.3	.618	.382	.236	.764
-3.6	.000159	.999841	0.4	.655	.345	.311	.689
-3.5	.000233	.999767	0.5	.691	.309	.383	.617
-3.4	.000337	.999663	0.6	.726	.274	.451	.549
-3.3	.000483	.999517	0.7	.758	.242	.516	.484
-3.2	.000687	.999313	0.8	.788	.212	.576	.424
-3.1	.000968	.999032	0.9	.816	.184	.632	.368
-3.0	.00135	.99865	1.0	.841	.159	.683	.317
-2.9	.00187	.99813	1.1	.864	.136	.729	.271
-2.8	.00256	.99744	1.2	.885	.115	.770	.230
-2.7	.00347	.99653	1.3	.9032	.0968	.806	.194
-2.6	.00466	.99534	1.4	.9192	.0808	.838	.162
-2.5	.00621	.99379	1.5	.9332	.0668	.866	.134
-2.4	.00820	.99180	1.6	.9452	.0548	.890	.110
-2.3	.0107	.9893	1.7	.9554	.0446	.9109	.0891
-2.2	.0139	.9861	1.8	.9641	.0359	.9281	.0719
-2.1	.0179	.9821	1.9	.9713	.0287	.9425	.0575
-2.0	.0228	.9772	2.0	.9772	.0228	.9545	.0455
-1.9	.0287	.9713	2.1	.9821	.0179	.9643	.0357
-1.8	.0359	.9641	2.2	.9861	.0139	.9722	.0278
-1.7	.0446	.9554	2.3	.9893	.0107	.9786	.0214
-1.6	.0548	.9452	2.4	.99180	.00820	.9836	.0164
-1.5	.0668	.9332	2.5	.99379	.00621	.9876	.0124
-1.4	.0808	.9192	2.6	.99534	.00466	.99068	.00932
-1.3	.0968	.9032	2.7	.99653	.00347	.99307	.00693
-1.2	.115	.885	2.8	.99744	.00256	.99489	.00511
-1.1	.136	.864	2.9	.99813	.00187	.99627	.00373
-1.0	.159	.841	3.0	.99865	.00135	.99730	.00270
-0.9	.184	.816	3.1	.999032	.000968	.99806	.00194
-0.8	.212	.788	3.2	.999313	.000687	.99863	.00137
-0.7	.242	.758	3.3	.999517	.000483	.999033	.000967
-0.6	.274	.726	3.4	.999663	.000337	.999326	.000674
-0.5	.309	.691	3.5	.999767	.000233	.999535	.000465
-0.4	.345	.655	3.6	.999841	.000159	.999682	.000318
-0.3	.382	.618	3.7	.999892	.000108	.999784	.000216
-0.2	.421	.579	3.8	.9999277	.0000723	.999855	.000145
-0.1	.460	.540	3.9	.9999519	.0000481	.9999038	.0000962
0.0	.500	.500	4.0	.9999683	.0000317	.9999367	.0000633

GAU (∗ | 0, 1)
Ordinates

	.0	.1	.2	.3	.4	.5	.6	.7	.8	.9
0	.3989	.3970	.3910	.3814	.3683	.3521	.3332	.3123	.2897	.2661
1	.2420	.2179	.1942	.1714	.1497	.1295	.1109	.0940	.0790	.0656
2	.0540	.0440	.0355	.0283	.0224	.0175	.0136	.0104	.0079	.0060
3	.0044	.0033	.0024	.0017	.0012	.0009	.0006	.0004	.0003	.0002

GAU (∗ | 0, 1)
Special Areas for Center and both tails

Center	Both tails	Z
.5	.5	.674
.6	.4	.842
.7	.3	1.036
.75	.25	1.150
.8	.2	1.282
.9	.1	1.645
.95	.05	1.960
.98	.02	2.326
.99	.01	2.576
.995	.005	2.810
.998	.002	3.091
.999	.001	3.291
.9999	.0001	3.891
.99999	.00001	4.417
.999999	.000001	4.892

IGAM: The inverse gamma distribution

$$\operatorname{igam}(\sigma \mid q, \nu) = \frac{2\left(\dfrac{\nu+1}{2}\right)^{\nu/2} q^{\nu} \exp\left(-\dfrac{\nu+1}{2}\dfrac{q^2}{\sigma^2}\right)}{\left(\dfrac{\nu-2}{2}\right)! \, \sigma^{\nu+1}} \quad \text{for} \quad \begin{cases} 0 < \sigma < \infty, \\ 0 < q < \infty, \\ 1 \leq \nu, \end{cases}$$

$$E[\operatorname{igam}(* \mid q, \nu)] = \sqrt{\dfrac{\nu+1}{2}} \, \dfrac{\left(\dfrac{\nu-3}{2}\right)!}{\left(\dfrac{\nu-2}{2}\right)!} \, q,$$

$$D^2[\operatorname{igam}(* \mid q, \nu)] = \left(\dfrac{\nu+1}{\nu-2}\right) q^2 - E^2,$$

$$\operatorname{mode}[\operatorname{igam}(* \mid q, \nu)] = q.$$

If $w \sim \operatorname{IGAM}(* \mid 1, \nu)$, then

$$\dfrac{1}{w^2} \sim \operatorname{GAMMA}\left(* \mid \dfrac{\nu}{2}, \dfrac{\nu+1}{2}\right).$$

Two tables of IGAM appear on the following page.

1. *Highest density regions:* intervals containing a specified amount of probability and having the property that the density inside is larger than the density outside.

Example. $\nu = 14$. 90% of the probability is contained in (.75, 1.43), and the density inside this interval is larger than the density outside.

In univariate Gaussian analysis, the posterior distribution of σ is

$$\operatorname{IGAM}(* \mid \sqrt{S_{\bar{x}\bar{x}}/n}, \, n-1).$$

The highest density regions from the table must be multiplied by $\sqrt{S_{\bar{x}\bar{x}}/n}$ to get the highest density region for σ.

Example. $n = 15$, $\nu = 14$, $S_{\bar{x}\bar{x}} = 375$, $\sqrt{S_{\bar{x}\bar{x}}/n} = 5$. The 90% HDR is

$$(5 \times .75, \, 5 \times 1.43) = (3.75, 7.15).$$

2. *Expected value and variance:* given because of the awkward formulas.

IGAM (∗ | 1, ν)
Highest density regions

ν	.80		.90		.95		.99	
1	.44	5.62	.38	11.29	?		?	
2	.55	2.65	.49	3.81	.45	5.44	.38	12.25
4	.65	1.80	.59	2.21	.55	2.69	.48	4.13
9	.75	1.42	.70	1.59	.66	1.77	.59	2.22
14	.79	1.31	.75	1.43	.71	1.55	.65	1.82
19	.82	1.26	.78	1.35	.75	1.44	.69	1.64
24	.84	1.22	.80	1.30	.77	1.37	.71	1.54
29	.85	1.20	.82	1.27	.79	1.33	.73	1.47
34	.86	1.18	.83	1.24	.80	1.30	.75	1.42
39	.87	1.17	.84	1.22	.81	1.27	.76	1.39
44	.88	1.16	.85	1.21	.82	1.25	.78	1.36
49	.88	1.15	.85	1.19	.83	1.24	.79	1.33
74	.90	1.12	.88	1.15	.86	1.19	.82	1.26
99	.91	1.10	.89	1.13	.87	1.16	.84	1.22
149	.93	1.08	.91	1.10	.90	1.13	.87	1.17
199	.94	1.07	.92	1.09	.91	1.11	.88	1.14

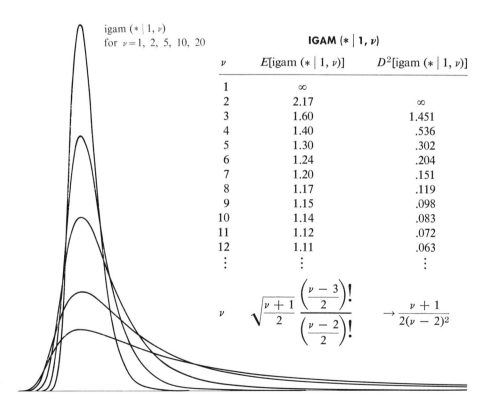

igam (∗ | 1, ν) for ν = 1, 2, 5, 10, 20

IGAM (∗ | 1, ν)

ν	$E[\text{igam}(* \mid 1, \nu)]$	$D^2[\text{igam}(* \mid 1, \nu)]$
1	∞	
2	2.17	∞
3	1.60	1.451
4	1.40	.536
5	1.30	.302
6	1.24	.204
7	1.20	.151
8	1.17	.119
9	1.15	.098
10	1.14	.083
11	1.12	.072
12	1.11	.063
⋮	⋮	⋮
ν	$\sqrt{\dfrac{\nu+1}{2}} \dfrac{\left(\dfrac{\nu-3}{2}\right)!}{\left(\dfrac{\nu-2}{2}\right)!}$	$\to \dfrac{\nu+1}{2(\nu-2)^2}$

NEGB: The negative binomial distribution

$$\text{negb}(n \mid s, p) = \binom{n-1}{s-1} p^s (1-p)^{n-s} \quad \text{for} \quad \begin{cases} n = s, s+1, \ldots, \\ 0 < p \leq 1, \end{cases}$$

$$E[\text{negb}(* \mid s, p)] = s/p,$$
$$D^2[\text{negb}(* \mid s, p)] = s(1-p)/p^2,$$
$$\text{mode}[\text{negb}(* \mid s, p)] \text{ is near } (s - \tfrac{1}{2})/p.$$

negb $(n \mid s, p)$ is the probability that it takes exactly n trials to achieve s successes when the probability of success on each trial is p and the trials are independent. This is a "waiting-time" distribution. The continuous relative is GAMMA.

POIS: The Poisson distribution

$$\text{pois}(k \mid \lambda) = \frac{e^{-\lambda} \lambda^k}{k!} \quad \text{for} \quad \begin{cases} k = 0, 1, \ldots, \\ 0 < \lambda, \end{cases}$$

$$E[\text{pois}(* \mid \lambda)] = \lambda,$$
$$D^2[\text{pois}(* \mid \lambda)] = \lambda,$$
$$\text{mode}[\text{pois}(* \mid \lambda)] \text{ is near } \lambda - \tfrac{1}{2}.$$

This distribution is named for a man, not a fish, but it does have a long tail!

It sometimes is called a "rare-event" distribution, since it is the limit of BIN when

$$n \to \infty, \quad p \to 0, \quad np \to \lambda.$$

BIN is to NEGB as POIS is to GAMMA.

If $x \sim \text{POIS}(* \mid \lambda_x)$ and $y \sim \text{POIS}(* \mid \lambda_y)$ independently, then

$$x + y \sim \text{POIS}(* \mid \lambda_x + \lambda_y).$$

RECT: The rectangular distribution

$$\text{rect}(x \mid \mu, \sigma^2) = \frac{1}{2\sigma} \quad \text{for} \quad \begin{cases} \mu - \sigma < x < \mu + \sigma, \\ -\infty < \mu < +\infty, \\ 0 < \sigma < \infty, \end{cases}$$

$$E[\text{rect}(* \mid \mu, \sigma^2)] = \mu,$$
$$D^2[\text{rect}(* \mid \mu, \sigma^2)] = \sigma^2/3,$$
the mode is not unique.

The graph of this distribution looks like a rectangle, hence the name. It is symmetric about the mean.

STU: The Student or "t" distribution

$$\text{stu}(x \mid \mu, \sigma^2, \nu) = \frac{\left(\frac{\nu-1}{2}\right)!}{\left(\frac{\nu-2}{2}\right)! \sqrt{\pi \nu}\, \sigma} \frac{1}{\left(1 + \frac{t^2}{\nu}\right)^{(\nu+1)/2}} \quad \text{where} \quad t = \frac{x-\mu}{\sigma},$$

for
$$\begin{cases} -\infty < x < +\infty, \\ -\infty < \mu < +\infty, \\ 0 < \sigma < \infty, \\ 1 \leq \nu, \end{cases}$$

$$E[\text{stu}(* \mid \mu, \sigma^2, \nu)] = \mu,$$
$$D^2[\text{stu}(* \mid \mu, \sigma^2, \nu)] = \sigma^2 \nu/(\nu - 2),$$
$$\text{mode}[\text{stu}(* \mid \mu, \sigma^2, \nu)] = \mu.$$

The distribution is symmetric about the mean. ν is the number of degrees of freedom. If $\nu \to \infty$, then

$$\text{stu}(x \mid \mu, \sigma^2, \nu) \to \text{gau}(x \mid \mu, \sigma^2).$$

A good approximation to the numerical coefficient is

$$\frac{\left(\frac{\nu-1}{2}\right)!}{\left(\frac{\nu-2}{2}\right)! \sqrt{\pi \nu}} \doteq .3989 - \frac{.1009}{\nu} + \frac{.0203}{\nu^2}.$$

STU$(* \mid 0, 1, \nu)$
Percentiles

LEFT TAIL	.0005	.001	.005	.01	.025	.05	.1	.15	.25	.4	.5
RIGHT TAIL	.9995	.999	.995	.99	.975	.95	.9	.85	.75	.6	.5
ν											
1	-637	-318	-63.7	-31.8	-12.7	-6.31	-3.08	-1.96	-1.00	-.32	0
2	-31.6	-22.3	-9.92	-6.96	-4.30	-2.92	-1.89	-1.39	-.82	-.29	0
3	-12.9	-10.2	-5.84	-4.54	-3.18	-2.35	-1.64	-1.25	-.76	-.28	0
4	-8.61	-7.17	-4.60	-3.75	-2.78	-2.13	-1.53	-1.19	-.74	-.27	0
5	-6.87	-5.89	-4.03	-3.36	-2.57	-2.02	-1.48	-1.16	-.73	-.27	0
6	-5.96	-5.21	-3.71	-3.14	-2.45	-1.94	-1.44	-1.13	-.72	-.26	0
7	-5.41	-4.79	-3.50	-3.00	-2.36	-1.90	-1.42	-1.12	-.71	-.26	0
8	-5.04	-4.50	-3.36	-2.90	-2.31	-1.86	-1.40	-1.11	-.71	-.26	0
10	-4.59	-4.14	-3.17	-2.76	-2.23	-1.81	-1.37	-1.09	-.70	-.26	0
20	-3.85	-3.55	-2.85	-2.53	-2.09	-1.72	-1.33	-1.06	-.69	-.26	0
30	-3.65	-3.39	-2.75	-2.46	-2.04	-1.70	-1.31	-1.06	-.68	-.26	0
50	-3.50	-3.26	-2.68	-2.40	-2.01	-1.68	-1.30	-1.05	-.68	-.25	0
∞	-3.29	-3.09	-2.58	-2.33	-1.96	-1.64	-1.28	-1.04	-.67	-.25	0

LEFT TAIL	.5	.6	.75	.85	.9	.95	.975	.99	.995	.999	.9995
RIGHT TAIL	.5	.4	.25	.15	.1	.05	.025	.01	.005	.001	.0005
CENTER	0	.2	.5	.7	.8	.9	.95	.98	.99	.998	.999
BOTH TAILS	1.0	.8	.5	.3	.2	.1	.05	.02	.01	.002	.001
ν											
1	0	.32	1.00	1.96	3.08	6.31	12.7	31.8	63.7	318	637
2	0	.29	.82	1.39	1.89	2.92	4.30	6.96	9.92	22.3	31.6
3	0	.28	.76	1.25	1.64	2.35	3.18	4.54	5.84	10.2	12.9
4	0	.27	.74	1.19	1.53	2.13	2.78	3.75	4.60	7.17	8.61
5	0	.27	.73	1.16	1.48	2.02	2.57	3.36	4.03	5.89	6.87
6	0	.26	.72	1.13	1.44	1.94	2.45	3.14	3.71	5.21	5.96
7	0	.26	.71	1.11	1.42	1.90	2.36	3.00	3.50	4.79	5.41
8	0	.26	.71	1.11	1.40	1.86	2.31	2.90	3.36	4.50	5.04
10	0	.26	.70	1.09	1.37	1.81	2.23	2.76	3.17	4.14	4.59
20	0	.26	.69	1.06	1.33	1.72	2.09	2.53	2.85	3.55	3.85
30	0	.26	.68	1.06	1.31	1.70	2.04	2.46	2.75	3.39	3.65
50	0	.25	.68	1.05	1.30	1.68	2.01	2.40	2.68	3.26	3.50
∞	0	.25	.67	1.04	1.28	1.64	1.96	2.33	2.58	3.09	3.29

WEIB: The Weibull distribution

$$\text{weib}(x \mid \mu, \sigma^2, c) = \left(\frac{c}{\sigma}\right)\left(\frac{x-\mu}{\sigma}\right)^{c-1} \exp\left[-\left(\frac{x-\mu}{\sigma}\right)^c\right]$$

for
$$\begin{cases} \mu \leq x < \infty, \\ -\infty < \mu < +\infty & \text{(location parameter)}, \\ 0 < \sigma < \infty & \text{(scale parameter)}, \\ 0 < c < \infty & \text{(shape parameter)}, \end{cases}$$

$$E[\text{weib}(* \mid \mu, \sigma^2, c)] = \mu + \sigma \left(\frac{1}{c}\right)!,$$

$$D^2[\text{weib}(* \mid \mu, \sigma^2, c)] = \sigma^2 \left\{\left(\frac{2}{c}\right)! - \left[\left(\frac{1}{c}\right)!\right]^2\right\},$$

$$\text{mode}[\text{weib}(* \mid \mu, \sigma^2, c)] = \begin{cases} \mu + \sigma\left(1 - \frac{1}{c}\right)^{1/c} & \text{for } c \geq 1, \\ \mu & \text{for } c < 1, \end{cases}$$

$$p\text{th percentile} = \mu + \sigma[-\ln(1-p)]^{1/c},$$

$$\text{median} = .50 \text{ percentile} = \mu + \sigma(\ln 2)^{1/c},$$

$$1 - \text{WEIB}(x \mid \mu, \sigma^2, c) = \exp\left[-\left(\frac{x-\mu}{\sigma}\right)^c\right],$$

$$\text{hazard function} = \frac{c}{\sigma}\left(\frac{x-\mu}{\sigma}\right)^{c-1}.$$

Index

accept hypothesis, 253–254
actions, 232
alternative hypothesis, 253
Anahuatl, 81, 83, 88
angles, 251, 318ff
Arabic, 15
area, 106, 161
autoregression, 308
average, 27, 179, 219, 352

band-limited signal, 332
Bayes' Theorem, 62ff, 65, 86, 259
Bayesian confidence interval = highest density region
Behrens-Fisher problem, 275
bell-shaped distribution, 204
Bernoulli trials, 115, 162, 261
Bessel function, 319
BETA = beta distribution
beta distribution, 115ff, 140, 158, 376ff
BIGAU = bivariate Gaussian distribution
bimodal distribution, 17, 33, 248
BIN = binomial distribution
binomial cofficient, 77, 357–359
binomial distribution, 20, 165, 220, 385
bit, 55, 237, 240
bivariate distribution, 35, 52, 58, 246, 291
bivariate Gaussian distribution, 246, 251, 384
blind experiment, 282
block, 201, 288

both tails, 141
bridge, 36, 39, 45
bulge, 171
burn-in, 325

Cauchy distribution, 205
center, 141
center of symmetry, 143
chances, 6
chess, 41, 46, 238
chi-square statistic, 260
circular normal distribution, 253, 318ff, 385
CIRG = circular normal distribution
cloudiness, 12, 14, 29
clustering, 247
composite alternative, 89ff, 130
compound probability, 21, 22, 24, 185, 301
conditional distribution, 33, 43ff, 51, 63
conditional probability, 2, 6, 8
consequence, 232
contaminated distribution, 217, 224
contingency table, 131–132
continuous variable, 100, 106, 114ff
contours, 37, 131, 191
control chart, 148
control limits, 148
controls, 9, 132, 280
correlation, 291, 300
correlation coefficient, 299
cost, 261

398 Index

credible region = highest density region
cross-product, 334, 336
cumulative distribution, 116, 136, 210

decision theory, 123, 231ff
decisions, 232ff
   sequential, 245, 260ff
   terminal, 260
degrees of freedom, 194, 273, 278
density, 106, 116
   two-dimensional, 190
dependence, 50, 131
direction finding, 35, 251
discrete variable, 12, 106
discrimination, 240ff
distribution-free = nonparametric
double-blind experiment, 282

e, 31, 359
E = expected value
efficiency, 31, 207ff, 283
English, 15, 60
entropy, 168ff, 369
error, of the first kind, 253
   of the second kind, 254
   of the third kind, 254
estimation, 121ff
exhaustive set of alternatives, 12, 21–22, 65
exp, 31, 140
EXPD = exponential distribution
expected income, 232ff
   under certainty, 235
   under uncertainty, 235
expected value, 28ff, 122
exponential distribution, 324ff, 386

factor, 84, 96–97, 131, 259
factor analysis, 247
factorial, 82, 131, 356–357, 368
fair coin, 5, 25, 29, 38, 79
false negatives and positives, 66
finite population, 59
fish, 78
FLAT, 180, 191, 294, 319
flat distribution, 124, 163
flat random digits, 57, 370
flat set of probabilities, 95, 167, 171

GAMMA = gamma distribution
gamma distribution, 311, 327, 387

gamma function, 357
GAU = Gaussian distribution
Gaussian distribution, 134, 140ff, 158, 177ff, 241, 294, 327, 331, 387ff
   sum or difference of, 272, 387
Geiger counter, 310, 311, 316
genetics, 36, 39, 46, 80, 100, 287
goodness of fit, 255ff

hazard function, 328
HDR = highest density region
highest density region, 119–120, 124
   joint, 198
histogram, 13

idealization, 4
IGAM = inverse gamma distribution
income, 233
independence, 50, 74, 87, 131, 155
infinite population, 59
information, 170
information theory, 168
integral, 107, 359
interpenetrating samples, 283
interval, 105
inventory, 237
inverse gamma distribution, 193, 295, 390–391

J-shaped distribution, 30, 119
joint density, 191, 204
joint distribution, 35, 51, 63–64

kernel, 115, 375

least squares, 292ff, 318
left tail, 120, 141
LFLAT, 180, 191, 294, 311, 325
life expectancy, 48, 49, 324
life on Mars, 95
life test, 324
likelihood, 83, 86, 115, 130, 177ff, 246
linear functional relationship, 291
location parameter, 144, 179
logarithms, 364ff
log-factor, 87, 133
log-likelihood, 87
log-odds, 87
loss, 123, 234
   due to uncertainty, 235
lump of probability, 107, 129–130

marginal density, 192–193
marginal distribution, 38, 51, 63
marginal probability, 51
Markov chain, 11
maximum entropy, 173
maximum likelihood, 123
mean, 30, 122, 136, 140, 206, 352
mean deviation, 222
median, 31, 123, 136, 140, 206
   distribution of, 220–221
midrange, 205
minks, 80, 82, 119
mixture of distributions, 217, 247ff, 315
mode, 123, 140
   of posterior distribution, 206
modulo 2 addition, 55
moments, 150ff
monotonic, 136
mortality table, 47
multimodal distribution, 30, 119
multiple comparisons, 278
multivariate distribution, 35, 271
mutually exclusive set of alternatives, 12, 21–22, 65

negative binomial distribution, 21, 165, 314, 392
NEGB = negative binomial distribution
noise, 331–332
noise tube, 55
noninformative prior, 115
nonparametric methods, 222
normal distribution = Gaussian distribution
normal equations, 307
normal scores, 226
normalization, 20, 44, 86, 106, 112, 123, 125
null hypothesis, 253
numerical integration, 107, 125, 359ff
numerical taxonomy, 247

odds, 84, 119
or, 21
order statistics, 202, 204, 212–214, 226
ordinate, 110, 118, 166, 177, 198
outlier, 215ff

P = probability
paired comparisons, 286, 289

parameter, 17, 197, 212
   location, 144, 179
   scale, 143, 179
pattern recognition, 247
pd = probability density
PD = cumulative distribution
phenyl thiocarbamide, 127
placebo, 280
point estimate, 124
POIS = Poisson distribution
Poisson distribution, 51, 310ff, 392
pooling of variances, 278
population, 7
   finite and infinite, 59
posterior density, 117
posterior distribution, 62, 118, 130, 183
posterior odds, 84
posterior probability, 65, 86
precise point hypothesis, 129ff
principal components, 247
prior density, 114–115, 124
prior distribution, 62, 124, 173, 179ff, 189
prior odds, 84
prior probability, 65, 86, 95, 242, 256
probability, 2, 6, 106
probability density, 106
probability distribution, 11, 14
probability paper, 210ff
probit, 226, 305
product notation, 131, 356
pseudo-random numbers, 55

quality control, 74, 148

radar, 330ff
radioactivity, 290, 311, 314
random numbers, 56
   flat, 56–57, 370
   Gaussian, 138, 218, 371–372
   pseudo-, 55
random sample, 57
randomization, 201, 282, 285
ranks, ranking, 14, 27, 226, 334
ratio, 314
RECT = rectangular distribution
rectangular distribution, 134, 204ff, 393
recursion, 316
regression, linear, 290ff
   stepwise multiple, 247
reject hypothesis, 253

relative factor, 86
relative frequency, 28
relative odds, 86
repeat rate, 171
replication, 307
residual, 132, 196, 200, 294, 353
right tail, 120, 141
robustness, 216
rough set of probabilities, 167

Salk vaccine, 9, 132, 282
sampling, from continuous distribution, 134ff
  from discrete distribution, 56ff
  with replacement, 59, 67, 92
  without replacement, 59, 67
  stratified, 26, 284, 288
sampling experiment, 101–105
scale, 14, 27, 226
scale factor or parameter, 143, 179
screening for disease, 65, 85
sequential analysis, sequential decisions, 72, 245, 260ff
sequential probability ratio test, 262, 270
sign test, 132
significance level, 257–259
significance test, 253ff, 258
simple alternative, 89
Simpson's rule, 112, 114, 181, 279, 327, 361ff
simulation, 56, 125
standard deviation, 152
  variation with n, 184
standard error = standard deviation
standardized variable, 145ff, 155
state of the world, 232
stationary process, 48
statistic, 27, 208, 334
  sufficient, 118, 184, 311

stepwise multiple regression, 247
stratified sampling, 26, 284, 288
STU = Student distribution
Student, 197
Student distribution, 193, 197, 217, 274, 295, 301, 393–394
sufficient statistics, 118, 184, 311
sum of squares, 179, 292–294
summation notation, 24, 351ff
symmetric distribution, 30, 140
systematic error, 185, 201

t-distribution = Student distribution
t-test, 258
terminal decision, 260
theory of games, 232
transformation of a variable, 153, 225ff
translation, 14, 15
trivariate distribution, 35, 53

U-shaped distribution, 30, 119, 204
unit normal density, 143, 387
univariate distribution, 35
uranium, 14, 17
utiles, 236
utility, 236

variance, 152, 207, 273ff
  addition of, 155
  variation with n, 184
velocity of light, 199

WEIB = Weibull distribution
Weibull distribution, 212, 327, 395
weighing design, 285
weighted least squares, 307
weights, 22
Welch approximation, 275ff
winsorized mean, 218